Biogeochemistry of estuarine sediments

# Biogeochemistry of estuarine sediments

Proceedings of a Unesco/SCOR workshop
held in Melreux, Belgium
29 November to 3 December 1976

The designations employed and the presentation of the material in this work do not imply the expression of any opinion whatsoever on the part of Unesco concerning the legal status of any country or territory, or of its authorities, or concerning the delimitations of the frontiers of any country or territory.

Published in 1978 by the United Nations
Educational, Scientific and Cultural Organization,
7 Place de Fontenoy, 75700 Paris
Printed by Imprimerie Floch, Mayenne

ISBN 92-3-101594-X



*Printed in France*

# Preface

The coastal zone represents one of the Earth's most complex and dynamic ecosystems. In these areas which fringe the continents of the globe, the intricate terrestrial and marine systems become even more complex as they respond to the pressures created by man. Development of coastal resources has obvious and profound effects on this environment which can be neither predicted nor controlled without full understanding of a particular coastal area.

Estuaries constitute a major area of concern to scientists and administrators as they focus their attention on the problems of coastal zone development and management. This interest is accentuated by the fact that many of the world's largest cities are located on or near estuaries.

Within the United Nations, particular emphasis has been placed on assisting Member States in planning their use of coastal areas, and each of the Specialized Agencies has devoted considerable time and expertise to the subject areas falling within its competence.

Unesco's contribution to this effort is being carried out in co-operation with the Scientific Committee on Oceanic Research (SCOR) along three major lines: mangrove ecology and productivity; coastal lagoons; and the biogeochemistry of estuarine sediments. *Ad hoc* advisory panels on each of these subjects were set up in 1974 to assist Unesco in drafting programmes based upon criteria which each panel was to identify as being essential for international research.

The history of the panel on the biogeochemistry of estuarine sediments is outlined by Edward D. Goldberg in the introduction to this volume. At its first meeting, panel members emphasized the need to develop an understanding of the natural chemical processes taking place in estuarine zones so that an assessment could be made of man's present and predicted impacts upon the environment. It was decided that the convening of an international workshop on the subject would be the most productive way of responding to this need.

This volume presents the proceedings of the workshop and reflects the diversity of approaches to a number of significant research problems. Within the four working group reports can be found succinct summaries of priority research areas as defined by workshop participants. Unesco hopes that scientists around the world, and particularly those in developing areas, will be able to use this information as a basis for their own programmes of research. The coastal area offers an unusual opportunity for the scientist and the administrator to work together and to ensure that vital decisions for economic growth can be made within the context of environmental conservation.

# Contents

# Introduction

There is a certain redundancy in the term "coastal marine chemistry". The numbers of chemical reactions and their intensities in the near shore zones far exceed those of the open ocean. Over 90 percent of photosynthetic activity takes place in coastal areas and the consequential production of organic matter produced triggers off the high level of activity. These compounds, in their original form or as modified by other levels of the marine food chain, fall to the sea floor taking with them chemical energy. The subsequent oxidation of these electron-rich substances, of time assisted by micro-organisms, is responsible for the wealth of chemical reactions in the coastal zone.

It was with this in mind that I accepted the invitation of Unesco and the Scientific Committee on Oceanic Research (SCOR) of the International Council of Scientific Unions (ICSU) to seek ways of formulating priority problems concerned with the biogeochemistry of estuarine sediments. The zone where the rivers meet the sea is a vital resource to many countries and an understanding of its properties can provide a basis for its management.

An international committee to consider this general problem was assembled and included J.D. Burton (Department of Oceanography, The University, Southampton, England), James Carpenter (School of Marine and Atmospheric Science, University of Miami, Florida, USA), M.J. Orren (Geochemistry Department, University of Capetown, South Africa), Erwin Suess (Geologisches-Paleontologisches Institüt der Universitat Kiel, Federal Republic of Germany), and R. Wollast (Institute of Industrial Chemistry, Université Libre de Bruxelles, Belgium).

The committee defined its aim as follows : to seek out a critical evaluation of the procedures for characterizing the nature of reactions in estuarine sediments with special reference to biological processes. The committee explored various ways to accomplish this goal. In many cases the statements of tractable problems in the biogeochemistry of estuarine sediments have not as yet been formulated. This situation has largely arisen because scientists with laboratory or field techniques, potentially applicable to estuarine science, are often unaware of the priority problems and hence cannot assist in proposing strategies for their solution. The committee thus decided that a workshop which brought together estuarine scientists with their colleagues from other areas of marine science would be most rewarding. Many of the problems as yet unsolved in the estuarine environment have counterparts in other oceanic or freshwater areas, such as the speciation of metals or diffusional processes in pore waters.

The committee met 8-9 September 1975 at the Department of Oceanography of the University of Southampton, England, and proposed that the workshop be held in Europe during 1976. An initial list of invitees was prepared and the workshop format developed. A set of eleven questions (Table 1) concerned with substantial aspects of estuarine sedimentation was posed. Each invitee was asked to consider one or more of these questions and to prepare a pre-conference presentation which would be sent to all the participants. A five-day meeting was arranged with both plenary and panel meetings. In order that the combined expertise of all participants might be brought to bear upon the problems, the number of parallel sessions was kept to a minimum. Twenty to thirty participants was proposed as an ideal workshop size.

The workshop was held at La Maison des Metallurgistes de Belgique,"La Reine Pedauque", in Melreux, Belgium 29 November - 3 December 1976 under the sponsorship of Unesco, SCOR and the U.S. NOAA and management of Professor R. Wollast (Belgium) and Dr. M. Steyaert (Unesco). There were thirty three participants from fourteen countries: Belgium (3); Federal Republic of Germany (3); France (2); India (1); Malaysia (1); Netherlands (3); Saudi Arabia (1); South Africa (2); Sweden (1); Thailand (1); Turkey (1); United Kingdom (6); United States (7); and Yugoslavia (1).

The workshop participants were initially sub-divided into five groups to intensively study the eleven questions :

Group 1. Forms and species of dissolved elements in estuarine systems (Questions 1, 2, 3, 4, 5 and 8);

Group 2. Forms and species of particulates in estuarine systems (Same questions considered by Group 1);

Group 3. Transfer processes between the water and the sediment (Question 9);

Group 4. The role of organisms in estuarine sedimentation processes (Question 2, 8 and 10); and

Group 5. The role of man in estuarine sedimentation processes (Question 11).

Five plenary sessions were held to consider the concerns of each of the five groups, followed by individual group discussions and the consolidation of their deliberations into written reports which were presented in plenary on the final day of the workshop. During the latter days of the workshop, the members of Group 5 were incorporated into the other groups, as were their concepts, and the four group reports which emerged form the basis of this volume (note: because there was movement of the workshop participants between groups, the group reports are presented anonymously).

In addition to the reports twenty-seven presentations are included in this volume nearly all of which were available in a somewhat modified form to the workshop participants, the original versions being edited in the months following the meeting. They have been ordered in this volume according to the four groups, rather than with respect to the eleven questions, and it is hoped that the volume will be considered as a set of working manuscripts, stepping stones to future research.

Edward D. Goldberg

# Questions put to invitees

1. What are the pertinent physical, chemical and biological characteristics of the input of sedimentary source materials entering the estuarine zone?

2. The most important reactions in estuarine sediments involve the biodegradation of organic matter. What are the products and reactants in such microbiological processes? How are the reactions mediated? What are the parametres that determine the rates of biological degradation at various levels in the sediments?

3. How can we effectively measure the vertical distribution and forms of dissolved chemical species in interstitial waters with adequate resolution?

4. What techniques are available to identify metallic species in the interstitial waters?

5. How can we identify and quantitatively determine highly disordered solid phases (i.e. poorly crystallized minerals).

6. How can we sample, with minimum disturbance, the water/sediment interface?

7. How can we determine the time frames of chemical reactions in the sediments?

8. What are the modes of association of chemical constituents (metals, phosphate, amino acids, etc.) with various sedimentary components (water, gases, surfaces, crystallinsphases, coatings, etc.)? What are the rational systems we might devise to define meaningful associations? Do existing systems such as those of Gibbs, Kaplan, etc. provide a basis to understand chemical and biochemical processes?

9. What are the dominant mechanisms of the transfer of chemical species within the sediments and to and from the overlying waters? What are the rates of transfer of materials to and from the sediments associated with advective and convective processes including bubbling, bioturbation, molecular diffusion, physical mixing caused by hurricanes, floods, storms, tides, etc?

10. How do the activities of organisms influence the chemical speciations within the sediment and influence the exchange of materials with overlying waters?

11. How can we distinguish between natural and anthropogenic materials in the sediments? Can we predict the effects of man's additions?

# List of workshop participants

| Name | Address |
|---|---|
| S. AKSORNKOAE | School of Forestry<br>Kasetsart University<br>BANGKOK 9, Thailand |
| A. BEHAIRY | Oceanography Institute<br>King Abdul-Aziz University<br>JEDDAH, Saudi Arabia |
| K. BERTINE | Department of Geological Sciences<br>San Diego State University<br>SAN DIEGO, California, USA |
| G. BILLEN | Laboratoire d'Environnement<br>Université Libre de Bruxelles<br>50 Ave. F. Roosevelt<br>1050 BRUSSELS, Belgium |
| M. BRANICA | "Ruder Boskovic" Institute<br>P.O.B. 1016<br>41.000 ZAGREB, Yugoslavia |
| O. BRICKER | Maryland Geological Survey<br>2100 Guilford Ave.<br>BALTIMORE, Md. 21218, USA |
| D. BURTON | Department of Oceanography<br>The University<br>SOUTHAMPTON, U.K. |
| J. DUINKER | Netherlands Institute for Sea Research<br>P.O.Box 59 den BURG-TEXEL<br>Netherlands |
| J. DAY | Department of Zoology<br>University of Cape Town<br>South Africa |
| H. ELDERFIELD | Graduate School of Oceanography<br>University of Rhode Island<br>Narrogonsett Bay Campus<br>KINGSTON, RI 02551, USA |

H. ERLENKEUSER — Institut für Reine und Angewandte Kernphysik
University of Kiel
Olshausenstr. 40-60
23 - KIEL, Germany (Federal Republic of)

U. FORSTNER — Laboratorium fur Sedimentforschung
University of Beidelberg
Im Nenenheimer Feld
D-69 HEIDELBERG, Germany (Federal Republic of)

R. GIBBS — College of Marine Studies
University of Delaware
LEWES, Delaware 19958, USA

E. GOLDBERG — University of California
Scripps Institution of Oceanography
LA JOLLA, California, USA

R. HALLBERG — University of Stockholm
Department of Geology
Box 6801
11386 STOCKHOLM, Sweden

Th. HOPNER — Universitat
Postfach 943
D 2S OLDENBURG

J. JEDWAB — Laboratoire de Géochimie - ULB
50 Ave. Roosevelt
1050 BRUSSELS, Belgium

D. LAL — Physical Research Laboratory
ARMEDABAD 380009, India

P. LISS — School of Environmental Sciences
University of East Anglia
NORWICH, U.K.

J-M. MARTIN — Laboratoire de Géologie
Ecole normale Supérieure
46, rue d'Ulm
75230 PARIS CEDEX 05, France

E. MYERS — MESA Program Office (RX-5)
NOAA/ERL
BOULDER, Colorado 80302, USA

M. ORREN — Geochemistry Department
University of Cape Town
South Africa

H. POSTMA — Netherlands Institute for Sea Research
P.O. Box 59
TEXEL, Netherlands

| | |
|---|---|
| N. PRICE | Grant Institute of Geology<br>University of Edinburgh<br>EDINBURGH, U.K. |
| W. REEBURGH | Institute of Marine Sciences<br>University of Alaska<br>FAIRBANKS, Alaska 99701, USA |
| A. SASEKUMAR | Department of Zoology<br>University of Malaya<br>KUALA LUMPUR 22-11, Malaysia |
| S. STANLEY | Dunstaffnage Marine Research Laboratory<br>P.O. Box n°. 3<br>OBAN, Argyll, Scotland, U.K. |
| M. STEYAERT | UNESCO Division of Marine Sciences<br>Place de Fontenoy<br>75007 PARIS, France |
| E. SUESS | Oregon State University<br>CORVALLIS, Oregon 97330, USA |
| TURGUT BALKAS | Department of Marine Science<br>Middle East Technical University<br>ANKARA, Turkey |
| P. WILLIAMS | University of Southampton<br>SOUTHAMPTON, U.K. |
| R. WOLLAST | Université de Bruxelles<br>50 Ave; F; Roosevelt<br>1050 BRUSSELS, Belgium |

# Group 1
# Forms and species of dissolved elements in estuarine systems

# Report

## 1. INTRODUCTION

An estuary can be viewed as a large chemical reaction vessel in which solutions having very different chemistries mix in the presence of chemically reactive particles. There are three main reasons for studying their chemistry. First, in such environments the flux of riverborne material to the oceans can be substantially altered both in form and magnitude, with implications for global geochemical budgets. In oceanic element budgets the riverine input term is often the largest of the known sources. Secondly, many estuaries receive significant anthropogenic inputs which may be profoundly affected in the estuarine zone. Thirdly, estuaries are important areas of biological activity, with strong interrelationships between biological and chemical processes.

Here we are mainly concerned with the solution chemistry of estuarine waters although, in view of the interactions between dissolved and particulate phases, and the potential importance of the dissolved material as a source of particulates, such a limitation is arbitrary. The frequently used analytical distinction between dissolved and particulate fractions is itself arbitrary depending, as it does, on filtration of water samples through a membrane having an average pore diameter of approximately 0.45µm. The so-called "dissolved fraction" will, in fact, consist of an electrolyte solution together with colloidal and fine particulate material. There are two major aqueous compartments, i.e. (i) the water above the bottom sediments which, depending on the mixing regime in the estuary, may be sub-divided into various water bodies, and (ii) the interstitial waters of the sediments. Due to the high concentration of both aqueous species and solid phases in the bottom deposits compared with the overlying waters, the sediments provide the more important sites for solid-water interactions.

There are large horizontal and vertical variations in chemical parameters in these compartments. In true estuaries the extreme boundary conditions for salinity are $< 0.1‰$ and $> 35‰$ leading to a convenient definition of an estuary - from the chemical viewpoint - as extending from where the freshwater enters to the area where marine processes are the dominant control on composition. Changes in concentrations of the major solutes are largely under the control of the marine-dominated salinity gradient. For some minor constituents, including the plant nutrients P, N, Si, the chemical composition of the inflowing riverwater can be of paramount importance. Since rivers show substantial differences in chemical composition (determined by such factors as climate, geology, soil and vegetation cover in their drainage basins), large inter-estuary differences occur. A major variable is pH, with a range of $<5$ to $>8$ . An important contrast is found between the acidic streams carrying high concentrations of humic material and associated trace metals, and alkaline rivers draining carbonate basins which have pH values $>7$ and lower concentrations of organic material and most trace metals. The situation in estuaries under substantial anthropogenic influence will be still more complex since such inputs (sewage, agricultural and industrial wastes variously enriched in P, N, trace metals and a wide range of organic compounds) may enter either with the river or directly into the estuarine zone.

Probably the most important chemical factor influencing vertical gradients in estuarine waters is the decomposition of organic material formed by biological fixation in surface waters or entering via riverwater and waste effluents. The magnitude and vertical variation of this breakdown depends particularly on the intensity of heterotrophic activity but also on many other factors including the amount of degradable material, water stratification and residence time, etc. The consequence is the consumption of oxygen and other oxidants in the water and pore waters, and the addition of phosphorus, nitrogen and inorganic carbon compounds formed during the degradation. This often leads to a decrease in oxygen concentration with increasing depth and at some point - generally in the sediments but sometimes in the water column - the water become anoxic. Due to the lowered pE, multi-valent elements will tend to change their speciation which in some cases (e.g. manganese, cobalt) will lead to their release from the solid to the aqueous phase. In other instances (e.g. cadmium, mercury, association with insoluble sulphides will occur when sulphate is reduced to sulphide. There is also the possibility of soluble metal-organic complexes being formed during the decomposition of organic material. In most cases these reactions will take place in the sediments so that concentration gradients, and hence fluxes of various chemicals will be established between the interstitial and overlying waters.

## 2. CONCENTRATION VARIATIONS OF DISSOLVED ELEMENTS IN ESTUARIES

2.1 Variations due to supply. In addition to the factors which determine the general composition of dissolved elements in riverwater, changes in the dissolved load carried by the river discharge also contribute to concentration variability. Temporal changes in river flow, which modify the riverine flux of elements to an estuary, can be important in that the effect of anthropogenic input assumes greater or lesser importance because of this shifting baseline of natural stream supply. Also, short-term variations (e.g. as a consequence of storms) alter the balance of surface to sub-surface supply and change the proportionate contribution of ground waters to the total supply; this fraction may have a very different chemical composition than that of the surface discharge. It can be assumed that the seawater component in an estuary is of relatively constant composition and in this sense variations in the riverwater component are the only ones to consider. However, one source of dissolved elements in estuaries is from dissolution of particles (e.g. during anoxic diagenesis) and these particles can be supplied by both river and sea. Thus an identification of the sources of estuarine sediments is relevant in assessing sources of and variations in the dissolved load.

2.2 Variations due to mixing processes. Because of the generally contrasting concentrations of constituents in the end-members of estuarine mixing series, the extent of mixing of fresh and saline waters is a major controlling factor of the distribution of constituents in an estuary. Mixing may, however, be accompanied by interactions affecting concentrations, e.g. removal into solid phases or addition by release from solid phases. The extent of such interactions has often been evaluated by examining the linearity of relationships between constituent concentrations and salinity, or some other conservative index of mixing. These studies may give an over-simplified view. Changes in distribution between aqueous and solid phases should be studied by examining the compositional variations in solution and in particulates along co-genetic pathways. This requires selection of estuaries with suitable sedimentological characteristics, e.g. river-dominated input of suspended matter, low resuspension of bottom sediments. The use of concentration/salinity relationships may not be sufficiently sensitive to detect small-scale removal which may be nonetheless significant in terms of sedimentary supply for the estuary as a whole. Budget studies designed from knowledge of hydrodynamics and sediment transportation in the system are essential to assessment of an estuary as a sink or source for specific constituents. Those estuaries possessing relatively simple and well-understood circulation and thus suitable for such work, will not necessarily be major in terms of global inputs. From the standpoint of flux modification, the very early stages of mixing, i.e. up to salinity of 1 - 2‰, warrant special attention because deviations from conservative behaviour often appear to be associated with the initial increase in ionic strength. Normally conservative ions may behave non-conservatively in some circumstances, e.g. under strongly reducing conditions sulphate might be affected by reduction to sulphide.

An important process in mixing is the flocculation of riverborne humic material as a result of increase in ionic strength, with concomitant removal of iron. It is necessary to establish whether such removal commences in the freshwater zone and to examine the associated behaviour of trace elements. The removal of iron is not necessarily confined to humic-rich inputs and colloid coagulation probably underlies both organic and inorganic removal mechanisms.

2.3 Variations due to biological processes in overlying waters. Biological processes are important in creating temporal variations in nutrients and other species in estuarine waters. The uptake of trace metals by living or detrital material can give rise to very high concentrations in the suspended particles. More information is needed on the eventual fates of these materials, i.e. whether sediments are formed directly from the enriched particles or whether there is release into the water column as the result of decomposition of the organic matrix.

2.4 Variations due to processes in bottom sediments. There is a general understanding of the relationship of the oxidising-reducing boundary to processes in estuarine sediments which modify the composition of interstitial waters. A commonly-cited example is the enrichment of manganese in surface sediments caused by diagenetic remobilization of dissolved manganese in the pore fluids under the mere reducing conditions deep in the sediments. In some cases it is possible, at least empirically, to see relations between pore-fluid compositions and the presence of authigenic solid phases (e.g. iron and manganese compounds) which are presumed to control the concentrations of dissolved elements. A major task is to seek authigenic phases within different zones of estuarine sediments and determine their relations to pore-fluid compositions. The identification of sinks for elements in sediments is important in understanding the concentration profiles of dissolved elements and in predicting the fate of anthropogenic inputs to estuaries. Complications are that the pore-fluid chemistry of the upper zone ($\sim$ 20 cm) of the sediments can respond to changes in the composition of the overlying water and is subject to the effects of bioturbation. Thus inhomogeneities in sediment composition and lateral changes in productivity must be taken into account in designing the temporal and spatial framework for sampling. Another important factor is that the time-scale of reactions involving pore-fluids may be independent of the deposition rate of the sediments. Also such variations as occur in pore-fluid composition and which are presumably caused by dissolution reactions, must be checked by budget calculations to show whether the postulated reactions are indeed adequate to account for them.

The decomposition of organic material is a primary control of the chemical composition of interstitial waters, especially through its relationship to the redox potential. An important approach to modelling these decomposition processes is through stoichiometric budgets for reactants and products. High concentrations of dissolved organic matter in pore-waters suggest that the role of complexation reactions may be increased relative to overlying waters. The presence of interstitial metal concentrations considerably above those predicted for equilibrium with solid sulphides, in the presence of $H_2S$, can be partially accounted for by the presence of organic ligands. Within the sediments, methylation reactions for Hg, Se, Sn, Te and As and trasmethylation reactions between metal ions have been observed. Useful information may be obtainable from environments where unusually high accumulation of organic matter in sediments occurs, e.g. mangrove regions.

Most fluxes of dissolved species from interstitial to overlying waters are probably minor in relation to other inputs to the estuarine regions but they may be significant in modifying speciation, particularly through the introduction of forms which are metastable in the oxic environment.

## 3. SPECIATION OF DISSOLVED MATERIALS

Given the redox potential of an aquatic environment and a knowledge of the major inorganic ligands present, it is possible to arrive at a description of the speciation of major constituents and trace elements through equilibrium modelling. An example for $Cd^{2+}$ and $Fe^{3+}$ - for conditions typically found along the length of an estuary - is shown in Fig. 1. The variations with change in redox conditions down a vertical water/sediment profile are shown for nitrogen species in Fig. 2. In evaluating these equilibria, all relevant competing reactions must be considered e.g. competing ligand associations with a metal in one oxidation state (reaction scheme A) or a redox couple (reaction scheme B) as in Fig. 3. For equilibrium modelling in general, specific interaction methods have considerable advantages over ion-association approaches.

An important question regarding speciation is the identification of the chemical forms specifically involved in uptake processes by organisms or on particulate sites. A simplified cycle of metal transfer involving speciation changes is shown in Fig. 4.

The applicability to real environments of descriptions of speciation, obtained by equilibrium modelling, is limited by inherent uncertainties in the specific assumptions required (e.g. values for pE and equilibrium constants, activities in solid phases and effects of surface coatings on particles) and by the importance of kinetic factors. Much of the information on the latter question is, in fact, obtained indirectly by comparison of equilibrium predictions and environmental data. The value of these comparisons is, however, limited by the uncertain resolution of species afforded by present empirical analytical differentiations. So far as dissolved species are concerned, the major uncertainties lie in the role of organic associations. The bulk of the organic material is not adequately characterized. It is possible, however, to measure empirical association constants for humic-metal interactions which can be introduced into equilibrium models. A more intractable problem lies in the existence of thermodynamically non-labile complexes, whether formed and released by organisms, introduced in riverwater or formed from the dissolved components in estuaries. There is a need for improved methods to determine the extent of organic association and the nature of the complexes, and for studies of the kinetics of their formation and exchange reactions. Pore-waters may afford a better entry here than overlying waters. Experimetnal studies, for example, on model solutions or comparing unperturbed and organic-free (photo-oxidised) aliquots of natural waters, can provide relevant information. It is important that such experimental systems have compositions consistent with the environmental concentration of all relevant constituents.

## 4. ANALYTICAL METHODS

Sampling procedures must be carefully tailored to the environment of the particular estuary. The temporal and spatial frequency of sampling intervals, for example, should be determined with due regard to the time-scale of the phenomena under examination. An effort should be made to develop in situ or field instrumentation to avoid any changes in speciation or concentration between sampling and analysis. Procedures for sampling overlying water are reasonably well developed, but problems of contamination and changes on storage remain. Interstitial water sampling presents more difficulties. Coring, and subsequent expression of interstitial waters, suffers from problems of disturbance of the sediments, channeling, and changes in speciation due to the effects of, for example, temperature change and atmospheric oxygen. Coring remains an art and there is not complete agreement as to the relative merits of box and gravity corers. In situ methods, using probes inserted in the sediment, are promising, but disturbance of the sediment still occurs, and the vertical resolution of these systems in uncertain as water is sampled from a volume of sediment around the sampling port whose sampling field is not well establ-shed. Furthermore, all procedures yield only small volumes of sample imposing limitations on the analytical procedures. The development of improved in situ samplers or measuring devices is required; these devices may have to remain in position for an extended period to avoid disturbance. For all types of samples, errors are likely to be introduced during sampling, filtration, storage and analysis, through contamination or by loss from solution due to adsorption or colloid formation. Minimum sample handling is desirable. There is also a need for new approaches to the separation of particulate material, e.g. by centrifugation. Comparisons of such separation procedures would enable use of filtration for the numerous applications for which it is probably appropriate and would define the problem areas where alternative methods are required.

Analytical procedures should be developed where possible to distinguish forms of elements involved. Assuming an adequate amount of samples available (rarely a practical limitation except for pore-waters) determination of the dissolved constituents can potentially draw on almost all analytical techniques. Estuarine samples are expected to have widely varying concentration of total salt and organic compounds, and analytical techniques must accommodate these large matrix variations without important loss in accuracy and precision. For example, some difficulties occur in determining phosphorus in interstitial waters due to interference by silicon. In view of their specificity, enzymatic methods may prove useful in such situations.

Knowledge of the redox state of a system is most important, particularly in interstitial waters, but in practice is most difficult to determine. The platinum electrode has recognized deficiencies, but probably gives a relative measure of the redox state in pore-waters. Development of improved techniques for obtaining meaningful information on redox potential and capacity is a major priority in methodology.

Another priority problem concerns the measurement of dissolved organic material. Photo-oxidation methods combined with various techniques for the measurement of $CO_2$ have been developed to the automated stage, for measurement of total dissolved organic carbon and, despite reservations expressed by some workers, there is considerable evidence for their accuracy. There is, however, a lack of suitable procedures for many specific organic substances or groups of related compounds both in the water column and in interstitial waters. In particular, the lack of characterization of the humic materials means that much analytical information on these substances is empirical.

Dissolved gases ($CO_2$, $O_2$, $H_2S$, $N_2O$, hydrocarbons) are fairly readily determined in the overlying water, but such measurements are more difficult in interstitial waters. These measurements are required both for a complete description of the speciation of several elements (e.g. S, C, N) and for calculating fluxes of these gases from the estuarine water to the atmosphere.

Various instrumental methods (e.g. atomic absorption spectrophotometry, electrochemical procedures, neutron activation analysis) are available for trace metal analyses of estuarine waters. Electrochemical determinations give information on the species measured. Anodic stripping voltammetry and potentiometric methods have sufficient sensitivity to allow direct measurement of trace metals (Cu, Zn, Cd, Pb, Bi, TI, Ag) in seawater at their natural levels. Chemical pretreatment to eliminate the matrix (e.g. ion exchange, solvent extraction, co-precipitation) may favour particular species and, until more is known about speciation, total analyses are preferable. Methods such as flameless atomic absorption spectrophotometry, spark source mass spectrometry, arc or plasma emission spectrography and neutron activation analysis, are capable of yielding values for the total element concentration. These techniques are also applicable to small samples of interstitial water. The most widely applicable of these techniques is probably flameless atomic absorption spectrophotometry and development of improved selectivity is needed to eliminate further matrix effects.

Cross-calibration of analytical procedures is essential in chemical studies of estuarine systems. The development of suitable reference solutions for this purpose is a priority requirement. Equal attention should be given to the interlaboratory differences which may arise in the collection and initial treatment of samples.

## 5. SUMMARY OF PRIORITY AREAS

1. There is a need to identify estuaries which are appropriate for specific biogeochemical studies due to their particular characteristics of water circulation and sediment transport. For purposes such as budgeting, simplicity of the system is an important criterion. In studying processes, extremity of conditions (strongly contrasting circulation patterns, major seasonal variations in salinity distribution) can enable valuable field experiments to be designed.

2. The most important dissolved substances of general significance in estuaries are the main micronutrients (N, P, Si); elements important in redox processes (particularly Fe, Mn, S, I); those which offer special advantages for budget studies (e.g. U); dissolved organic matter (particularly the humic materials); and specific pollutants.

3. Behaviour of dissolved species in early mixing (up to salinity of 1-2‰) requires critical attention.

4. Budget studies are needed to complement and extend the information available from studies of concentration variations along salinity gradients.

5. More information is required on the molecular characteristics of the humic material in estuaries, in relation to their sources, role in complexation and flocculation behaviour.

6. The role of heterotrophic processes in overlying waters in the release of metals from suspended particulates needs clarification.

7. Information is needed concerning the occurence of non-labile, dissolved complexes and the kinetics of their exchange reactions.

8. Movement of water and dissolved species in pore-waters - especially in the upper part of the sediment column - warrants more detailed study.

9. Improved sampling methods and *in situ* measurement techniques are needed for interstitial water investigations.

10. More satisfactory methods are required for studies of redox potential and capacity in natural waters.

11. The resolution of dissolved and particulate species in the colloidal region has still received insufficient practical attention.

12. Laboratory intercomparisons of measurements of dissolved elements are needed with particular attention to the compositional variability of estuarine waters and the errors which arise in sampling and during initial treatment.

13. Identification of these areas stresses the paramount importance of multidisciplinary effort in the study of chemical processes in estuarine sedimentary environments.

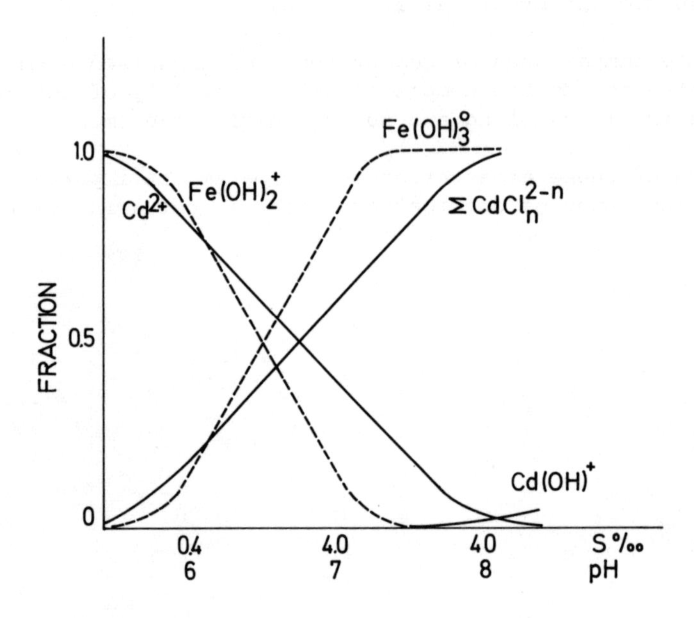


Fig. 1 Equilibrium specialism for Cd (———) and Fe (- - - -) for conditions of salinity and pH typically found along the length of an estuary.

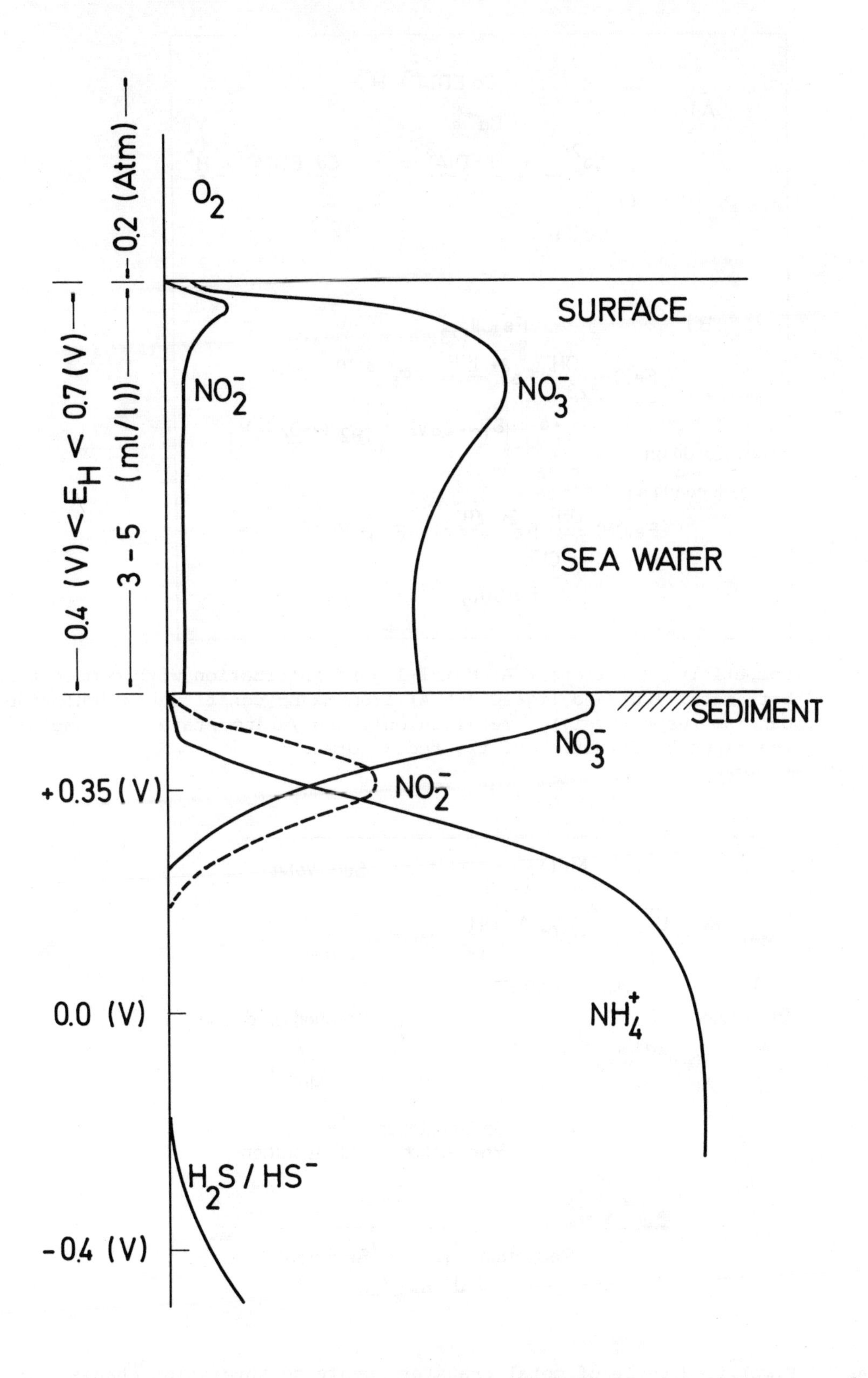


Fig. 2 Variation in equilibrium speciation of nitrogen with change in redox conditions down a vertical water/sediment profile.

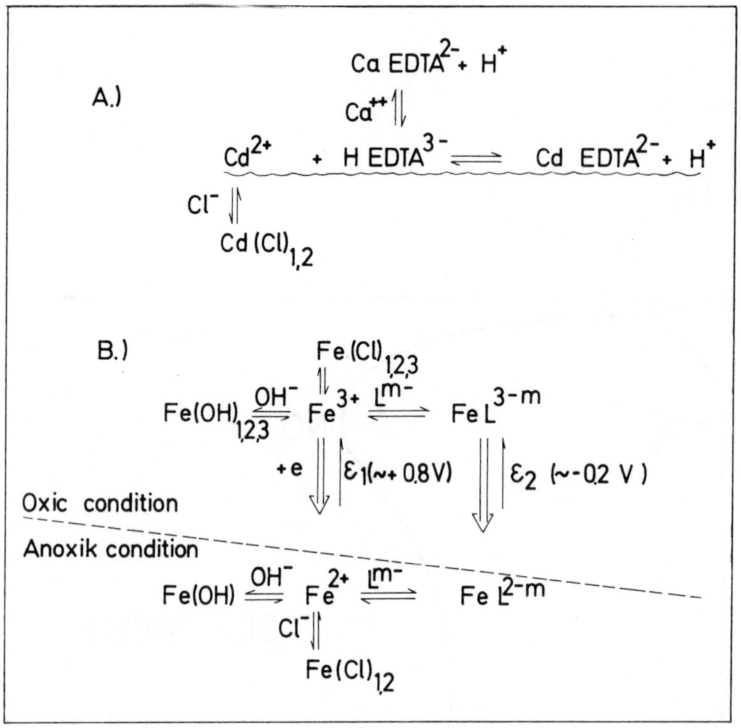

A.)

$$Ca\,EDTA^{2-} + H^{+}$$

$$Ca^{++} \rightleftharpoons$$

$$Cd^{2+} + H\,EDTA^{3-} \rightleftharpoons Cd\,EDTA^{2-} + H^{+}$$

$$Cl^{-} \rightleftharpoons$$

$$Cd(Cl)_{1,2}$$

B.)

$$Fe(Cl)_{1,2,3}$$

$$Fe(OH)_{1,2,3} \overset{OH^{-}}{\rightleftharpoons} Fe^{3+} \overset{L^{m-}}{\rightleftharpoons} FeL^{3-m}$$

$+e$ $\varepsilon_1(\sim +0.8V)$ $\varepsilon_2\ (\sim -0.2\ V)$

Oxic condition

Anoxik condition

$$Fe(OH) \overset{OH^{-}}{\rightleftharpoons} Fe^{2+} \overset{L^{m-}}{\rightleftharpoons} Fe\,L^{2-m}$$

$$Cl^{-} \rightleftharpoons$$

$$Fe(Cl)_{1,2}$$

Fig. 3 Competitive Reactions : A) Metal-ligand interaction with competition for both metal (Cd) and ligand (EDTA) from major constituents (chloride and calcium respectively) - metal in only one valency state; B) Complex redox reactions by formation of two redox forms.

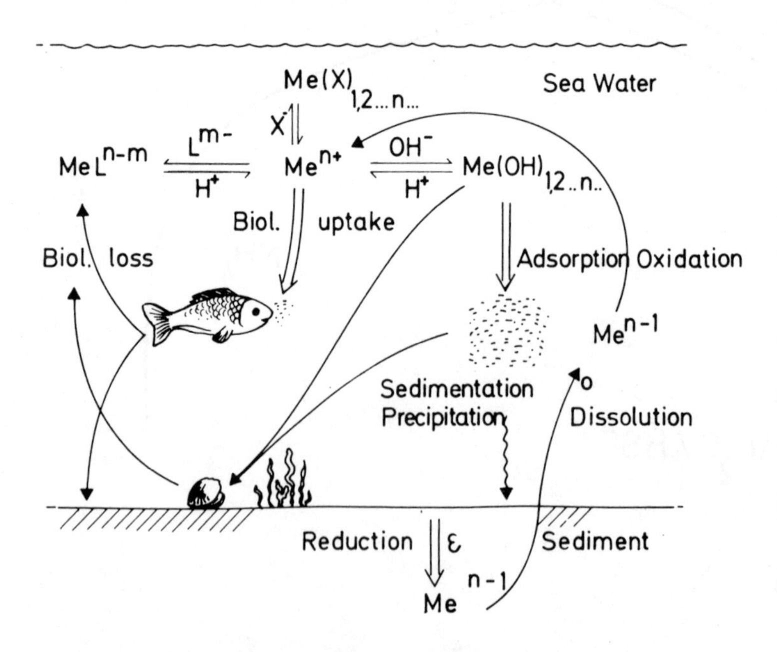


Fig. 4 Simplified cycle of metal transfer involving speciation change and incorporating the effects of uptake by organisms and on particulate sites.

# Techniques available to identify metallic species in the interstitial waters

Michael J. Orren*

The interstitial water sample volume obtained for analysis depends on several factors, including the resolution desired (i.e. the size of the "slice" to be squeezed), the water content and the particle size of the sediment, but will generally not exceed a few tens of millilitres. With the present techniques, high resolution studies will generate samples of only a few millilitres. Since samples deteriorate on storage (Riley, 1975), an analytical method should have a reasonably high throughput. This also allows for a more intensive sampling programme. Estuarine samples exhibit a widely varying range of composition both in inorganic salt and organic compound content (Burton and Liss, 1976). The analytical problem may initially be stated to require a reasonably rapid determination of as many metallic species - present at very low concentrations ($10^{-5}$ - $10^{-7}$M) - as possible, in a small sample of variable and complex matrix.

A brief discussion of instrumental techniques which may be of use is as follows :-

(1) Atomic absorption spectrophotometry (flame and flameless);

(2) Ultraviolet/visible spectrophotometry;

(3) Electrochemical methods (polarography, anodic stripping);

(4) Mass spectrography (spark source);

(5) Emission spectrography (arc and plasma); and

(6) Neutron activation.

## (1) Atomic Absorption (flame and flameless)

Flame atomic absorption spectrophotometry (AAS) is sensitive, simple, rapid, applicable to most elements and comparatively interference-free (Mavrodineanu, 1971). For most water analyses, a simple separation procedure, e.g. solvent extraction or co-precipitation, is used to separate the major interfering components of the matrix (Riley, 1975). A disadvantage is that the standard nebulizers consume several millilitres solution per minute and since only one element may be determined at a time, determination of several elements in a small sample is volume-limited.

Flameless atomic absorption, using a graphite furnace or heated metal ribbon atomizer, requires small volumes (5 - 100 µl), has very high absolute sensitivity

*National Research Institute for Oceanology, South Africa.

(pg level for, e.g. Cd) and is otherwise similar to flame AAS, except that interference is more marked (Woodward, 1975) (though not severe) and sample throughput is slower. Background correction, to reduce non-atomic absorption interference, is rarely required in flame AAS, but is usually necessary in flameless AAS (Woodward, 1975).

(2) Ultraviolet/visible spectrophotometry

Methods exist to determine most elements at trace concentrations (Sandell, 1959). While sensitivity in general is about the same as for AAS, more time-consuming chemical pre-treatment is required to avoid interference; only one constituent can usually be determined at a time and, even when using long path length and small volume optical cells, sample volume becomes a limiting factor.

(3) Electrochemical methods (polarography, anodic stripping)

For several elements (e.g. Pb, Cd, Zn) electrochemical techniques can give high sensitivity and yield additional useful information on the ionic species present (Whitfield, 1975). The techniques are not generally applicable to a wide range of elements at trace concentration and interference is a serious problem. Considerable technical skill is required and sample volume may be limiting.

(4) Mass spectrography (spark source)

This has the advantage of multi-element capability and high absolute sensitivity, and needs only small samples (about 10ml). The method is time-consuming as the sample is usually dried onto graphite before presenting to the instrument, and volatilization losses or contamination may occur; photographic spectral records must be read and computer processing is necessary (Jackson et al., 1976). In addition, the equipment is expensive and requires considerable skill in operation.

(5) Emission spectrography (arc and plasma)

Although high sensitivity is attainable with multi-element capability using an arc source emission spectrograph, interference is considerable and the major constituents of natural water (e.g. $Na^+$, $Ca^{2+}$, $K^+$, $Cl^-$, $SO_4^{2-}$) seriously affect the determination of trace constituents. The interference can be reduced by drying the sample onto standard powder and graphite mixtures (Brenner et al., 1975) but this effectively "dilutes" the sample, thus raising the detection threshold. Precision is usually lower than the other techniques above, and operator skill is important.

Inductively coupled plasma emission sources, in conjunction with direct reading spectrometers, have multi-element capability, high sensitivity and reasonable freedom from interference (Kniseley, 1975). Samples may be introduced by nebulization (about 20ml required) or, more recently, by atomizing small volume samples in a graphite furnace and sweeping the atoms formed into the plasma. This latter technique is very promising as it most nearly meets the requirements for interstitial water studies. The equipment is very expensive and much operator skill is required at this early stage of technical development of the method.

(6) Neutron Activation Analysis

This is again a multi-element technique of high absolute sensitivity and is reasonably free from interference, although some pre- or post-irradiation chemical treatment may be necessary (Robertson, 1975). The analyses are often time-consuming due both to user-pressure on the reactor facility and the need to allow some samples to "cool" for periods of days to weeks before counting. The equipment is costly, irradiation facilities are not generally available in smaller laboratories, and operator skill is important.

x x x x x x x x x

A concerted effort to understand the chemistry of estuarine interstitial water will require co-operation and comparison studies between many laboratories both large and small. The exchange of samples for inter-calibration - regarded as a high priority - would be facilitated if most laboratories used similar techniques. Larger, well-equipped laboratories should run samples by several different methods to establish accuracy criteria.

In principle, plasma emission spectrography probably best meets the analytical requirements. In practice, few laboratories have access to these complex and costly instruments. Flameless atomic absorption probably most closely approximates to the overall analytical requirements. This technique is simple and sensitive, reasonably rapid, many elements can be determined free from serious interference (if due care is taken), small volumes can be accepted often without pre-treatment (which may introduce errors), and due to its relatively low cost, most analytical laboratories are either equipped with, or have access to, such instrumentation.

With no chemical pre-treatment, atomic absorption, mass spectrography and emission spectrography generally yield numbers close to the "total" amount of element present, although there may be problems arising from differential rates of decomposition of compounds into atoms.

Chemical pre-treatment generally introduces a bias towards particular metallic species. For example, in solvent extraction of metalchelates, chelators react rapidly with ionic species, but metals strongly bound in natural organic complexes may be extracted with low efficiency or not al all. Similar arguments apply to ion exchange or co-precipitation separation and concentration techniques (Riley, 1975). Problems thus arise both in interpretation of the data and also because ionic species are almost universally employed in calibrating analytical techniques.

Many workers have used radio-active spikes to determine "chemical yield", but this implicitly assumes that the tracer reaches equilibrium with the naturally occurring species. Benes and Steinnes (1975), in a study of Norwegian fresh waters, pointed out that there were considerable differences between the behaviour of a radiotracer and its stable counterpart; ".... in particular, very slow isotope exchange was observed with colloidally or organically complexed forms of the natural trace elements".

In conclusion, the chemistry of any pre-treatment method should be carefully tested to ascertain what species are actually being measured. Since our knowledge of naturally occurring metallic complexes is meagre, it is suggested that "total" analyses be used until we better understand the natural complexes that exist in such samples.

## Reference

Benes, P.; Steinnes, E. 1975. Determination of the state of trace element in water. Paper delivered at the international conference on heavy metals in the environment, Toronto, Canada, October 1975.

Brenner, I.B.; Goldbart, Z.; Matthews-Bar, M.; Harel, A.; Halicz, Z. 1975. A versatile DC carbon arc spectrochemical procedure for the determination of 25 trace elements in natural water and industrial effluents. Paper delivered at the international symposium on the geochemistry of natural waters, Burlington, Canada, August 1975.

Burton, J.D.; Liss, P.S. (eds.), 1976. Estuarine chemistry. London, Academic Press.

Jackson, P.F.S.; Watling, H.R.; Watling, R.J. 1976. Mass spectrographic analysis of natural waters. Paper delivered at IUPAC international symposium on analytical chemistry in the exploration, mining and processing of materials, Rand Afrikaans University, Johannesburg, August 1976.

Kniseley, R.N. 1975. Inductively coupled plasma optical emission spectroscopy and its application to simultaneous multi-element determination of trace elements in bio-environmental samples. Paper delivered at the international conference on heavy metals in the environment, Toronto, Canada, October 1975.

Mavrodineanu, R. 1971. Analytical flame spectroscopy. London, Macmillan.

Riley, J.P. 1975. In : J.P. Riley and G. Skirrow (eds) Chemical oceanography (2nd ed.), Vol. 3, p. 193-514. London, Academic Press.

Robertson, D.E. 1975. In : J.P. Riley and G. Skirrow (eds) Chemical oceanography, vol. 3, p. 313-34, London, Academic Press.

Sandell, E.B., 1959. Colorimetric determination of traces of metals. New York, Interscience.

Whitfield, M. 1975. In : J.P. Riley and G. Skirrow, (eds.) Chemical oceanography (2nd ed.), vol. 4, p. 1-154, London, Academic Press.

Woodward, C. (ed.) 1975. Annual Reports on analytical atomic spectroscopy, p. 12-6., London, The Chemical Society.

# The modes of association of trace metals with certain components in the sedimentary cycle

J. D. Burton*

There has been much discussion of the interactions between dissolved and particulate material in estuaries, and the implications of such reactions for the supply of dissolved constituents to mere open waters and the composition of estuarine sediments, but clear evidence regarding the environmental significance of postulated processes is often scanty. Changes in the metal composition of sediments in polluted estuaries such as that of the Rhine were considered by de Groot (1973) to be attributable to release of metals from particulates, influences in part by complexation with decomposition products of organic material in the particulates. These changes may be accounted for, however, by an alternative mechanism of sediment mixing processes without invoking chemical exchanges (Müller and Förstner, 1975). Marked differences in the concentrations of mercury between suspended and deposited sediments have been diversely attributed to differences in particle size and to biological mechanisms involving mobilization of mercury by methylation (Aston and Chester, 1976).

While the environmental significance of desorption processes in estuaries is unclear, there is unequivocal evidence for the removal of readily hydrolysed elements, such as iron and rare earths (Coonley et al., 1971; Holliday and Liss, 1976; Martin et al., 1976); the removal mechanisms and sites are not, however, well understood. Existing particles influence removal processes (Aston and Chester, 1973) but solid phases are formed also in the absence of particles (Sholkovitz, 1976). Among other factors, differences in behaviour between inorganic and organically associated dissolved metal species are not known.

An understanding of the sources and cycles of metals in estuarine sediments requires studies of the changes, during estuarine mixing, in total composition of both dissolved and particulate phases along cogenetic pathways and also differentiation of associations within these materials - ideally a determination of the chemical speciation, although current techniques do not achieve this. These comments are addressed to aspects of this latter question.

## Dissolved trace metals in natural waters

The question of the speciation of dissolved trace metals has been approached by equilibrium calculations on the basis of either ion-association models as, for

* Department of Oceanography, The University, Southampton, United Kingdom.

example, in the work of Zirino and Yamamoto (1972) and Ahrland (1975), or ionic interaction models (Whitfield, 1975). Organic associations cannot be readily allowed for in such models because the bulk of the organic matter is not properly characterized and thermodynamically non-labile complexes may be significant. The determination of stability constants for the associations formed by metals with separated fractions of humic material (Mantoura and Riley, 1975) enables some progress to be made in this respect. Potentially important geochemical roles of organic associations include the entry of complexes into flocculation processes (Sholkovitz, 1976) and the stabilization of lower redox states of metals (Theis and Singer, 1974). In interstitial waters organic complexes may modify solubility equilibria and influence the migration of dissolved forms in the pore-fluids. Biologically, effects on accumulation and toxicity may be significant. Experimental approaches to assessing organic association have been of two kinds: (1) experimental models; and (2) analytical differentiation.

The capacity of humic materials to associate cadmium and mercury in model solutions has been shown by their effects on availability of the metals in electrochemical reactions (Gardiner, 1974; Millward and Burton, 1975). Changes in electroactivity of metals added to natural waters have also been demonstrated (Siegel, 1971; Chau et al., 1974; Millward and Burton, 1975). Matharu (1975) has shown that additions of humic material or algal extracts to organic-free seawater restores the phenomenon of chloroform extractability of a fraction of dissolved copper. Such experiments demonstrate conclusively that the presence of organic materials in solution affects the behaviour of some trace metals by speciation changes but the nature of the interactions remains uncertain.

Analytical differentiation methods have been particularly applied to the study of dissolved copper. In coastal and estuarine waters an additional fraction of this element is made available to analytical concentration techniques when organic matter is destroyed by photo-oxidation (see Table 1) and a fraction is extractable directly into chloroform (Slowey et al., 1967; Matharu, 1975). The comparability and interpretation of such data is difficult since the former fraction excludes labile associations and lability will vary with the nature of the competing analytical ligands and with pH. Data given in Table 2 confirm the non-equivalence of the operationally defined fractions; it is probable, moreover, that the larger fraction does not represent the total organically associated copper. Measurements on surface and intermediate open oceanic samples show these fractions to be less significant in these waters, each accounting for 10% or less of the dissolved copper (Moore and Burton, 1976). In estuarine waters ultrafiltration experiments (Smith, 1976) indicated that the main associating capacity is shown by material in the molecular weight range less than 1000 Daltons and that the capacity tends to increase with increasing salinity. In contrast in river water, similar associating capacity is shown by fractions from more than 100,000 to less than 1000 in molecular weight. It appears that differences in the sources of associating entities or of copper, which is already associated, exist over the range of environments from freshwaters to the open sea. Processes which may lead to such differences include the removal of terrestrially derived humic material in the estuarine region (Gardner and Menzel, 1974) and the increased competition for complexing sites by the abundant cations, such as calcium.

In addition to the obvious limitations of analytically defining fractions by the above methods or by changes in electroactivity with pH, as has been employed particularly in studies on zinc (Bernhard et al., 1975), there is uncertainty as to whether the effects observed are due to the existence of organic complexes in true solution. Significant fractions of elements such as iron and aluminium occur in dispersed states in the filtrates passing filters of ca. 0.5 - 1 μm average pore diameter and can be removed by finer filters (Burton and Head, 1970; Hydes, 1974). Most environemental observations are compatible with the hypothesis that non-labile fractions of metals may be associated with colloidal material stabilized by organic material. This was recognized by Williams (1969) but the point has received little further consideration. Destruction of organic material or acidification could make such fractions more freely available. Solvent extraction may be a consequence of interfacial reactions and need not necessarily reflect the initial presence of dissolved complexes - Matharu (1975) showed that copper can be removed from seawater by ion-flotation. The data of Smith (1976) are, however, more difficult to explain on this hypothesis, since they show an increasing complexing capacity of the low molecular weight fraction with increasing salinity.

The question as to whether or not these and similar phenomena are attributable to associations in true solution is one which can be explored by present techniques. An examination is required of the amounts of trace metals associated with the various size fractions in the sub-micron range down to about $10^{-3}$ μm, which are included in the operationally defined dissolved material by present conventional methods of filtration. The characterization of any dissolved organic complexes is beyond present capabilities unless non-modifying concentration methods can be applied to bring the levels within the range of appropriate spectroscopic methods (Campbell and Dwek, 1975). Ultrafiltration is perhaps the most promising technique for this purpose, though its value is limited in dealing with complexes with low molecular weight material. The implications of speciation in solution are important not only with respect to the roles of different pools of individual elements in river and seawaters which act as sources for authigenic sedimentary phases but also in relation to interstitial waters. The higher concentrations of trace metals in pore waters may facilitate more direct approaches.

## Metals in suspended and deposited estuarine sediments

Several methods have been devised to differentially extract metals from suspended and deposited solids.

The relatively straightforward technique described by Chester and Hughes (1967) was originally developed for use with pelagic sediments. It uses a solution of hydroxylamine hydrochloride and acetic acid to segregate lattice held and non-lattice held material. The former is often described as lithogenous but this may be misleading, particularly in the present context, in that continentally derived components may occur in non-lattice positions. The method has been applied to sediments in the polluted Bristol Channel (Chester and Stoner, 1975); their results suggested that the enhancement of zinc and lead was mainly by uptake from overlying waters, whereas tin was significantly enriched in the non-leachable fraction.

Young (unpublished data) has used the same method on suspended particulates in the Beaulieu Estuary (southern England). A marked removal of "dissolved" iron with increasing salinity in this system has been reported by Holliday and Liss (1976) suggesting that there might be variations in the authigenic iron in associated particulate material. By contrast, both studies showed the largely conservative behaviour of dissolved manganese over a wide range of salinities. Four metals were studied by Young (Fe, Mn, Zn and Cu) under conditions of minimal biological activity. The lattice held concentrations in suspended material were virtually constant over the same salinity range, averaging (dry wt. basis) 3.2 per cent for Fe, 120 ppm for Mn, 97 ppm for Zn and 16 ppm for Cu. In contrast, the contents of non-lattice held Fe and Mn decreased markedly with salinity (see Figs. 1 and 2); this trend was less clear for Cu and Zn. It appears that a primary control on the differential distribution of Fe and Mn in particulates in this system is the mixing of river-derived particulates high in non-lattice held iron and manganese, with marine particulates low in these components. Whereas the decrease in non-lattice held manganese conforms approximately to a single linear relationship, the relationship for iron is more complex with a more pronounced decrease occurring in the lower range of salinity (up to S = 5‰). The variation in non-lattice held particulate iron with salinity unexpectedly resembles that of "dissolved" iron as found by Holliday and Liss (1976).

These data illustrate that a coarse differentiation method can provide some insight into the processes affecting solid phases. Techniques involving successive leachings include those of Nissenbaum (1972), Gibbs (1973) and Gupta and Chen (1975). While these techniques give more refined information they raise problems of interpretation and comparability. Gupta and Chen (1975) distinguished eight sedimentary fractions, for some of which alternative extractants were examined. The agreement between total sediment concentrations and the sum of the fractions was generally very good. It is uncertain, however, what materials in the sediments are actually attacked by the successive treatments. The data emphasize the dilemma that mild leaching solutions may give incomplete attack on components such as hydrous iron oxide, organic material, and sulphides, while more effective extractants are probably not selective. Work in this laboratory (Armannsson, unpublished data) has shown that while a solution of sodium pyrophosphate and sodium hydroxide powerfully attacks humic materials, the reagent as generally employed also attack silicates to an appreciable degree; for a modified reagent in which the pH of the sodium pyrophosphate solution was adjusted to 9.5 with ammonia, the attack on silicates was reduced.

Coarse differentiation, as discussed above, can give envrionmentally useful data, particularly if the effects of carbonates are taken into account where relevant. Truly selective attacks giving differentiation of organic and other fractions would be of great value but present methods appear limited. At this stage more information is required from experimetnal studies; for example, of the changes in availability of added radiotracers to extraction techniques and the biological availability of different sedimentary fractions. In the absence of this information a multiplicity of empirical techniques is potentially confusing. Studies on properties of bulk sediments and fractions need to be complemented by detailed examination of element associations in particles and at their surfaces.

Table 1

Organically associated fractions of dissolved copper in natural waters as defined by difference in availability following photo-oxidation

| Region | Initially available copper $\mu g\ l^{-1}$ | Fraction of total made available by photo-oxidation percent | Analytical method | Comments | Reference |
|---|---|---|---|---|---|
| Californian coast | 0.4-4.3 | 0-22 | Accidified to pH 4. SDDC/$CCL_4$extraction; molecular absorption spectrophotometry. | Iron gives some interference in SDDC method. | Williams (1969) |
| Menai Straits, Wales | 1.8-3.3 | 9-30 | As above, but no pH alteration. | Some evidence of seasonal variation in seawater related to planktonic cycle. | Foster and Morris (1971) |
| Rivers adjacent to Menai Straits | 1-1600 | 0-11 | | | |
| Southampton Water system | | | Concentration by imino-diacetic chelating resin at natural pH (river water) or pH 5-6 (sea water); atomic absorption spectrophotometry. | Cadland Creek concentrations directly affected by industrial discharge; difference fraction minimal. | Matharu (1975) |
| Surface estuarine waters | | | | | |
| Marchwood (S = 21-32‰) | 1.3-3.7 | n.d.-24 (mean 10) | | | |
| N.W.Netley (S = 26 - 33‰) | 1.2-3.6 | n.d.-16 (mean 6) | | | |
| Cadland Creek (S = 26-33‰) | 1.3-17 | n.d.-17 (mean 4) | | | |
| Hillhead (S = 28 - 35‰) | 1.1-3.6 | n.d.-28 (mean 9) | Larger numbers of samples than were included in comparative study (Table 2) | | |
| Riverwaters | | | | | |
| Test | 0.8-2.3 | n.d.-41 (mean 9) | | | |

Table 2

Fractions of dissolved copper in river and estuarine waters in the Solent region[a]

| Location | Range of total concentrations of dissolved copper $\mu g\ l^{-1}$ | Average per cent extractable into chloroform | Average per cent made available to chelating resin[b] by photo-oxidation |
|---|---|---|---|
| River Test (6 samples) | 0.8-1.8 | 30 | 4 |
| Hillhead (average S *ca.* 33‰) (18 samples) | 1.3-2.7 | 25 | 7 |

[a] Data from Matharu (1975). Samples covered a range of seasonal conditions.

[b] Uptake on resin was at natural pH for riverwater and at pH 5-6 for saline waters.

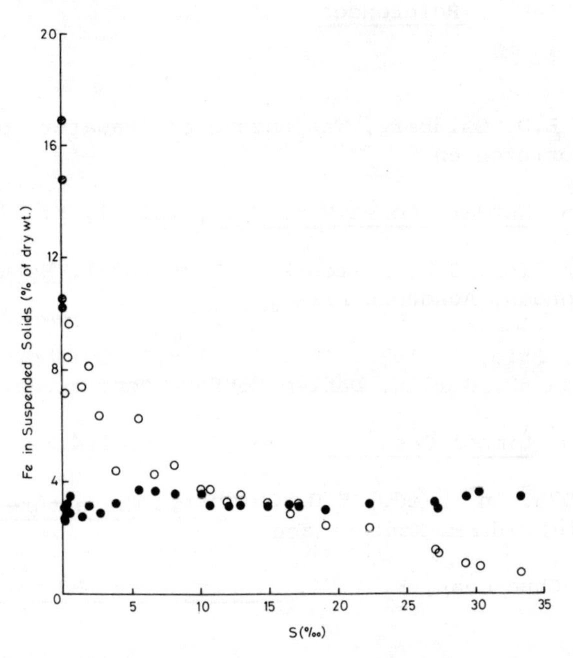


Fig. 1. The leachable (o) and non-leachable (●) iron content of suspended material in the Beaulieu Estuary and its variation with salinity (Young, unpublished data). Cross-hatched points are for leachable contents of river particulates sampled on four days prior to the estuarine survey. Fractions were resolved by the technique of Chester and Hughes (1967).

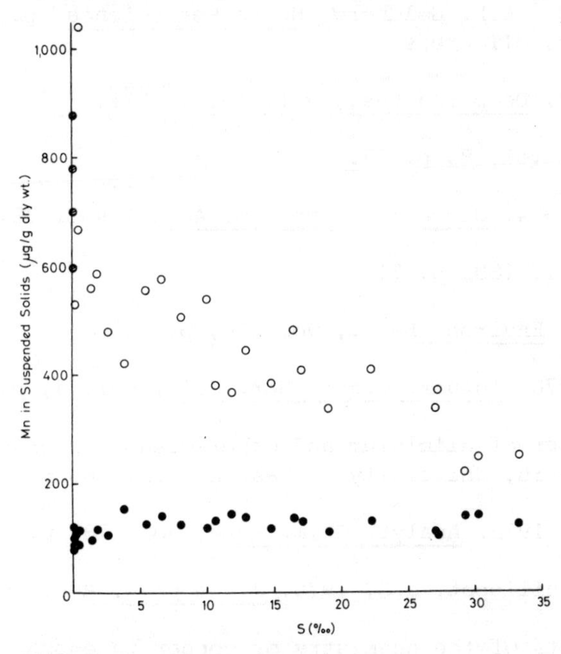


Fig. 2. The leachable (o) and non-leachable (●) manganese content of suspended material in the Beaulieu Estuary and its variation with salinity (Young, unpublished data). Cross-hatched points are for leachable contents of river particulates sampled on four days prior to the estuarine survey. Fractions were resolved by the technique of Chester and Hughes (1967).

References

Ahrland, S. 1975. In : (ed.) E.D. Goldberg, The nature of Seawater, p. 219-44. Berlin, Dahlem Konferenzen.

Aston, S.R.; Chester, R. 1973. Estuar. coast Mar; Sci., vol. 1, p. 225.

Aston, S.R.; Chester, R. 1976. In : J.D. Burton;P.S. Liss (eds). Estuarine chemistry, p. 37-52. London, Academic Press.

Bernhard, M.; Goldberg, E.D.; Prio, A. 1975. In : (ed.) E.D. Goldberg, The nature of seawater, p. 43-68. Berlin, Dahlem Konferenzen.

Burton, J.D.; Head, P.C. 1970. Limnol Oceanogr., vol. 15, p. 164.

Campbell, I.D.; Dwek, R.A. 1975. In : (ed.) E.D. Goldberg, The nature of seawater, p. 165-89, Berlin, Dahlem Konferenzen.

Chau, Y.K.; Gächter, R., Lum-Shue-Chan, K. 1974. J. Fish. Res. Bd Can., vol. 31, p. 1515.

Chester, R.; Hughes, M.J. 1967. Chem. Geol., vol. 2, p. 249.

Chester, R.; Stoner, J.H. 1975. Mar. Pollut. Bull., vol. 6, p. 92.

Coonley, L.S., Jr.; Baker, E.B.; Holland, H.D. 1971. Chem. Geol. vol. 7, p. 51.

de Groot, A.J. 1973. In : (ed.) E.D. Goldberg, North Sea science, p. 308-25, Cambridge Massachusetts, MIT Press.

Foster, P.; Morris, A.W. 1971. Deep-Sea Res., vol. 18, p. 231.

Gardiner, J. 1974. Water Res. vol. 8, p. 23.

Gardner, W.S.; Menzel, D.W. 1974. Geochim. Cosmochim. Acta., vol. 38, p. 813.

Gibbs, R.J. 1973. Science, vol. 180, p. 71

Gupta, S.K.; Chen, K.Y. 1975. Environ. Lett., vol. 10, p. 129.

Holliday, L.M.; Liss, P.S. 1976. Estuar. Coast. Mar. Sci., vol. 4, p. 349.

Hydes, D.J. 1974. The behaviour of aliminium and other dissolved species in estuarine waters, Thesis, University of East Anglia, 168 p.

Mantoura, R.F.C.; Riley, J.P. 1975. Analyt. Chim. Acta, vol. 78, p. 193.

Martin, J.-M.; Høgdahl, O.; Philippot, J.C. 1976. J. Geophys. Res., vol. 81, p.3119.

Matharu, H.S. 1975 Some aspects of the chemistry of copper in estuarine and marine environments. Thesis, Southampton University, 186 p.

Millward, G.E.; Burton, J.D. 1975. Mar. Sci. Comm., vol. 1, p. 15.

Moore, R.M.; Burton, J.D. 1976. Nature, vol. 264, p. 241.

Müller, G.; Förstner, U. 1975. Environ. Geol., vol. 1, p. 33.

Nissenbaum, A. 1972. Israel J. Earth-Sci. vol. 21, p. 143.

Sholkovitz, E.R. 1976. Geochim. Cosmochim. Acta, vol. 40, p. 831.

Siegel, A. 1971. In : (eds.) S.D. Faust, J.V. Hunter, Organic compounds in aquatic environments. p. 265-95. New York, Dekker.

Slowey, J.F.; Jeffrey, L.M.; Hood, D.W. 1967. Nature, vol. 214, p. 377.

Smith, R.G., Jr. 1976. Analyt. Chem., vol. 48, p. 74.

Theis, T.L.; Singer, P.C. 1974. Environ. Sci. Technol., vol. 8, p. 569.

Whitfield, M. 1975. Geochim. Cosmochim. Acta, vol. 39, p. 1545.

Williams, P.M. 1969. Limnol. Oceanogr., vol. 14, p. 156.

Zirino, A. and Yamamoto, S. 1972. Limnol. Oceanogr., vol. 17, p. 661.

# The analysis of estuarine interstitial waters for trace metals

J. C. Duinker*

A number of the problems encountered in the analysis of trace metals at the $\mu g\ l^{-1}$ level and below in sea water are illustrated by a comparison of several intercalibration exercises (Brewer & Spencer, 1972; Anon, 1974; Duinker et al., 1975; see also review by Tölg, 1972). Errors are likely to be introduced during sampling, filtration, storing and analysing the samples due to the introduction of metals from dust or the equipment used (Robertson, 1968) or to their removal from the solution due to adsorption or colloid formation (Struempler, 1975; Issaq and Zielinski, 1974). Minimum sample handling, including the addition of chemicals, is essential for obtaining reliable data. A number of analytical techniques for trace metal determinations are available (Mancy, 1972; Minear and Murray, 1973). Three analytical methods, neutron activation analysis, flameless atomic absorption spectrometry and electrochemical methods (anodic stripping voltammetry (ASV) and potentiometric methods) possess sufficient sensitivity to allow direct measurement of trace metals in seawater at their natural levels (Ariel and Eisner, 1962; Ariel et al., 1964; Branica, 1970; Allen et al., 1970; Jasinski et al., 1974; Segar and Cantillo, 1975). ASV has been shown to be applicable for Cu, Zn, Cd, Pb, Bi, Tl and Ag analyses; with special modifications, it may also work for a number of other elements (Neeb, 1969).

Concentration and separation techniques, such as coprecipitation, ion-exchange and solvent extraction of metal complexes formed after addition of a complexing agent, have all been used to concentrate these metals, leaving major elements and other interfering substances behind.

These procedures allow the application of a number of techniques for an accurate analysis of a large number of trace elements (Anon, 1975). In most cases, it is not yet certain whether the total amount or just a certain fraction of any one trace element in seawater is being analysed by any one particular procedure (Riley, 1965; Goldberg, 1965; Lindgren, 1975; Röndell, 1975; Ekedahl, 1975; Burton, 1976).

Theoretical models for the mainly inorganic speciation of elements in seawater are available in the literature (Sillen, 1961; Garrels and Thompson, 1962; Goldberg, 1965; Zirino and Yamamoto, 1972; Dyrssen and Wedborg, 1974; Millero, 1975). Experimental tests (mainly electrochemical methods) of these models describing the complexation of Cd, Zn, Cu and Pb by inorganic ligands ($Cl^-$, $SO_4^{2-}$, $CO_3^{2-}$, $OH^-$) are found in studies by Duursma and Sevenhuysen (1966), Baric and Branica (1967), Zirino and Yamamoto (1972), Bradford (1973) and Duinker and Kramer (1977).

---

* Netherlands Institute for Sea Research, P.O. Box 59, Texel, the Netherlands

It is well established that dissolved organic complexes are important for a wide range of trace metals (Lerman and Childs, 1973; Gamble and Schnitzer, 1973; Theis and Singer, 1973; Fukai, 1974; Andren and Harriss, 1975). Experimental evidence for the occurrence of organically bound fractions has been derived (1) from extraction of organic matter (Schindler & Alberts, 1974), (2) from the differences - analytically detected - in metal concentrations before and after decomposition of organic matter (Laevestu and Thompson, 1958; Alexander and Corcoran, 1967), (3) from the application of separation techniques such as dialysis (Rona et al., 1962) and (4) from the basis of ASV studies at different pH values (Branica et al., 1969; Zirino and Healy, 1972; Duinker and Kramer, 1977). Although evidence for the presence of both inorganic and organic complexes is available for a number of metals, no general conclusions can yet be drawn as to the relative importance of organic and inorganic ligands in metal complexes in seawater.

Neutron activation analysis and flameless atomic absorption spectrometry, although having sufficient sensitivity to measure metal concentrations dissolved in seawater, cannot discriminate between their different chemical forms. The only techniques presently available to analyze directly different forms of a number of metals in solution are electrochemical ones. Whitnack (1961) was the first to apply one of these direct methods (cathode ray polarography) to the field of oceanography. He was able to show that Cu, Pb, Cd, Zn, Ni, Co and Mn all responded in an useful way. The use of polarographic methods has already resulted in considerable information on speciation (Whitnack, 1961, 1964; Branica, 1970; Branica et al., 1969; Maljkovic and Branica, 1971; Odier and Plichon, 1971). The levels in ocean water are generally lower, however, than the detection limits reported. ASV is, in principle, ideally suited for measurements of very low metal concentrations in high ionic strength solutions like seawater and although not being without its problems, the method has sufficient analytical sensitivity for quantitative analysis of coastal and ocean waters for a number of elements, of which Zn, Cd, Pb and Cu have been most studied. The electrochemistry of seawater has been recently reviewed by Whitfield (1975).

Trace metals in estuarine interstitial waters may occur in concentrations well above the detection limit of suitable analytical instrumentation. This leads to more reliable data, not the least because the amount of sample handling and the need for concentration techniques are reduced. The techniques that are used for the analysis of dissolved metals in seawater are therefore also suitable for the analysis of interstitial waters. However, the preferred techniques for any one element may be different for the two compartments. A number of observations may support this and are summarized below.

(1) The relative concentrations of trace metals with respect to those of the major elements interfering with their analytical determination in the analysis of water can be quite considerably larger in interstitial waters. This may allow direct determination - without the need for further concentration if the major elements do not interfere in the analytical procedure. Where such interference does occur, dilution of the sample may reduce it to a negligible level, e.g. Fe (II) by colorimetry (Matisoff et al., 1975) and Cr, Mn and Fe by flameless atomic absorption spectrometry (Segar and Cantillo, 1975).

(2) Decaying organic matter in the sediment causes an increase in the concentration of organic molecules that are capable of complexing metals (Gamble and Schnitzer, 1973; Morel et al., 1973; Theis and Singer, 1973; Schindler and Alberts, 1974). The possibility of much interaction between metals due to the presence of organic ligands has been suggested by Morel et al. (1973). The organic matter may cause serious interference with trace metal determination in AAS and spectrometric methods. In addition, the differences in relative concentrations of trace metal species with respect to major elements such as $Ca^{2+}$ and

$Mg^{2+}$, and the expected differences in dissolved organic matter between estuarine interstitial waters and overlying water, may result in a larger organically bound fraction of particular trace elements in interstitial waters than has been suggested for seawater (Kester et al., 1975). Finally, inorganic polynuclear complexes have been suggested as playing a significant role in interstitial waters (Kester et al., 1975). The presence and relative contributions of organically bound metal fractions are a function of Eh and pH. The apparent complexing capacity of seawater is also dependent on pH (Duinker and Kramer, 1977) for inorganic metal species. Electrochemical techniques (polarography and ASV and gas-liquid chromatography (Jensen and Jernelöv, 1969; Moshier and Sievers, 1965) also in combination with a mass-spectrometer (GC-MS - Skinner and Schnitzer, 1975) seem to be promising tools in tackling the problems associated with the organic matter-trace metal associations.

(3) The actual oxidation state of an element may be strongly dependent on Eh and pH, a critical factor in selecting the preferred analytical procedure, including sample treatment. Alternatively, different oxidation states of a particular element being present may be analytically detected, Fe(II) by polarography and not Fe(III), distinction of Cu(I) and Cu(II) (Stiff, 1971), Cr(III) - Cr(VI) (Fukai, 1969; Elderfeld, 1970). "Free" ions in solution may be analyzed by potentiometry: $Cu^{2+}$, $Cd^{2+}$, $Pb^{2+}$, $Ag^{+}$.

(4) Within the sediment, transformations between inorganic and organic species have been found to occur for Hg, Se, Te, Sn and As. Also, transmethylation reactions between heavy metal ions have been observed (Jewett et al., 1975). Methylated species can be analyzed with the aid of gas-liquid chromatrography (Jensen and Jernelöv, 1969; Moshier and Sievers, 1965) and possibly be electroanalytical techniques (Heaton and Laitinen, 1974).

(5) Despite the presence of $H_2S$ in reducing sediments, concentrations of dissolved metal species may be larger than the equilibrium concentration of solid sulphides. It has been suggested that organic complexes and polysulphide species may account for this observation (Lindberg and Harris, 1974). Also, some sulphides (Cu, Ni, Fe and Hg) may be solubilized by aerobial bacterial activity (Silverman and Ehrlichm 1964; Duncan and Trussell, 1964; Fagerström and Jernelöv, 1971).

References


Alexander, J.E.; Corcoran, E.F. 1967. The distribution of copper in tropical sea water. -- Limnol. Oceanogr. vol. 12, p. 236-42.

Allen, H.E.; Matson, W.R.; Mancy, K.H. 1970. Trace metal characterization in aquatic environments by anodic stripping voltammetry. -- J. Water-Poll. Control Fed. vol. 42, p. 573-81.

Andren, A.W.; Harriss, R.C. 1975. Observations on the association between mercury and organic matter dissolved in natural waters. -- Geochim. Cosmochim. Acta, vol. 39, p. 1253-7.

Anon, 1974. Interlab. Pb analyses of standardized samples of sea water. -- Mar. Chem. p. 69-84.

Anon, 1975. Methods for detection, measurement, and monitoring of water pollution.-- FAO Fish. Techn. Pap. N° 137.

Ariel, M.; Eisner, U. 1962. Trace analysis by anodic stripping voltammetry. I. Trace metals in Dead Sea Brine 1. Zn + Cd. -- J. Electro-anal. Chem., vol. 5, p. 362-74.

Ariel, M.; Eisner, U.; Gottesfeld, S. 1964. Trace analysis by anodic stripping voltammetry. II. The method of medium exchange. -- J. Electro-anal. Chem., vol. 7, p. 307-14.

Baric, A.; Branica, M. 1967. Polarography of sea water. I. Ionic state of Cd and Zn in sea water. -- J. Pol. Soc., vol. 13 (1), p. 4-8.

Bradford, W.L. 1973. The determination of a stability constant for the aqueous complex $Zn(OH)_2^0$ using anodic stripping voltammetry. -- Limnol. Oceanogr., vol. 18, p. 757-62.

Branica, M. 1970. Determination of zinc in the marine environment. -- Vienna, IAEA report, N° 118, p. 243.

Branica, M.; Petek, M.; Baric, A.; Jeftic, L. 1969. Polarographic characterization of some trace elements in seawater. -- Rapp. Comm. Int. Mer. Medit., vol. 19, p. 929-933.

Brewer, P.G.; Spencer, D.W. 1972. Trace element intercalibration study. -- Woods Hole unpubl. rep.

Burton, J.D. 1976. Basic properties and processes in estuarine chemistry. -- In : (ed.) J.D. Burton; P.S. Liss, Estuarine chemistry, London, Acad. Press, p. 1-36.

Duinker, J.C.; Elskens, I.; Jones, P.G.W. 1975. An intercalibration exercise for analysing dissolved trace metals in sea water by marine laboratories.-- Int. Council Expl. Sea E27, 15 p.

Duinker, J.C.; Kramer, C.J.M. An experimental study on the speciation of dissolved Zn, Cd, Pb and Cu in river Rhine and North Sea water by differential pulsed anodic stripping voltammetry. -- Mar. Chem. (in press).

Duncan, D.W.; Trussell, P.C. 1964. -- Can. Metal. Qtrly., vol. 3, p. 43-55.

Duursma, E.K.; Sevenhuysen, W. 1966. Note on chelation and solubility of metals in sea water at different pH values. -- Neth. J. Sea. Res., vol. 3 (1), p. 95-106.

Dyrssen, D.; Wedborg, M. 1974. Equilibrium calculations of the speciation of elements in sea water. -- In : E.D. Goldberg, (ed.), The Sea; Ideas and observations on progress in the study of the sea, vol. 5, p. 181-95, New York, Wiley Intersci.

Ekedahl, G. 1975. Introduction to analyses of heavy metals. -- FAO Fish. Techn. Pap., N° 137, p. 47-54.

Elderfeld, M. 1970. Chromium speciation in sea water. -- Earth Planet Sci. Letters, vol. 9, p. 10.

Fagerström, T.; Jernelöv, A. 1971. -- Water Res., vol. 5 (3), p. 121-2.

Fukai, R. 1969. -- J. Oceanogr. Soc. Japan, vol. 25, p. 47.

Fukai, R. 1974. A contribution to the stability aspect of metalorganic complexes in sea water. -- Tokyo, Proc. Symp. Biogeochem. p. 562-70.

Gamble, D.S.; Schnitzer, M. 1973. The chemistry of fulvic acid and its reactions with metal ions. -- In : (ed.) P.C. Singer; Ann Arbor Sci. Publ., Trace metals and metalorganic interactions in natural water, p. 265-302.

Garrels, R.M.; Thompson, M.E. 1962. -- Am. J. Sci., vol. 206, p. 57.

Goldberg, E.D. 1965. Minor elements in sea water. -- In : (eds.) J.P. Riley; G. Skirrow, Chemical Oceanography, vol. 1, New York, Academic Press.

Heaton, R.C.; Laitinen, H.A. 1974. Electroanalytical studies of methylmercury species in aqueous solution. -- Analyt. Chem., vol. 46, p. 547-53.

Issaq, H.J.; Zielinski, W.L. 1974. Loss of Pb from aqueous solutions during storage. -- Analyt. Chem., vol 46 N° 9, p. 1326-7.

Jasinski, R.; Trachtenberg, I.; Andrychuk, D. 1974. Potentiometric measurement of copper in sea water with ion selective electrodes. Analyt. Chem., vol. 46, P. 364.

Jensen, S.; Jernelöv, A. 1969. -- Nature, vol. 223, p. 753.

Jewett, K.L.; Brinckman, F.E.; Bellama, J.M. 1975. Chemical factors influencing methyl alkylation in water. In : (ed.) T.M. Church, Marine chemistry in the coastal environment, Am. Chem. Soc., p. 304-18.

Kester, D.; Byrne, R.H.; Yu-Jean Liang. 1975. Redox reactions and solution complexes of iron in marine systems. -- In : (ed.) T.M. Church, Marine chemistry in the coastal environment. Am. Chem. Soc., p. 56-80.

Laevastu, T.; Thompson, T.G. 1958. Soluble iron in coastal waters. J. Mar. Res., vol. 16, p. 192.

Lerman, A.; Childs, C.W. 1973. In: (ed.) P.C. Singer. Trace metals and metal organic interactions in natural waters, Ann Arbor Sci. Publ., p. 201-35.

Lindberg, S.E.; Harris, R.C. 1974. -- Environ. Sci. Technol., vol. 8, p. 459-62.

Lindgren, O. 1975. Heavy metal analysis by atomic absorption spectroscopy (AAS) and flame emission spectroscopy (FES). FAO Fish. Techn. Paper, N° 137, p. 55-61.

Maljkovic, D.; Branica, M. 1971. Polarography of sea water. II. Complex formation of cadmium with EDTA. -- Rap. Comm. Int. Mer. Medit., vol. 16, N° 5, p. 779-85.

Mancy, K.H. 1972. Analytical problems in water pollution control. -- NBS Spec. Publ., N° 351.

Matisoff, G.; Bricker, O.P.; Holdren, G.R., Jr.; Kaerk, P. 1975. Spatial and temporal variations in the interstitial water chemistry of Chesapeake Bay sediment. -- In : (ed.) T.M. Church, Marine Chemistry in coastal environment. ACS Symposium series, N° 18, p. 343-63.

Millero, F.J. 1975. The physical chemistry of estuaries. -- In T.M. Church, (ed.), Marine chemistry in the coastal environment, ACS Symposium series, N° 18, p. 25-55.

Minear, R.A.; Murray, B.B. 1973. Methods of trace metal analysis in aquatic systems. -- In P.C. Singer, (ed.) Trace metals and metal-organic interactions in natural water, Ann Arbor Sci. Publ. p. 1-42.

Morel, F.; McDuff, R.E.; Morgan, J.J. 1973. Interactions and chemostasis in aquatic chemical systems: role of pH, solubility and complexation. -- In : P.C. Singer (ed.), Trace metals and metal-organic interactions in natural water, Ann Arbor Sci. Publ. p. 157-201.

Moshier, R.W.; Sievers, E.R. 1965. Gaschromatography of metal chelates. (Int. series of monographs in analytical chemistry, vol. 23), p. 163, Oxford, Pergamon Press.

Neeb, R. 1969. Inverse Polarographie and Voltammetrie. -- Weinheim: Verlag Chemie 1969.

Odier, M.; Plichon, V. 1971. Le cuivre en solution dans l'eau de mer: forme chimique et dosage. -- Analytica Chim. Acta, vol. 55, p. 209-20.

Riley, J.P.. 1965. Analytical chemistry of sea water. -- In : J.P. Riley, G. Skirrow, (eds.), Chemical oceanography, vol. 2, London, Academic Press, p. 295-424.

Robertson, D.E. 1968. Role of contamination in trace element analysis of sea water. Anal. Chem., vol. 40, p. 1067.

Rona, E.; Hood, D.W.; Muse, L.; Buglio, B. 1962. Activation analysis of Mn and Zn in sea water. -- Limnol. Oceanogr., vol. 7, p. 201-6.

Röndell, B. 1975. Heavy metal analysis in water by colorimetric methods. -- FAO Fish. Techn. Paper, N° 137, p. 61-8.

Schindler, J.E., Alberts, J.J. 1974. Analysis of organic-inorganic associations of four Georgia reservoirs. -- Arch. Hydrobiol., vol. 74, p. 429-40.

Segar, D.A., Cantillo, A.Y. 1975. Direct determination of trace metals in sea water by flameless atomic absorption spectrophotometry. -- In : T.R.P. Gibb Jr. (ed.): Analytical methods in oceanography, chapter 7, p. 56-81. Advances in chemistry series, N° 147, Am. Chem. Soc.

Sillen, L.G. 1961. The physical chemistry of sea water. -- Publ. Am. Ass. Adv. Sci., N° 67, p. 549-82.

Silverman, M.P.; Ehrlich, H.L. 1964. -- Adv. Appl. Microbiol., vol. 6, p. 153-206.

Skinner, S.I.M.; Schnitzer, M. 1975. Rapid identification by gc-ms computer of organic compounds resulting from the degradation of humic substances. -- Anal. Chem. Acta, vol. 75, p. 207-11.

Stiff, M.J. 1971. The chemical states of copper in polluted fresh water and a scheme of analysis to differentiate them. -- Water Res., vol. 5, p. 585-99.

Struempler, A.W. 1975. Trace element composition in atmospheric particulates during 1973 and the summer of 1974 at Chadron, Neb. -- Env. Sci. Techn. vol. 9 (13), p. 1164-1168.

Theis, T.L.; Singer, P.C. 1973. The stabilization of ferrous ion by organic compounds in natural waters. -- In : P.C. Singer (ed.), Trace metals and metal-organic interactions in natural water, Ann Arbor Sci. Publ., p. 305-20.

Tölg, G., 1972. Extreme trace analysis of the elements. I. Methods and problems of sample treatment, separation and enrichment. -- Talanta, vol. 19, p. 1489-521.

Whitfield, M. 1975. The electroanalytical chemistry of sea of sea water. -- In : J.P. Riley and G. Skirrow, (ed.): Chemical Oceanography (2nd ed.), London, Academic Press, p. 1-154.

Whitnack, G.C. 1961. Applications of cathode-ray polarography in the field of oceanography. -- J. Electroanal. Chem., vol. 2, p. 110.

Whitnack, G.C. 1964. Application of single sweep polarography to the analysis of trace elements in sea water. -- Proc. of the 3rd Int. Conf., Southamption, G.J. Hills, (ed.), London, McMillan, p. 110.

Zirino, A.; Healy, M.L. 1972. pH controlled differential voltammetry of certain trace transition elements in natural waters. -- Env. Sci. Technol., vol. 6, p. 243-9.

Zirino, A.; Yamamoto, S. 1972. A pH dependent model for the chemical speciation of Cu, Zn, Cd and Pb in sea water. -- Limnol. Oceanogr., vol. 17, p. 661-71.

# Heavy metal contamination in estuarine and coastal sediments: sources, chemical association and diagenetic effects

Ulrich Förstner, German Müller and Peter Stoffers*

By virtue of their composition, sediments conserve heavy metal contamination and "express the state of a water body" (Züllig, 1956). Vertical sediment profiles (cores) often uniquely preserve the historical sequence of pollution intensities; lateral distributions (quality profiles) are used to determine and evaluate local sources of pollution (Förstner and Müller, 1974). This review will further deal with the different forms of sediment associations of heavy metals as well as with the methods used in their determination.

## I. Sources and distribution of heavy metals in coastal sediments.

The contamination of coastal regions, including estuaries and marginal seas, can be attributed to the following sources (Weichart, 1972): (1) direct input of effluents from industries and communities situated near the coasts; (2) dumping of waste from ships and effluents from communities (and, more seldom, from industries) via sewer outfalls; (3) contamination as a result of atmospheric fallout; (4) input of soluble and suspended loads from rivers; (5) waste derived from the extraction of materials from the sea; and (6) pollution by the shipping industry. High metal concentrations in sediments can come from all these sources; the most common pollution factors are, however, those listed from (1) to (4).

### 1. Direct input of heavy metals from effluents

The most tragic example of industrial metal pollution is that at Minamata, Japan, where 46 people died and many others still suffer from permanent damage to their health due to mercury discharges into the sea (Tokuomi, 1969). According to Kitamura's (1968) investigations, more than 2000 ppm mercury were detected in the sediment at the channel inflow to Minamata Bay; these values decreased to around 130 ppm a few hundred metres from the source of input. The sediments of the central part of the bay contained mercury concentrations between 40 and 60 ppm. In the estuary leading to the open sea, sediment samples taken 1 - 1.5 km from the discharge point contained 22 ppm and 12 ppm, respectively. In the coastal zone adjacent to Minamata Bay, 0.4 to 3.4 ppm Hg were measured (Takeuchi, 1972).

Severe mercury pollution in coastal waters has also been reported in Europe. A chemical factory in Ravenna, near the Adriatic Sea, produced acetaldehyde applying the same process as was used in Minamata (Ui and Kitamura, 1971). Heavy mercury en-

* Institut für Sedimentforschung der Universität Im Neuenheimer Feld 236, D-69 Heidelberg, Federal Republic of West Germany.

richment was observed in fish from the Kragerö Fjord in Norway (Underdal, 1971) and in sediments from the coastal area of the Bothnian Bay (northern Baltic Sea) (Jernelöv et al., 1975; Olsson, 1974). In both the latter examples, waste from the paper industry is regarded as the source of the pollution.

Unregulated effluents from communities into harbour areas, estuaries and marine coastal regions are most widespread. Applequist et al. (1972) were able to trace a number of sewage inflows in New Haven Harbour, Conn., from the distribution of mercury in sediments. Similar tests on bottom sediments were performed for various metals in Tokyo Bay (Goto, 1973; Goldberg, 1976; Matsumoto and Yokota, 1976), in lagoons and bays of Hokkaido (Japan) (Saroma, 1972), in Kaoshung Harbour (Taiwan) (Hung et al., 1975), in Port Phillip Bay, Victoria (Talbot et al., 1976) and Halifax Bay, Queensland (Australia) (Knauer, 1976). Metal studies from the western coast of North America include sediment analyses from Puget Sound (Washington) (Schell and Nevissi, 1975) and from the Southern California Bight (Young et al., 1973; Bruland et al., 1974). A large number of metal investigations have been undertaken on the eastern coast of North America (from north to south): in bays in Newfoundland (Cranston, 1974; Slatt, 1974; Willey, 1976), Great Bay (N.H.) (Capuzzo and Anderson, 1973; Armstrong et al., 1976), East Haven (Conn.) (Siccama and Porter, 1972), Delaware Bay (Bopp and Biggs, 1973), Chesapeake Bay (Brinckman et al., 1975; Carpenter et al., 1975) with Baltimore Harbour (Villa and Johnson, 1974; Helz, 1976), in the Florida Everglades (Horvath et al., 1972; Segar and Pellenbarg, 1973), and in Mobile Bay (Lindberg et al., 1975) and Corpus Christi (Holmes et al., 1974) on the Gulf of Mexico.

The case of severe metal pollution in a marine coastal environment was investigated by Stoffers et al. (submitted) in New Bedford Habour (Mass.). In Fig. 1 is summarized the metal data from four selected cores, taken at distances of a few hundred metres (N° 9), 2 km (N° 8), 6 km (N° 4), and 12 km (N° 19) from the presumed industiral waste effluents. Marked increases in the concentrations of heavy metals occur at burial depths of 100 cm in core 9, 50 cm in core 4, and at about 20 cm in core 19. The maximum amount of metal enrichment in the bottom sediments of New Bedford Harbour is 150 times for Cu, 100 for Cd, 30 for Pb and Cr, and 10 times for Zn.

Other areas of intensive sediment studies are the bays and harbours of Great Britain: deposits from Liverpool Bay (Abdullah and Royle, 1974a), Conway Bay (Elderfield et al., 1971), Swansea Bay (Bloxam et al., 1972) and Poole Harbour (Boyden, 1975) are enriched in copper, lead, cadmium and zinc from municipal and industrial sources.

Very little published data is available on the metal levels in recent sediments in the Mediterranean which, according to newspaper reports, is particularly endangered by pollution (Proc, 1974). Investigations were undertaken on the mercury enrichments in the surface sediments of the Golfe de Lion (Arnoux et al., 1975) and the metal pollution in the Lake of Tunis (Harbridge et al., 1976). Studies made on sediment samples from the Adriatic Sea have shown contradictory results with respect to anthropogenic pollution (Paul and Meishner, 1976; Stirn et al., 1974).

When dealing with the problem of direct inputs of pollutants into the sea, mention must be made of the disposal of tailings from mines and smelters. One of the main places where this is done at present is on the British Columbian coasts (Ellis, 1975; Littlepage, 1975; Thompson and McComas, 1974), but high metal enrichment from mine waste has also been observed in the coastal waters off south-western England (Bryan and Hummerstone, 1971; Thornton et al., 1975; Yim, 1976), as well as in the Philippines (Lesaca, 1975), New Guinea (Brown, 1974) and Tasmania (Thrower and Eustace, 1973).

Particular problems may arise when mine and other metalliferous wastes are released into fjords, where exchange with open ocean waters is often limited: the Sagueney Fjord on the St. Lawrence Estuary, is injected with mercury from industrial effluents (Loring, 1975). Compared to a century ago, the mercury level in the Skagerrak has increased one hundredfold (Olausson, 1975). The Sörfjord in western Norway receives daily 45 tons of lead and 60 tons of zinc from a smelting plant and levels of 7% Pb and 12% Zn (by weight) are attained in some parts of the fjord (Skei et al., 1972). Even in western Greenland, deposits of mining wastes contribute considerably to metal enrichment in fjord sediments (Bondam, unpublished).

Diagenetic processes in coastal basins and fjords (some of which only became anoxic in recent times), can lead to further metal enrichment in surface sediments (Price, 1973; Presley, 1972), as well as occasionally in water, such as in bottom waters of the Agfardlikavsa Fjord in Greenland where approximately 1 mg/l of lead was detected (Thompson, 1975). These processes, however, seem to be controlled to a somewhat greater extent by bacterial activity and subsequent complex formation than by ion migration (Cline, 1973), as the calculated stability for the sulphide compounds is strongly exceeded by the concentration of Cu, Cd, Zn, Pb and other metals in the pore-water (Brooks et al., 1968; Hallberg, 1974; Elderfield and Hopworth, 1975).

## 2. Dumping of waste from ships and from sewer outfalls

Metal analyses from bottom sediments of an area in the Firth of Clyde (Scotland) (MacKay et al., 1972; Halcrow et al., 1973), where sewage sludge from the city of Glasgow is dumped into the sea, confirmed the findings of Applequist et al. (1972), that the contents of mercury may serve as an indicator of sewage discharges: based on a background value of approximately 0.1 ppm in the less polluted parts of the Firth of Clyde, an increase in the mercury concentrations by a factor of 20 or more could be detected in the centre of the disposal area (Fig. 2A). Similar investigations on sewage sludge disposals and their effects on the composition of marine sediments have been carried out in the Liverpool Bay (Wood et al., 1973/1974), Swansea Bay (Clifton and Vivian, 1975), Tor Bay (Taylor, 1974), Thames Estuary (Shelton, 1971), in the southern California Bight (Young, 1973), in the Back River Estuary Maryland (Helz et al., 1975) and Delaware Bay (Szucs and Oostdam, 1975). From several studies in the New York Bight (Pearce, 1970; Gross, 1972; Carmody et al., 1973; Ali et al., 1975), it was shown that in an area of approximately 60 $km^2$ of sediment, copper, chromium, lead and zinc concentrations were ten times above normal. General aspects of the metal content of sediments with respect to dumping and discharge controls are discussed by Andrulewica and Portmann (Andrulewic and Portmann). In the southern Baltic and on the east coast of the United States, several million tons of acidic iron waste is dumped from special barges annually ( Weichart, 1972; Proc, 1974; Peschiera and Freiherr, 1968); the question of dumping alkaline "red mud" from bauxite extraction is still unresolved owing to the harmful effects it could have on various filter and suspensions feeders (Rosenthal and Stelzer, 1973).

Further evidence for "descrete association of mercury with sewage sludge" (Halcrow et al., 1973) can be found in sediment analyses near sewer outfalls. An example from the Californian coast which was studied by Klein and Goldberg is shown in Fig. 2B (Klein and Goldberg, 1970); near the outfall, the mercury content of the sediment is as much as fifty times higher than the presumably uncontaminated sediments further away. A similar study in the upper Saronikos Gulf (Greece) revealed an 8 to 200-fold increase of the concentrations of Sb, As, Cr, Hg, Ag and Zn in the bottom sediments in the neighbourhood of the Athens sewage outfall (Papakostidis et al., 1975).

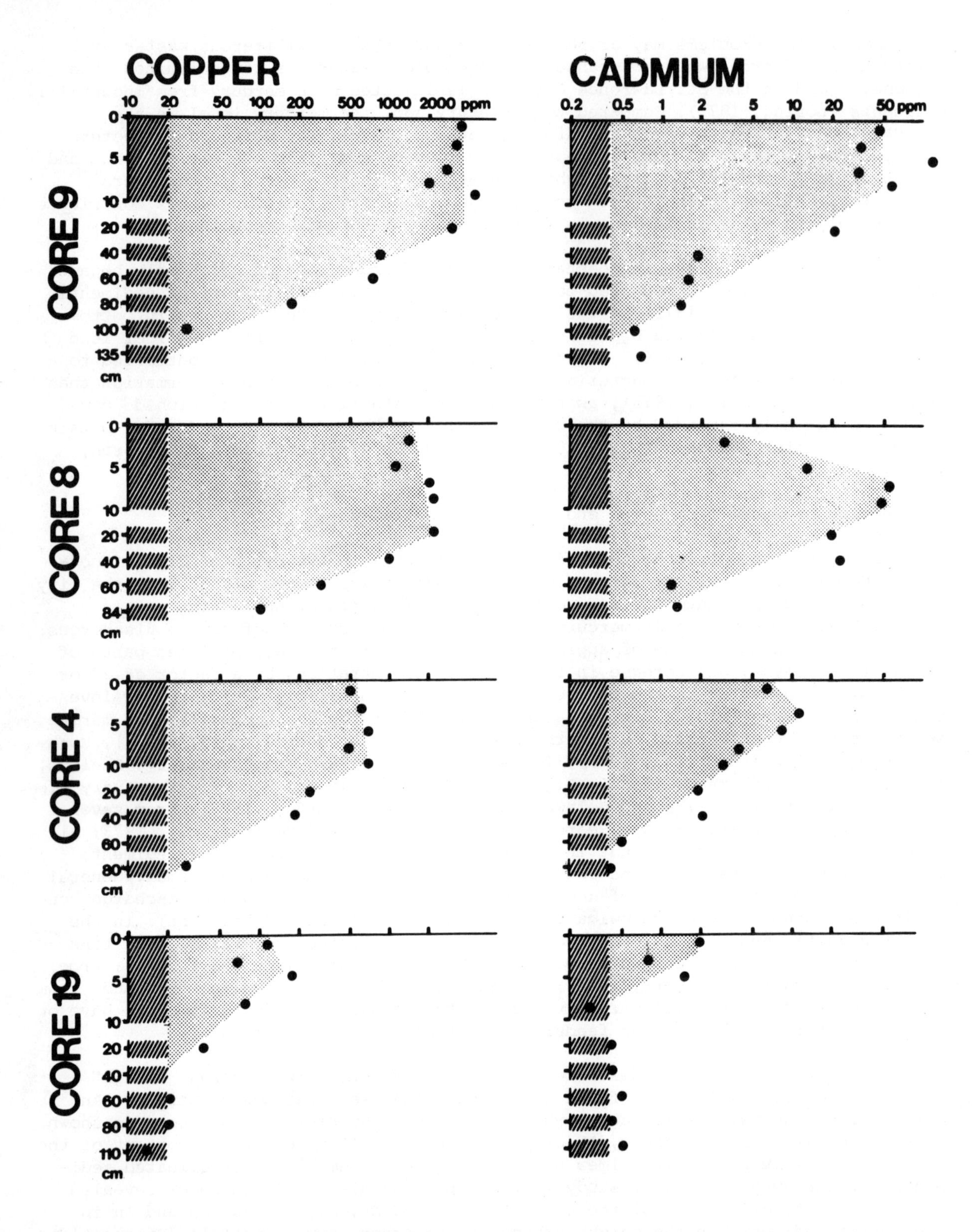


Fig. 1 Heavy-metal concentrations in sediment cores from New Bedford Harbour (Mass.) Data from Summerhayes et al. (in preparation).

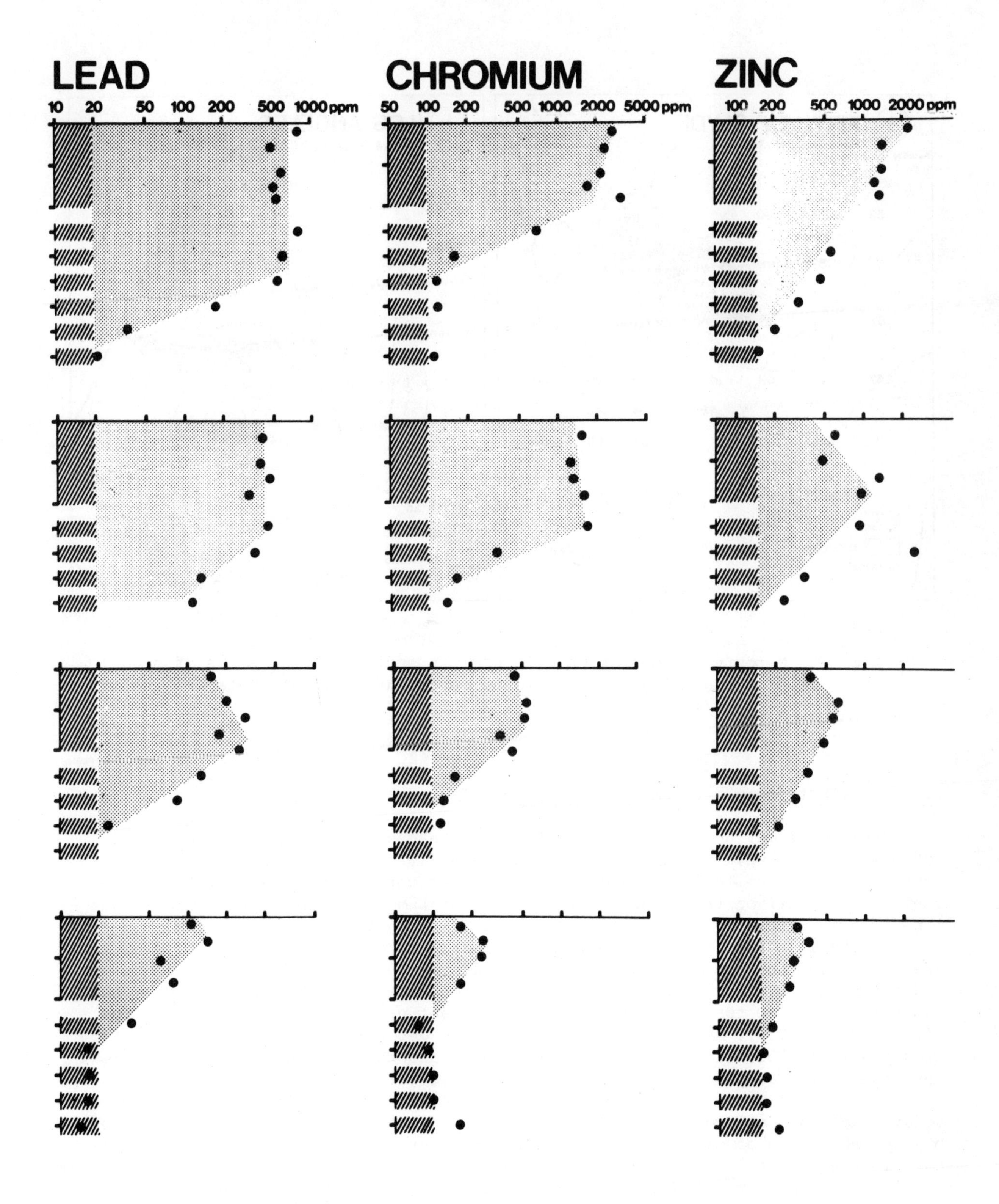
LEAD
10 20 50 100 200 500 1000 ppm
CHROMIUM
50 100 200 500 1000 2000 5000 ppm
ZINC
100 200 500 1000 2000 ppm

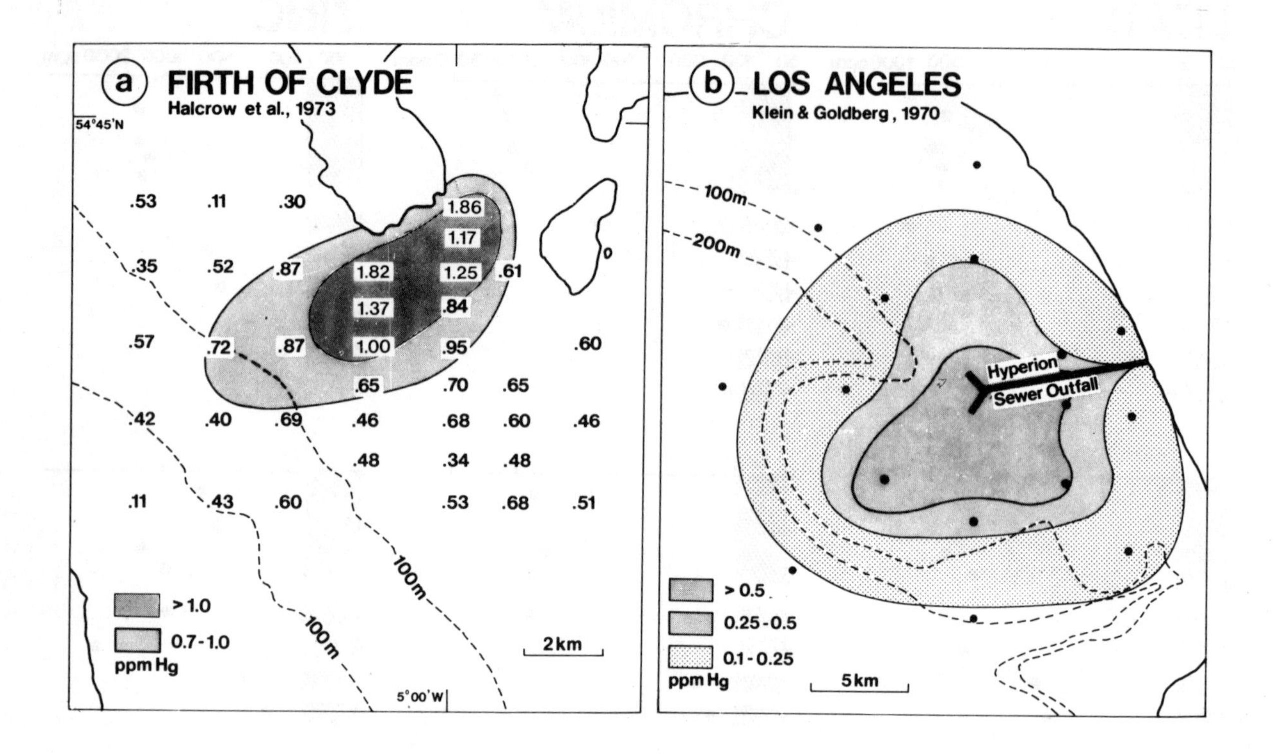


Fig. 2. Mercury concentrations (in ppm) in sediments from areas of sewage sludge dumping (A) and around sewer outfalls (B). Data from Halcrow et al. (1973) and Klein and Goldberg (1970).

## 3. Contamination as a result of atmospheric fallout

Low-level marine atmospheric particulates are strongly enriched in lead (Chester and Stoner, 1973). Chow and co-workers (1973), showed that, due to their isotope ratios, the lead accumulation in sediment cores from Baja California province can be derived from lead additives such as those present in the type of gasoline sold in Southern California; it is estimated that, on a daily basis, 0.5 tons of lead are injected into the zone off California by atmospheric fallout, 0.55 tons are discharged through municipal waste outfalls and 0.24 tons from the average storm and river runoff. The total world lead pollution is estimated to be 430,000 tons annually - 250,000 tons of which stem from tetraethyl lead and 180,000 tons from rivers (Chow, 1973). About 1000 tons of lead are deposited into the North Sea from atmospheric fallout (for comparison - the Rhine discharges about 2000 tons of lead into the North Sea per annum - Weichart, 1972). Studies by Erlenkeuser et al (1974) in the Baltic Sea show that a number of other metals released by the burning of fossil fuels are enriched in marine sediments via the atmosphere. This could possibly be the source of zinc enrichments which Hallberg (1974b) observed in a sediment core in the Baltic Sea. Goldberg (1976) has shown that after fossil fuel buring, lower but similar types of emissions are a consequ-nce of cement production.

## 4. Metal from rivers

A certain amount of the metal from atmospheric fallout reaches the sea via soil erosion and municipal and industrial effluents into rivers. In estuarine areas, a characteristic reduction of metal levels in particulates and sediments can generally be observed seawards, particularly among those metal species which are strongly enriched by anthropogenic influences. Müller and Förstner (1975 - see Fig. 3) concluded that mixing of heavily polluted river sediments with relatively "clean" marine sediments is the main cause of the dilution of heavy metals in the Elbe and Rhine estuaries - rather than the mechanisms of adsorption or solubilization by complexation as was assumed by De Groot and co-workers (1973) for the Rhine estuary.

Heavy metals can be transported over great distances from the mainland into the sea via the sediment load of the rivers. Förstner and Reineck (1974) found enrichments of zinc, cadmium, and mercury in sediment cores off Helgoland (North Sea) which were 7 - 10 times those of the precultural level of that area, probably caused by the metal load carried by the River Elbe. In general, sediments from the southern North Sea clearly showed metal enrichment from the heavily polluted rivers of central Europe (Müller and Förstner, 1976; Degens and Valeton, 1975; Nauke, 1975; Gadow and Schafer, 1974; De Groot *et al.*, 1971; Wollast *et al.*, 1973).

In Great Britain, considerable metal pollution was observed in the Mersey estuary (Craig and Morton, 1976), Bristol Channel and Severn estuary (Butterworth *et al.*, 1972; Abdullah and Royle, 1974b; Chester and Stoner, 1975), and Thames estuaries (Smith *et al.*, 1971). In contrast, the sediments from the Solway Firth (Perkins *et al.*, 1973), the Solent (Burton and Leatherland, 1971), the Humber estuary (Jaffe and Walters, 1975) and the Firth of Forth (Covill, 1972) are relatively clean. Examples of metal studies in estuarine sediments from areas in North America are those done at the mouth of the Columbia River (Cutshall *et al.*, 1973) and the Frazer River (Hall and Fletcher, 1974); on the east coast, the estuaries of the St. Lawrence River (Bewers *et al.*, 1974) were studied as well as the La Have River (Nova Scotia), various tributaries of the Chesapeake Bay (Potomac River) (Cranston and Buckley, 1972), Rappahannock, York and James River (Huggett *et al.*, 1972), Newport estuary (Cross *et al.*, 1971; Wolfe *et al.*, 1973), the Pamlico River estuary (Harding and Brown, 1975), the southeastern Atlantic salt marsh estuaries (Windom, 1975) and the estuaries of the Gulf of Mexico (Andren, 1973). The high enrichment rates, which were observed in examples of direct waste input or dumping from ships or via sewer outfalls, were not reached. In the past two decades, there has been a general trend towards increased levels of heavy metals in river sediments, particularly of the most toxic heavy metals - mercury and cadmium; consequently, higher amounts have also been found in estuarine and coastal sediments. De Groot *et al.* (1973) observed a 29% increase in the mercury in sediments from the mouth of the River Rhine over the decade from 1960 to 1970; for cadmium the increase was as high as 94%. Müller and Förstner (1976), measured a tripling of the mercury content of clay sized sediments in the Elbe estuary over a three-year period (Jan. 1973 - Oct. 1975). This correlates with the heavy mercury contamination in fish caught in the Elbe River estuary during the 1974-1975 period (Krüger *et al.*, 1975).

## II. Associations of Metals in Estuarine Sediments

The study of the associations of chemical constituents in natural waters has its major goal in the evaluation of availability of toxic substances to metabolic processes under certain environmental conditions. Our compilation in this section indicates some of the aspects of the bonding and release of trace metals in aquatic sediments which may be particularly relevant with respect tot the estuarine environment.

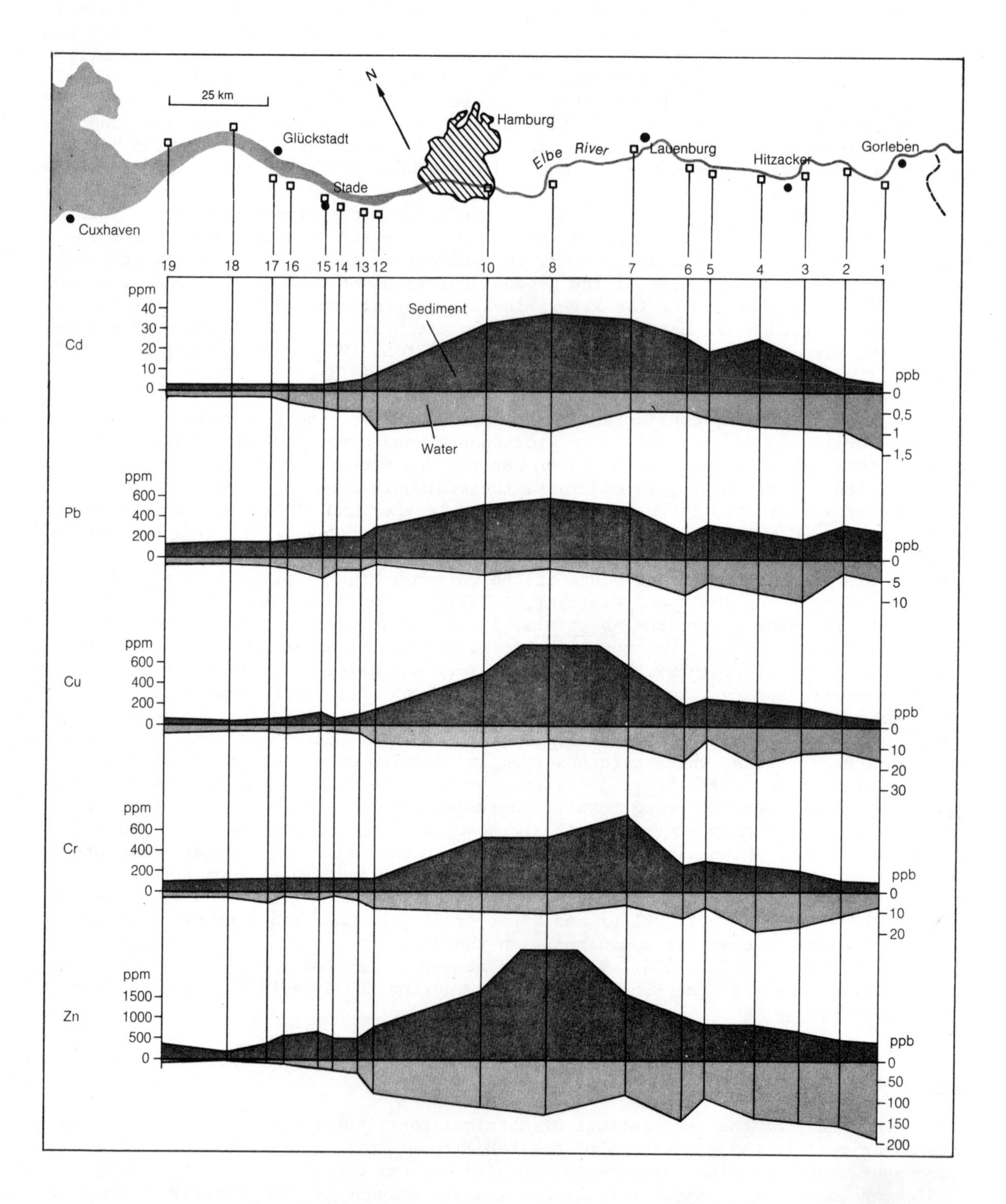


Fig. 3 Heavy-metal concentrations in the clay-sized Fraction of Elbe sediments and waters in N. Germany (Müller and Förstner, 1975 reprinted from "Environmental Geology", Springer-Verlag, New York).

## 1. Metal transport in rivers

The source of trace metals significantly determines their distribution ratio between the aqueous and solid phases. For example, the bulk of the detrital trace element particulates never leaves the solid phase from initial weathering to ultimate deposition. Similarly, metal dust particles (e.g. from smelters) and effluents containing heavy metals associated with inorganic and organic matter undergo little or no change after being discharged into riverwater (Wittmann and Förstner, 1976).

There are characteristic differences in the metal distribution of the aqueous and solid phases. Examples from U.S. rivers (Kopp and Kroner, 1968), in West Germany (Heinrichs, 1975) and from the Rhine River in the Netherlands (De Groot et al., 1973) indicated the following ordered sequence for the "mobility" of selected metals (Förstner and Patchineelam, 1976a): alkali and earth-alkali metals are predominantly present in a dissolved form; metals like boron, zinc and cadmium are associated to particles 30-45% with respect to the total discharge of the respective metals; metals like copper, mercury, chromium and lead are even more fixed to the solid phase, 60% to more than 80% being transported by the particulate material; iron, aluminum and manganese (the latter only under normal $E_h$ conditions) are almost totally transported in the solid phase.

## 2. Changes of metal associations at the river/sea interface

The ratios of metal distributions in the aqueous and solid phases undergo particular changes when riverwater reaches the marine environment. It is evident that the ratio of natural metal content in sediment particles to that in water can vary over seven orders of magnitude in the oceanic environment (Turekean and Wedepohl, 1961; Brewer, 1975), whereas the values for the continental regions show greater uniformity (Arbeitsgreppe "Metalle", 1975). A group of metals, such as Na, Ca, Sr, Li, B, Sn, Sb, Se and U, all tend to be more closely associated with the aqueous phase under oceanic conditions than under the less saline conditions prevailing in rivers and lakes; on the other hand, Al, Be, Mn, Fe, W, and Pb and, to a lesser extent, Cd, Cr, Co, Cu and Zn seem to be effectively eliminated from the aqueous phase when introduced into the ocean in a dissolved state.

## 3. Chemical forms of metals in aquatic sediments

The first step in the study of trace metals in sediments is usually the correction for the grain size as there is a marked decrease in the level of metal content as sediment particle size increases. Reduction of grain size effects can be achieved (Förstner, 1977) by : (1) separation of characteristic grain fractions, e.g. the pelitic fraction of less than 2 µm or the clay-silt fraction of less than 63 µm; (2) extrapolation from grain size distribution; (3) correction for the specific surface area; (4) selective chemical treatment (see Tables 1 and 2); (5) mineral separation, e.g. quartz and feldspar contents; and (6) by comparison with conservative elements, e.g. aluminum and silicon.

In spite of their different origin, environmentally related types of metal bonding are comparable with the natural forms. Table 1 is a compilation of the most common associations of heavy metals in particulate substances (Patchineelam, 1975; Förstner and Patchineelam, 1976b): (1) Heavy metals are transported and deposited as major, minor or trace constituents in the natural rock detritus, mostly in inert forms; (2) slight alkalinity prevailing in normal surface waters results in the formation and precipitation of hydroxides, carbonates, sulphides and phosphates of heavy metals; (3) sorption and cation exchange particularly take place on fine-grain substances with large surface areas. A generalized sequence of the capacity of the solids to sorb heavy metals was established (Guy and Chakkrabarti, 1975) in the order $MNO_2$ > humic acid > hydrous iron oxides > clay minerals; (4) dissolved organics are increasingly becoming regarded as the carrier material in the transfer of metals from the mineral phase into the organic material (Saxby, 1969; Nissenbaum and Swaine, 1976). The young, less-condensed humic acids (fulvic acids) play an especially important role in the bonding of heavy metals in aquatic systems because of their large number of functional groups (Rashid, 1974); and (5) co-precipitation of heavy metals with hydroxides and carbonates appears to be a very important means of controlling the metal concentrations in the aquatic environment (Groth, 1971; Lee, 1975). Experiments on calcium carbonate co-precipitation (Popova, 1961) indicated that heavy metals, which have carbonates with low solubilities, e.g. cadmium and lead, are completely eliminated from solution as a result of $CaCO_3$ precipitation.

## 4. Selective extraction procedures

Several attempts have been made to selectively extract different metal compounds from sediments. Hirst and Nicholls (1958) investigated the partitioning of trace metals in the detrital and non-detrital phases of carbonate rocks by acetic acid treatment, using organic precipitates for spectro-chemical analysis (Scott and Mitchell, 1943; Mitchell and Scot, 1947; Heggen and Strock, 1953). Additional methods for the diffentiation of metals in sorbed, calcareous, hydroxidic, organic and detrital phases are shown in Table 2 (after Förstner, 1977).

A combination of extraction procedures was first used by Nissenabum (1972) for sediments from the Okhotsk Sea (method Nos. 5, 8 and 12 in Table 2), then by Gibbs (1973) for analyses of trace metals associated with different carrier substances in suspended material samples from the Amazon and Yukon Rivers (1b, 6, 11, 12b). Engler and co-workers (1974) developed a separation technique, particularly suitable for dredged materials, which precluded atmospheric oxidation at sensitive steps (1c, 7, 8a, 6, 12a); sediments from Mobile Bay (Alabama), Ashtabula (Lake Erie) and Bridgeport (Long Island Sound) were investigated (Brannon et al., 1976). Further modifications of that scheme were performed by Gupta and Chen (1975) (1c, 7, 8, 5, 12a) and applied to sediment samples from Los Angeles Harbour. Patchineelam (1975) differentiated the metal contents associated with cation exchange positions (1a), humic acids (10), carbonate minerals (4a) and hydrous Fe/Mn oxides (5). The first three examples compiled in Fig. 4 represent relatively unpolluted systems (e.g. Dead Sea (Nissenbaum, 1974): 14 ppm Cu, 13 ppm Pb, 36 ppm Zn), while both the latter areas are strongly contaminated with heavy metals (e.g. Los Angeles Harbour (Gupta and Chen, 1975): 568 ppm Cu, 342 ppm Pb, 632 ppm Zn). In the example of the Lower Rhine sediment (Patchineelam, 1975), two groups of metals can be distinguished from a comparison of metal data on recent Rhine deposits with the analyses of ancient Rhine sediments obtained from drilling cores in the Cologne area (Förstner and Müller, 1973): high percentages of detrital fractions are found for those elements which are less affected by man's activities, e.g. iron (3.0% in fossil Rhine deposits versus 3.4% in recent pelitic sediments of the Rhine), nickel (85 ppm cf. 165 ppm), and cobalt (14 ppm cf. 35 ppm); on the other hand, lattice held fractions are very low for metals such as lead (30 ppm cf. 333 ppm), zinc (115 ppm cf. 1558 ppm) and cadmium (0.3 ppm cf. 28.4 ppm). Lead, copper and chromium in the Rhine sediments are particularly associated with the hydroxide phase. These findings can be explained by the specific sorption of lead to iron-hydroxide (Jenne, 1968), by the effective co-precipitation of copper together with hydrous Fe/Mn oxides (Groth, 1971), and by the absence of a carbonate phase of chromium in natural aquatic systems. The preferential carbonate binding of zinc and cadmium, on the other hand, can be attributed to the relatively high stability of zinc carbonate and cadmium carbonate under the pH-$E_h$ conditions of common inland waters (Hem, 1972) as well as to the characteristic co-precipitation of these components with calcium carbonate. Copper and cadmium contents are found to be closely associated to organic substances or sulphides. The metal fractions in cation exchange positions represent less than 10% of the total metal content associated with the sediment.

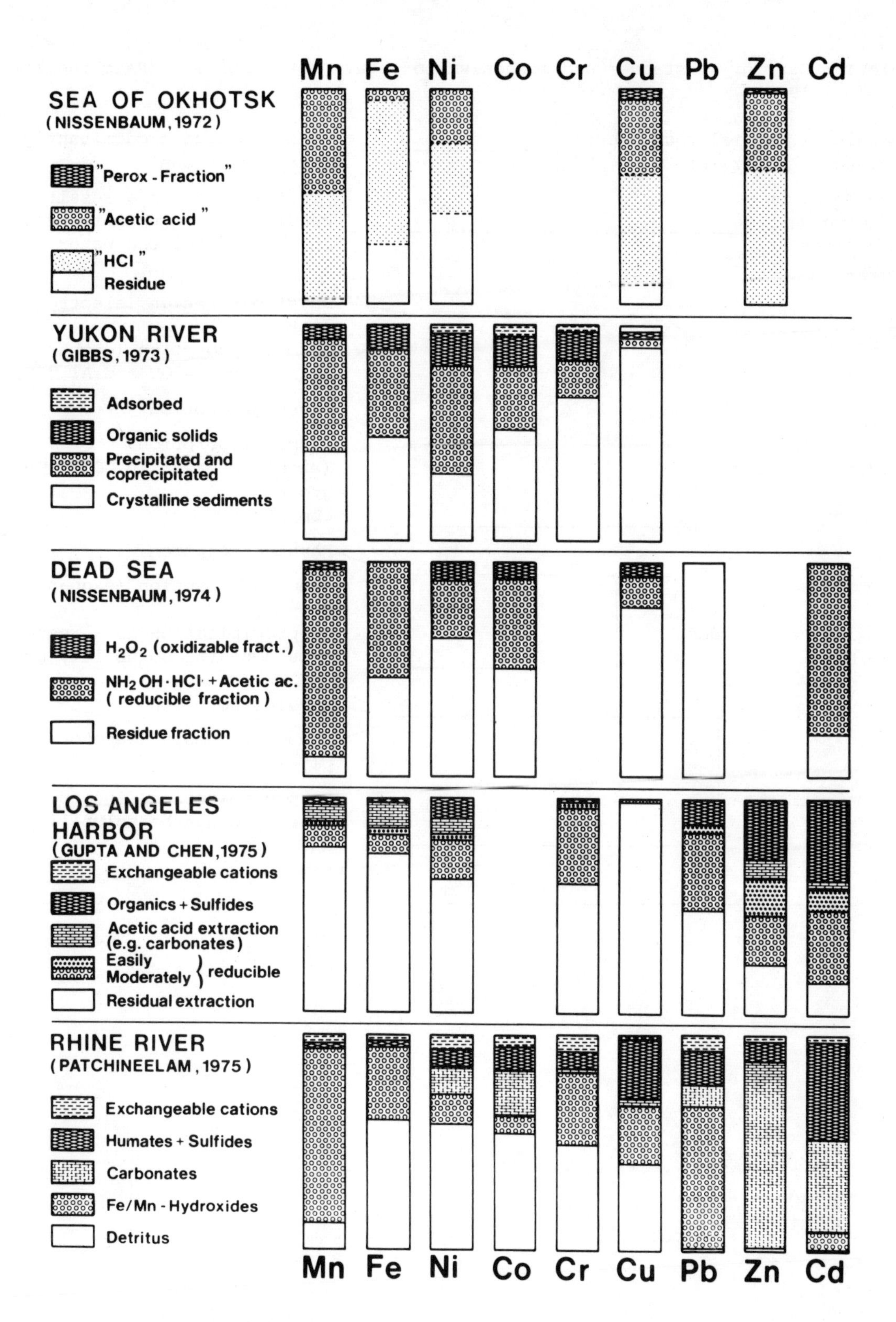


Fig. 4 Evaluation of different Forms of chemical associations in aquatic sediments. Data from Welte (1969), Weichart (1972), Rosenthal and Stelzer (1973) and Thrower and Eustace (1973).

Table 1 : Carrier substances and mechanisms of heavy metal bonding (Patchineelam, 1975).

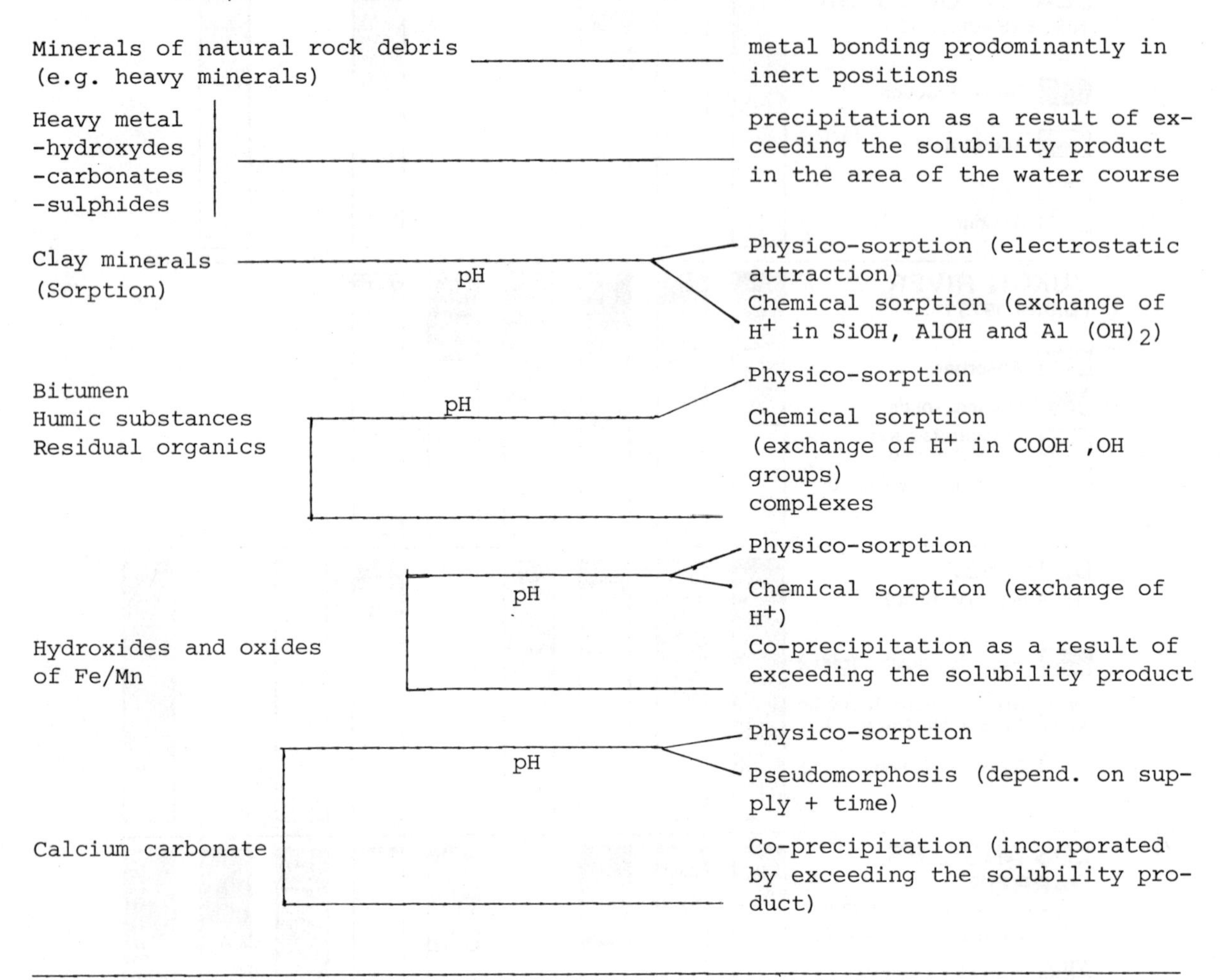

| Carrier substance | | Mechanism |
|---|---|---|
| Minerals of natural rock debris (e.g. heavy minerals) | | metal bonding prodominantly in inert positions |
| Heavy metal -hydroxydes -carbonates -sulphides | | precipitation as a result of exceeding the solubility product in the area of the water course |
| Clay minerals (Sorption) | pH | Physico-sorption (electrostatic attraction)<br>Chemical sorption (exchange of $H^+$ in SiOH, AlOH and Al $(OH)_2$) |
| Bitumen<br>Humic substances<br>Residual organics | pH | Physico-sorption<br>Chemical sorption (exchange of $H^+$ in COOH ,OH groups)<br>complexes |
| Hydroxides and oxides of Fe/Mn | pH | Physico-sorption<br>Chemical sorption (exchange of $H^+$)<br>Co-precipitation as a result of exceeding the solubility product |
| Calcium carbonate | pH | Physico-sorption<br>Pseudomorphosis (depend. on supply + time)<br>Co-precipitation (incorporated by exceeding the solubility product) |

Table 2 : Extraction procedures for metals in sediments (Förstner, 1977).

| Chemical phase | Extraction procedure | Reference |
|---|---|---|
| 1. Adsorption + cation exchange | (a) $BaCl_2$, (b) $MgCl_2$, (c) $NH_4OAc$ | Jackson, 1958 |
| 2. Detrital/authigenic phases in pelagic sediments | EDTA treatment | Goldberg & Arrhenius 1958. |
| 3. Hydrogenous/lithogenous | 0.1 M HCl (b); 0.1 M $HNO_3$; $HCl/HNO_3$ | Piper, 1971;Jones, 1973;Literathy and Laszlo, 1975. |
| 4. Carbonate fraction | (a) $CO_2$; exchange column | Lloyd, 1954. |
| 5. Reducible phases | 1M $NH_2OH \cdot HCl$; 25% v/v acetic acid | Chester and Hughes 1967. |
| 6. Moderately reducible phases (hydrous Fe-oxides) | Reduction with sodium dithionite complexing with sodium citrate | Holmgren, 1967. |
| 7. Easily reducible phase (Mn-oxide + amorph. Fe-oxide) | 0.1 M $NH_2OH \cdot HCl$; 0.01 M nitric acid | Chao, 1972 |
| 8. Organics, sulphides | 30% $H_2O_2$ at 95°C, extract with (a) 1N $NH_4OAc$ or (b) 0.01 M $HNO_3$ | Jackson, 1958; Engler <u>et al</u>., 1974; Gupta & Chen, 1975. |
| 9. Lipids, bitumen | fat solvents : e g. chloroform, ether, gasoline, benzene, carbon disulphide | Bergmann, 1963; Welte, 1969. |
| 10. Humic/fulvic acids | 0.5 N NaOH; 0.1 N $NaOH/H_2SO_4$ | Raschid, 1971; Volkov and Fomina, 1974. |
| 11. Solid organic material | Na-hypochlorite, dithionate/citrate | Gibbs, 1973. |
| 12. Detrital silicates | Digestion with $HF/HClO_4$; $HNO_3$ (a)<br>Lithium metaborate (1000°C), $HNO_3$ (b) | Smith and Windom, 1972.<br>Gibbs, 1973 |

## 5. Remobilization of metals from aquatic sediments

Heavy metals which are "immobilized" in the bottom sediments of rivers and estuaries constitute a potential hazard to water quality since they may be released as a result of chemical changes in the aquatic milieu (Förstner and Patchineelam, 1976b):

(1) Increased salinity of the water body leads to competition between dissolved cations and adsorbed heavy metal ions and results in partial replacement of the latter. Such effects should be particularly expected in the estuarine environment;

(2) A lowering of pH leads to the dissolution of carbonate and hydroxide minerals and also - as a result of hydrogen ion competition - to an increased desorption of metal cations. Long-term changes of the pH conditions have been observed from highly industrialized areas by atmospheric $SO_2$ emissions and acid mine drainage, mainly in waters poor in bicarbonate ions (Förstner and Wittmann, 1976). Since, however, the marine interstitial waters are well-buffered (Ben-Yaakov, 1973), the effects arising from acid inflows should be less detrimental than in continental waters;

(3) The growing input of organic degradation products as well as of synthetic complexing agents (e.g. nitriltriacetic acid in detergents to replace polyphosphates) increases the solubilization of heavy metals from aquatic sediments;

(4) A change in the redox conditions is usually caused by the increased input of nutrients. Oxygen deficiency in the sediments leads to an initial dissolution of hydrated manganese oxide, followed by that of iron compounds. Since these metals are readily soluble in their divalent states, any coprecipitates with metallic coatings become partially remobilized;

(5) Microbial activity enhances the release of metals in three major ways: (i) by the formation of inorganic compounds capable of complexing metal ions; (ii) by influencing the pH-$E_h$ conditions of the environment; and (iii) by converting inorganic metal compounds to organic molecules; and

(6) The latter two mechanisms (4) and (5), which seem to be particularly significant in the estuarine environment, may be enhanced by such physical effects as erosion, dredging and bioturbation.

Acknowledgements:

We are indebted to the Deutsche Forschungsgemeinschaft for supporting our research on heavy metals in the environment.

## References

Abdullah M.I.; Royle, L.G. 1974a. Symposium "Problems of the contamination of man and his environment by mercury and cadmium", Luxemburg, 1973, p. 69-81.

Abdullah, L.I.; Royle, L.G. 1974b. J. mar. biol. Ass. U.K. vol. 54, p. 581-97.

Ali, S.A. et al. 1975. Environ. Geol., vol. 1, p. 143-8.

Andren, A.W. 1973. Dissertation, Florida State Univ. 139 p.

Andrulewica; E., Portmann, J.E. ICES Papers and Reports, NO. E-22

Applequist, M.D. et al. 1972. Environm. Sci. Techn., vol. 6, p. 1123-4.

Arbeitsgruppe "Metalle" im DFG-Schwerpunktprogram "Wasserforschung-Schadstoffe im Wasser" - Interim Report, 110 p. (1975).

Armstrong, P.B. et al., 1976. Environm. Geol., vol. 1, p. 207-14.

Arnoux, L. et al., 1975. C.R. Acad. Sci. Paris (D), vol. 281, p. 743-6.

Ben-Yaakov, S. 1973. Limnol. Oceanogr., vol. 18, p. 86-94.

Bergmann, W. 1963. In I.A. Breger, Organic geochemistry, New York, Pergamon Press.

Bewers, J.M. et al. 1974. Can. J. Earth Sci., vol. 11, p. 939-50.

Bloxam, T.W. et al. 1972. Nature Phys. Sci., vol. 239, p. 158-9.

Bondam, J. (personnel communication).

Bopp, F.; Biggs, R.B. 1973. Delaware Bay Rept. Ser., vol. 3, p. 25-69.

Bothner, M.H.; Carpenter, R. 1973. IAEA - SMI58/5, p. 73-87.

Boyden, C.R. 1975. Mar. Pollut. Bull., vol. 6, p. 180-2.

Brannon, J.M. et al. 1976. U.S. Waterways Experim. Station Misc. Paper D-76-18.

Brewer, P.G. 1975. In : J.P. Riley, G. Skirrow (eds.), Chemical Oceanography, vol. 1, London, Academic Press, p. 415-96.

Brinckman, F.E. et al., 1975. In : P.A. Krenkel (ed.) Heavy metals in the aquatic environment, p. 251-2.

Brinckman, F.E.; Iverson, W.P. 1975. In : T.M. Church (ed.), Marine Chemistry in the coastal environment, Amer. Chem. Soc. Ser. 18, p. 319-37.

Brooks, R.F. et al., 1968. Geochim. Cosmochim. Acta, vol. 32, p. 397-444.

Brown, M.J.F. 1974. Geoforum, vol. 18, p. 19-27.

Bruland, K.W. et al., 1974. Environ. Sci. Techn., vol. 8, p. 425-32.

Bryan, G.W.; Hummerstone, L.G. 1971. J. mar. Biol. Ass. U.K., vol. 53, p. 721-39.

Burton, J.D.; Leatherland, T.M. 1971. Nature, vol. 231, p. 440-1.

Butterworth, J. et al., 1972. Mar. Pollut. Bull., vol. 3, p. 72-4.

Capuzzo, J.M.; Anderson, F.E. 1973. Mar. Geol., vol. 14, p. 225-35.

Carmody, D.J. et al., 1973. Mar. Pollut. Bull., vol. 4, p. 132-5.

Carpenter, J.H. et al. 1975. In : L.E. Cronin (ed.), Estuarine research, vol. 1, p. 188-214.

Chao, L.L. 1972. Soil Sci; Soc. Amer. Proc., vol. 36, p. 764-8.

Chester, R.; Hughes, M.J. 1967. Chem. Geol., vol. 2, p. 249-62.

Chester, R.; Stoner, J.H. 1973. Nature, vol. 246, p. 138-9.

Chester, R.; Stoner, J.H. 1975. Mar. Pollut. Bull., vol. 6, p. 92-6.

Chow, T.J. et al. 1973. Science, vol. 181, p. 551-2.

Chow, T.J. 1973. Chemistry in Britain, vol. 9, p. 258-63.

Clifton, A.P.; Vivian, C.M.G. 1975. Nature, vol. 253, p. 621-2.

Cline, J.T.; Upchurch, G.B. 1973. Proc. 16th Conf. Great Lakes Res., p. 349-56.

Covill, R.W. 1972. Proc. Roy. Soc. Edinburgh, vol. 71, p. 143-56.

Craig, P.G.; Morton, S.F. 1976. Nature, vol. 261, p. 125-6.

Cranston, R.E. 1974. Conf. transp. persist. chem. aqu. ecosyst., I-59.

Cranston, R.E.; Buckley, D.E. 1972. Environ. Sci. Techn., vol. 6, p. 274-8.

Cross, F.A. et al., 1971. Chesapeake Sci., vol. 12, p. 280-2.

Cutshall, N. et al. 1973. Rept. Oregon State Univ. 15 p.

Degens, E.T.; Valeton, I. 1975. Lebensraum Alster, Hamburg, Presse-stelle Univ., 71 p.

De Groot, A.J. et al. 1971. Geologie en Mijnbouw, vol. 50, p. 292-8.

De Groot, A.J. et al. 1973. Sympos. Waterloopkundein dienst van industrie en milieu. Publ. N° 110N, 27 p.

De Groot, A.J. 1973. In : E.D. Goldberg, (ed.) North Sea science, MIT Press, p. 308-25.

Dörjes, J.; Little-Gadom, S.; Schäfer, A. 1976. Senckenberg. marit., vol. 8, p. 103-9.

Elderfield, H. et al. 1971. Mar. Pollut. Bull., vol. 2, p. 44-7.

Elderfield, H.; Hepworth, A. 1975. Mar. Pollut. Bull., vol. 6, p. 85-7.

Ellis, D.V. 1975. Abstr. Conf. Heavy Metals Environm. Toronto, A-21.

Engler, R.M. et al., 1974. 168th ACS Natl. Meeting, Atlantic City, N.J.

R.M. et al. 1974. 168th ACS Natl. Meeting, Atlantic City, N.J.

Erlenkeuser, H.; Suess, E.; Willkomm, H. 1974. Geochim. Cosmochim. Acta., vol. 38, p. 823-42.

Förstner, U.; Müller, G. 1973. Geoforum, vol. 14, p. 53-61.

Förstner, U.; Müller, G. 1974. Schwermetalle in Flüssen und Seen als Ausdruck der Umweltverschmutzung. Berlin, Springer-Verlag, 225 p.

Förstner, U.; Reineck, H.-E. 1974. Senckenberg. marit., vol.6, p. 175-84.

Förstner, U.; Patchineelam, S.R. 1976a. Workshop on fluvial transport of sediment-associated nutrients and contaminants. Kitchener/Ont.

Förstner, U.; Patchineelam, S.R. 1976b. Chemikerzeitung, vol. 100, p. 49-57.

Förstner, U.; Wittmann, G.T.W. 1976. Geoforum, vol. 7, p. 41-9.

Förstner, U., SIL/UNESCO Symp. H.L. Golterman (ed.) "Interactions between Sediments and Freshwater", Amsterdam, 6-10 Sept., 1976, 1977, in press.

Gadow, S.; Schäfer, A. 1974. Senckenberg. marit., vol. 5, p. 165-78, ibid, vol. 6, p. 161-74.

Gibbs, R.J. 1973. Science, vol. 180, p. 71-3.

Goldberg, E.D. et al. 1976. Geochemical Journal, vol. 10, p. 165-74.

Goldberg, E.D. 1976. The health of the oceans, Paris, the Unesco Press, 172 p.

Goldberg, E.D.; Arrhenius, G.O.S. 1958. Geochim. Cosmoschim. Acta., vol. 13, p. 153-212.

Goto, M. 1973. Environm. Qual. Safety, vol. 2, p. 72-7.

Gross, M.G. 1972. Geol. Soc. Ameri. Bull., vol. 83, p. 2163-176.

Groth, P. 1971. Arch. Hydrobiol., vol. 68, p. 305-74.

Gupta, S.K.; Chen, K.Y. 1975. Environm. Lett., vol. 10, p. 129-58.

Guy, R.D.; Chakkrabarti, C.L. 1975. Abst. int. conf. heavy metals environment, Toronto, D-29.

Halcrow, W. et al. 1973. J. Mar. biol. Ass. U.K., vol. 53, p. 721-39.

Hall, K.J.; Fletcher, K. 1974. Proc. int. conf. transport of persistent chemicals in aquatic ecosystems, Ottawa, p. 83-8.

Hallberg, R.O. 1974a. Oikos Suppl. 15., p. 51-62: Estuar. Coast. Mar. Sci., vol. 2, p. 153-70.

Hallberg, R.O. 1974b. Merentutkimuslait Julk./Havsforkningsinst. Skr., N° 238, p. 3-16.

Harbridge, W. et al. 1976. Environ. Geol., vol. 1, p. 215-25.

Harding, S.C.; Brown, H.S. 1975. Environ. Geol., vol. 1, p. 181-91.

Heggen, G.E.; Strock, L.W. 1953. Anal. Chem., vol. 24, p. 859-63.

Heinrichs, H. 1975. Diss. Göttingen, 97 p.

Helz, G.R. 1976. Geochim. Cosmochim. Acta., vol. 40, p. 573-90.

Helz, G.R.; Huggett, R.H.; Hill, J.M. 1975. Water Res., vol. 9, p. 631-6.

Hem, J.D. 1972. Water Resour. Res., 8, p. 661-79.

Hirst, D.M.; Nichools, G.D. 1958. J. Sediment. Petrol., vol. 24, p. 461-8.

Holmes, C.W. et al. 1974. Environm. Sci. Techn., vol. 8, p. 255-9.

Holmgren, G.S. 1967. Soil Sci. Soc. Amer. Proc., vol. 31, p. 210-1.

Horvath, G.J. et al. 1972. Mar. Pollut. Bull., vol. 3, p. 182-4.

Huggett, R.J. et al. 1972. Chesapeake Sci., vol. 12, p. 280-2.

Hung, T.C.; Li, Y.H.; Wu, D.C. 1975. Abstr. int. conf. on heavy metals in the environment, Toronto, C-267.

Jackson, M.L. 1958. Soil Chemical Analysis, Englewood Cliffs, N.J., Prentice Hall Inc. 498 p.

Jaffe, D.; Walters, J.K. 1975. Abstr. conf. heavy metals in the environment, Toronto, C-270.

Jenne, E.A. 1968. Am. Chem. Soc. Adv. Ser., vol. 73, p. 337-87.

Jernelöv, A.; Landner, L.; Larsson, T. 1975. J. Water Pollut. Control Fed., vol. 47, p. 810-22.

Jones, A.S.G. 1973. Mar. Geol., vol. 14, M1-M9.

Kitamura, S. 1968. In : Minamata Disease. Shuhan and Co., p. 257-66.

Klein, D.H.; Goldberg, E.D. 1970. Environ. Sci. Techn., vol. 4, p. 765-8.

Knauer, G.A. 1976. Mar. Pollut. Bull., vol. 7, p. 112-5.

Kopp, J.F.; Kroner, R.C. 1968. Trace metals in waters of the Unites States., U.S. Department Interior, FWPCA, Cincinnati.

Krüger, U.E. et al. 1975. Arch. F. Lebenzmittelhygiene, vol. 26, p. 201-7.

Leatherland, T.M.; Burton, J.D. 1974. J. mar. biol. Ass. U.K., vol. 54, p.457-68.

Lee, J.F. 1975. In : P.A. Krenkel, (ed.) Heavy metals in the aquatic environment, p. 137-47.

Lesaca, R.M. 1975. Abstr. Conf. Heavy Metals Environ., Toronto, C-105.

Lindberg, S.E. et al. 1975. Estuarine research, vol. I, p. 64-107.

Literathy, P.; Laszlo, F. 1975. Int. symp. geochem. natural water, Burlington.

Littlepage, J. 1975. Abstr. conf. heavy metals environm., Toronto, C-159.

Lloyd, R.M. 1954. Jour. Sediment. Petrol., vol. 24, p. 218-20.

Loring, D.H. 1975. Can. J. Earth Sci., vol. 12, p. 1219-37.

MacKay, D.A. et al. 1972. Mar. Pollut. Bull., vol. 3, p. 7-11.

Matsumoto, E.; Yokota, S. 1976. Kaguku, vol. 46, p. 182-4.

Mitchell, R.L.; Scott, R.O. 1947. Soc. Chem. Ind. Jour., vol. 66, p. 330-6.

Müller, G.; Förstner, U. 1975. Environ. Geol., vol. 1, p. 33-9.

Müller, G.; Förstner, U. 1976. Naturwissenschaften, vol. 6, p. 175-84.

Nauke, M. 1975. In : H. Caspers, (ed.), Pollution of coastal waters, H. Boldt Verlag, Boppard, p. 25-30.

Nissenbaum, A. 1972. Israel J. Earth Sci., vol. 21, p. 143-54.

Nissenbaum, A. 1974. Israel J. Earth Sci., vol. 23, p. 111-6.

Nissenbaum, A.; Swaine, D.J. 1976. Geochim. Cosmochim. Acta., vol. 40, p.809-916.

Olausson, E. 1975. Geol. Fören. Stockholm Förhand, vol. 97, p. 3-12.

Olsson, M. 1974. Conf. transp. persist. chem. aqu. ecosystems. III-49-60.

Papakostidis, G. et al. 1975. Mar. Pollut. Bull., vol. 6, p. 136-8.

Patchineelam, S.R. : Unpubl. Diss. Heidelberg, 137 p. 1975.

Paul, J.; Meischner, D. 1976. Senckenberg, Marit., vol. 8, p. 91-102.

Pearce, J.B. 1970. FAO conf. on marine poll., Dec. 9-18, Rome, Italy.

Perkins, E.J. et al. 1973. Mar. Pollut. Bull., vol. 4, p. 59-61.

Peschiera, L.; Freiherr, F.J. 1968. J. Water Pollut. Control Fed., vol. 40, p. 127-31.

Pheiffer, T.H. 1972. Annapolis Field Office, Techn. Rept., N° 49, 22 p.

Piper, D.Z. 1971. Geochim. Cosmochim. Acta, vol. 35, p. 531-50.

Popova, T.P. 1961. Geochemistry, vol. 12, p. 1256-60.

Presley, B.J. et al. 1972. Geochim. Cosmochim. Acta, vol. 36, p. 1073-90.

Price, N.B. 1973. Woods Hole Oceanogr. Inst. Techn. Rept., N° 39, 73 p.

Proc, P.I. 1974. Umwelthygiene, vol. 25, p. 7-10.

Rashid, M.A. 1971. Soil. Sci., vol. 111, p. 298-306.

Rashid, M.A. 1974. Chem. Geol., vol. 13, p. 112-5.

Rosenthal, H.; Stelzer, R. 1973. Umschau Wiss. Techn., vol. 73, p. 118-21.

Saroma, A.M. 1972. Chikyu Kaguku, vol. 25, p. 149-58.

Saxby, J.D. 1969. Rev. Pure Appl. Chem., vol. 19, p. 131-50.

Schell, W.R.; Nevissi, A. 1975. Abstr. int. Conf. on heavy metals in the environment, Toronto, C-43.

Scott, R.O.; Mitchell, R.L. 1943. Soc. Chem. ind. Jour., vol. 66, p. 4-8.

Segar, D.A.; Pellenbarg, R.E. 1973. Mar. Pollut. Bull., vol. 4, p. 138-42.

Shelton, R.G.J. 1971. Mar. Pollut. Bull., vol. 2, p. 24-7.

Siccama, T.G.; Porter, E. 1972. BioScience, vol. 22, p. 232-4.

Skei, J.N. et al. 1972. Water Air Soil Poll., vol. 1, p. 452-61.

Slatt, R.M. 1974. Can. J. Earth Sci., vol. 11, p. 768-84.

Smith, R.G.; Windom, H.L. 1972. Georgia Mar. Sci. Center Techn. Rept., p. 72-76.

Smith, J.D.; Nicholson, R.A.; Moore, P.J. 1971. Nature, vol. 232, p. 393-4.

Stirn, J. et al. 1974. Rev. Int. Océanogr. Méd., vol. 25/26, p. 21-78.

Stoffers, P. et al. paper submitted to Environ. Sci. Techn.

Suess, E.; Erlenkeuser, H. 1975. Meyniana, vol. 27, p. 63-75.

Szucs, F.K.; Oostdam, B.L. 1975. Abstr. conf. heavy metals environment, Toronto, C-52.

Takeuchi, T. 1972. In : R. Hartung and B.D. Dirman (eds.), Environmental mercury contamination. Ann Arbor Sci. Publ., p. 79-81.

Talbot, V.; Magee, R.J.; Hussain, M. 1976. Mar. Pollut. Bull., vol. 7, p. 53-5.

Taylor, D. 1974. Estuar. Coast. Mar. Sci., vol. 2, p. 417-24.

Thompson, J.A.J. 1975. Abstr. conf. heavy metals environment, Toronto, C-165.

Thompson, J.A.J.; McComas, F.T. 1974. Fish. Res. Bd Canada Tech. Rept., N° 473, p. 33.

Thompson, J.A.J.; Paton, D.W. 1975. Techn. Rept. Fish. Mar. Serv. Canada, N°506, 24 p.

Thornton, L. et al. 1975. Sci. Total Environ., vol. 4, p. 325-45.

Thrower, S.J.; Eustace, I.J. 1973. Fd. Techn. Australia, vol. 25, p. 546-54.

Tokuomi, H. 1969. Rev. int. Océanogr. Med., vol. 13, p. 5-35.

Turekian, K.K.; Wedepohl, K.H. 1961. Bull. Geol. Soc. Amer., vol. 72, p. 175-92.

Ui, J.; Kitamura, S. 1971. Mar. Pollut. Bull., vol. 2, p. 56-8.

Underdal, B. 1971. Oikos 22, 101-105. Berlin, Springer-Verlach.

Villa, O.; Johnson, P.G. 1974. Annapolis Field Office, Techn. Rept., N° 59, 71p.

Volkov, I.I.; Fomina, L.S. 1974. AAPG Mem. 20, p. 456-76.

Weichart, G. 1972. Naturwissenschaften, vol. 60, in : M. Ruivo (ed.), Marine pollution and sea life, p. 186-8.

Weichart, G. 1973. Ambio, vol. 2, p. 99-106.

Welte, D. 1969. In : K.H. Wedepohl (ed.), Handbook of geochemistry, New York, Springler Verlag, 11-6-L.

Willey, J.D. 1976. Can. J. Earth Sci., vol. 13, p. 1393-410.

Windom, H.L. 1975. Estuarine Research, vol. I, p. 137-52.

Wittmann, G.T.W.; Förstner, U. 1976. Water S.A., vol. 2, p. 67-72.

Wolfe, D.A. et al. 1973. IAEA (Vienna), p. 159-75.

Wollast, R. et al. 1973. Rapport de synthèse. Bruxelles, Ministère des travaux Publics.

Wood, P.C. et al. 1973/74. In : Out of sight, out of mind, vol. IV, Appendix D, p. 57-67.

Yim, W.W.-S. 1976. Mar. Pollut. Bull., vol. 8, p. 147-50.

Young, D.R. et al. 1973. Nature, vol. 244, p. 273-4.

Young, D.R. 1973. In : Cycling and control of metals, EPA Cincinnati, p. 21-39., In : T.M. Church (ed.) Marine Chemistry in the coastal environment, Am.Chem. Soc. Symp. Ser. 18, p. 424-39.

Züllig, H. 1956. Schweiz. Zeitschr. Hydrologie, vol. 18, p. 7-143.

# Humic matter as a component of sediments in estuaries

Thomas Höpner and Christoph Orliczek*

## Introduction

Humic matter is a "natural material". Under the conditions in Northern Europe, however, the rate of release of humic matter into rivers and estuaries is anthropogenically influenced. Efflux of dissolved, "colloidal" and suspended humic matter has increased with increasing draining, cultivation and the improvement of marshy and boggy soils. Occurrence of humic matter in estuaries has risen over the past centuries - further increasing, perhaps, even today.

This report is based on the observation that dissolved and colloidal humic matter entering an estuary is precipitated during the mixing process with seawater where it contributes to the organic constituents of the sediment. From the nature and the known reactivity of humic matter, conclusions are drawn as to its further biodegradation.

## Origin and Properties

In Northern Germany, bogs and bog-draining are prevailing sources of humic matter, where it is formed by the anaerobic degradation of plant material. Two main lines of reactions are discussed : (1) the biodegradation of lignin; and (2) the biodegradation of other macromolecules involving synthetic reactions (see Kononova, 1966). The concerted action of both types of reactions leads to an enrichment of carbon with respect to hydrogen, oxygen and nitrogen and to the transformation of up to 50% of all carbon into aromatic compounds (Wagner, 1968). The whole, very slow process seems to be governed by the transduction of all non-volatile degradation products into a low energy state, thereby gaining energy for the microorganisms which catalyse the reactions (for a review, see Haider *et al.*, 1975). The result is a macromolecular, three-dimensional network of aromatic rings linked together directly, or by ether or methene bridges. The composition of C, O, H and N seems to be variable. For soil humic matter, approximately 40% oxygen is assumed, of which from 50 to 100% seems to be present in functional groups (Haider *et al.*, 1975). Molecular weights are assumed to range from very low up to 200,000 daltons. The high content of aromatic carbon compounds suggests hydrophobic behaviour, but humic matter tends to be hydrophilic. This seems to be due to a structure with a hydrophobic "core" and with hydrophilic groups (carboxylic and phenolic) being turned out to the surface like for the native configurations of some proteins.

---

* Universität Oldenburg, Postfach 943, D 29 Oldenburg, Federal Republic of Germany

Humic matter is characterized by its acid nature, its colour, its intense light absorbance in the UV, its fluorescence (both the latter being due to the aromatic rings), its ability to form complexes and precipitates with di- and polyvalent metal ions, its macromolecular nature, and its high affinity for anion exchange resins (best to DEAE-cellulose). Another outstanding property is its resistance to further biodegradation; this will be discussed at a later stage.

A sensitive analysis of humic matter in natural samples (freshwater, seawater) is based on its affinity for DEAE-cellulose, which can bind up to 1 g/g (dry weight). The elution with an alkaline gradient (0 - 2N NaOH) produces characteristic, origin-specific patterns. Fractions of the elution may be characterized by photometry in the visible or UV range or by fluorescence-spectrophotometry.

Transport to the estuary

Transport humic matter by brooks and rivers proceeds under mainly anaerobic conditions. A brown "bog water", before having been diluted or polluted, is acid (pH 3.5 - 5.5) and almost sterile. The outer, highly oxidized groups are present even at this stage which apparently originated during a former, aerobic state. Even later, at high pH values and in the presence of microorganisms, significant oxidative biodegradation is not detectable. We found that humic matter does not contribute to the "biochemical oxygen demand assay" (Deutsche Einheitsverfahren 1966; also, Rüffer <u>et al.</u>, 1973) even if the assay is prolonged to 20 days and under conditions of added microbial inoculum and nutrients. Humic matter nevertheless is rapidly oxidized by permanganate under the conditions of the "determination of oxidability" (Deutsche Einheitsverfahren 1968).

A part of humic matter is removed by precipitation and flocculation before entering an estuary. Inflow of iron-rich waters causes precipitation and the formation of sediments with colours varying from dark brown to yellow - according to the oxidation state of the iron. Formation and sedimentation of these fine materials proceeds very slowly (see later).

Behaviour in an estuary

What happens when freshwater, which contains humic matter, mixes with seawater in the estuary? The figure shows an experimental answer to this question.

Filtrated brown freshwater out of a bog drainage was used to make two dilution series with distilled water and seawater (35 °/oo salinity) respectively. Both were prepared with the original filtrate as well as with a filtrate centrifuges for 1 hour at 30.000 x g. Both resultant bogwate samples were clear, the uncentrifuged sample showing a just visible opalescence.

The mixtures containing seawater developed a turbidity within one to two days. After four days the turbidity had settled completely leaving the supernatant absolutely clear. The light absorption of the supernantants was recorded at 405 nm. The experiment shows :

(1) The 50% mixture (17.5 °/oo salinity) contained only one third of the initial brown colour;

(2) The time required for formation and sedimentation of the precipitates was in the range of some days;

(3) The formation of the precipitates was due to the seawater constituents and not simply to the dilution; and

(4) In the range of zero to 3.5 °/oo salinity, exactly that part of humic matter

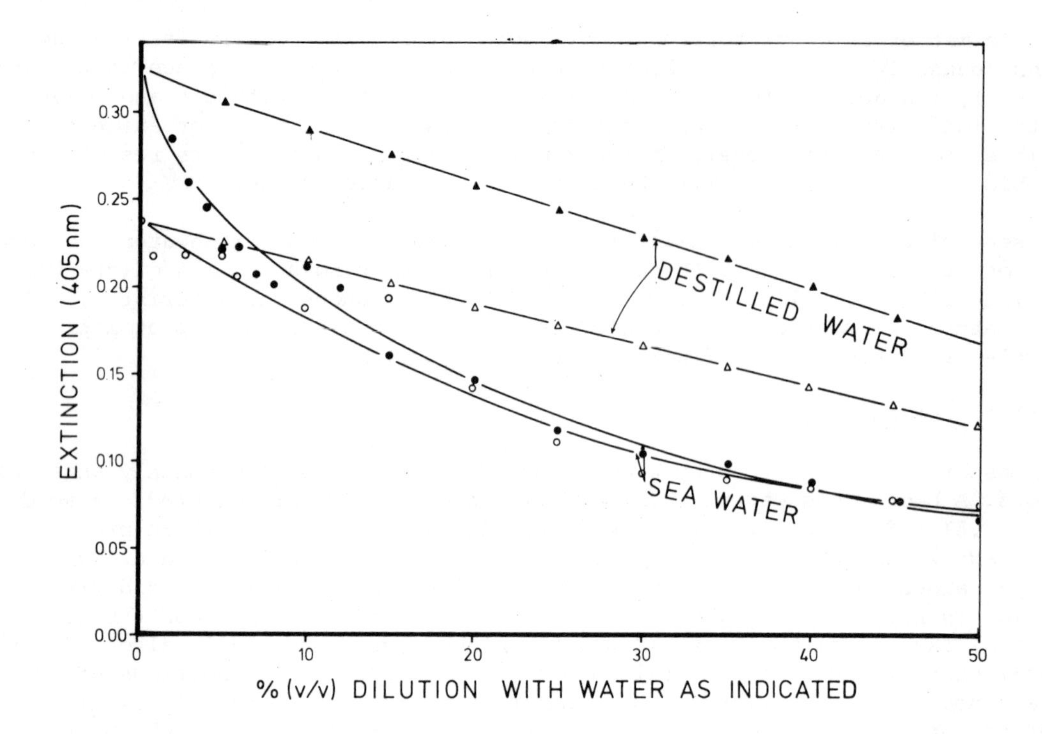


Figure

Elimination of humic matter out of freshwater-seawater mixtures. Dilution series of filtrated (closed symbols) and centrifuged (1 h, 30.000 x g, open symbols) humic freshwater with seawater (35 °/oo salinity, circles) and distilled water (triangles). Mixtures were allowed to stand for 4 days before the light extinction of the clear supernatants was recorded at 405 nm.

was precipitated which can be removed by centrifugation.

In addition it was found that in the presence of 1 mM $Ca^{2+}$ or $Mg^{2+}$ and in the absence of seawater, 80% of the brown colour was precipitated, while 1 mM $Fe^{3+}$ removed 98% of this colouring.

These results strongly suggest a contribution of humic matter to the formation of organic sediments in estuaries. Such a contribution was postulated earlier by Gardner and Menzel (1974) and by Burton and Liss (1976).

The question of further biodegradation in a sediment has hardly been examined and many facts point against easy degradation.

(1) Unlike other biomacromolecules, humic matter is not built up from chemically identical or similar constituents with simple, linear repetition of an unique bond principle. It has many different bond types. Thus different enzymes and, perhaps, microorganisms would be necessary for efficient degradation in contrast to many other biomacromolecules.

(2) As well for other biomacromolecules, extracellular or cell-wall bond enzymes are responsible for degradation because a macromolecule cannot enter the cell (see Williams, 1977). Extracellular enzymes are well known but only with hydrolytic catalytic activity, and there seem to be no bounds in humic matter which can be

cleaved by hydrolysis.

(3) Bond hydrolysis by extracellular enzymes involve thermodynamically favoured, cleavage reactions i.e. hydrolysis of ester, peptide or glycosidic bonds which belong to the "energy-rich bond type". There are no "energy-rich bonds" in humic matter and thus no thermodynamically favoured cleavage.

(4) Humic acids seem to act on proteins like tannins, i.e. adsorbing, denaturing and inactivating them. However, the effects seem to be complex. Stimulations of enzymic activities occur, inactivations can often be related to reversible metal ion chelation, and some other inhibitions seem to be reversible (for a review on humic matter protein interactions, see Ladd and Butler, 1975). It cannot be excluded that the presence of humic matter prevents the biodegradation of other substrates in a sediment.

These observations do not argue for a rapid or significant biodegradation of humic matter. Even its utilization as an electron acceptor is unlikely. Nevertheless, humic matter in a sediment may have important functions in the ecology of an estuary and its sediment, binding toxic heavy metals on its hydrophilic shell, or hydrophobic pollutants in its hydrophobic core. Apart from this, humic matter seems to be one of the final forms of carbon deposition in principle re-utilizable - but if so, only extremely slowly.

## References

Burton, J.D.; Liss, P.S. 1976. Estuarine chemistry, London, Academic Press.

Deutsche Einheitsverfahren. 1966. Bestimmung des biochemischen Sauerstoffbedarfs. Fachgruppe Wasserschemie in der GDCh., Weinheim. Verlag Chemie.

Deutsche Einheitsverfahren. 1968. Bestimmung der Oxidierbarkeit. Fachgruppe Wasserchemie in der GDCh., Weinheim., Verlag Chemie.

Gardner, W.S.; Menzel, D.W. 1974. Geochim. Cosmochim. Acta, vol. 38, p. 813-22.

Haider, K.; Martin, J.P.; Filip, Z. 1975. Humus biochemistry. In E.A. Paul; A.D. McLaren, (eds.), Soil biochemistry, New York, Marcel Dekker, p. 195-244.

Kononova, M.M. 1966. Soil organic matter, Oxford, Pergamon.

Ladd, J.N.; Butler, J.H.A. 1975. Humus enzyme systems and synthetic organic polymer enzyme analogs. In : Soil biochemistry, vol. 4, New York, Marcel Dekker.

Rüffer, H.; Möhle, K.A.; Schilling, J. 1973. Versuche zur Aufbereitung huminsäuerhaltigen Oberflächenwassers. In : Vom Wasser 41, Weinheim, Verlag Chemie, p. 243-76.

Wagner, G.H. 1968. In : Isotopes and radiation in soil organic matter studies, Vienna, IAEA, p. 197.

# Sampling the distribution and speciation of dissolved chemical substances in interstitial waters

Owen P. Bricker*

The composition of interstitial waters in sediments is perhaps the most sensitive indicator of the types and the extent of reactions that take place between sediment particles and the aqueous phase which bathes them. This is particularly true of the estuarine environment in which fine-grained, organic-rich sediment commonly accumulates. The large surface area of the sediment with respect to the small volume of its contained interstitial water insures that minor reactions with the solid phases will produce major changes in the composition of the aqueous phase. Identification of the reactions that take place in the estuarine sediment environment, and the chemical species produced by these reactions, is crucial to our understanding of the chemical bahaviour of estuaries. The sediment can behave either as a source or a sink for natural and man-made materials depending upon the reactions that take place in the sediment and the direction of transfer of dissolved species across the sediment-water interface.

The first step in understanding the chemistry of the sediment-interstitial water system is the collection of appropriate data that accurately reflect the vertical, horizontal, and temporal distribution of dissolved chemical species in the system. This requires :

(1) Vertical and horizontal sampling grids on a scale that will resolve changes in interstitial water chemistry;

(2) Frequency of sampling adequate to resolve temporal changes in interstitial water chemistry;

(3) Methods for separating the interstitial water from the sediment without changing the concentrations or the speciation of dissolved species or introducing contaminants; and

(4) Analytical methods of sufficient sensitivity and selectivity to detect dissolved species of interest in the interstitial waters.

In conjunction with this data, detailed investigation of the mineralogy and the chemical composition of the solids is necessary to develop a realistic model of the behaviour of the system.

* Maryland Geological Survey, The Johns Hopkins University, Baltimore, Maryland 21218. U.S.A.

Sampling Grids

The chemical composition of intersitial waters in estuarine sediments is usually quite different from the composition of the water above the sediment-water interface. This difference arises because of reactions between the sediment and the interstitital waters and in conjunction with bacterially mediated oxidation of organic matter. As a consequence of these reactions, the sedimentary environment is usually anoxic only a short distance beneath the interface and the interstitial waters contain much higher concentrations of nutrients, trace metals and other dissolved species than do the open estuarine waters. The interstitial waters are not completely isolated from the overlying estuarine waters, and exchange of dissolved species take place across the sediment-water interface by diffusion, advection, bioturbation, and othe physical processes that disturb the sediment. Steep gradients in the concentrations of dissolved species may be created in the upper part of the sediment column by these processes. In order to determine the concentrations of dissolved species in interstitial waters as a function of depth in the sediment, it is necessary to choose a sample spacing that will define these gradients. Investigations of the interstitial waters of Chesapeake Bay sediments using a 2 cm sample spacing over the upper 10 cm oc sediment disclosed remarkably steep gradients in some dissolved species and suggest that closer sample spacing would be desirable over this interval (Bricker and Troup, 1975). The resolution of the gradients in the interstitial water of the upper centimetre of sediment, where conditions are most rapidly changing, may require a sample spacing of sub-centimetre intervals. There are very serious problems associated with obtaining undisturbed samples of the sediment-water interface, and these are compounded many times by trying to subsample the upper centimentre of sediment. Thus better sampling techniques must be developed for the water-rich upper layer of the sediment column.

Temporal changes in interstitial water chemistry have been observed in Chesapeake Bay sediments on a time scale of less than a month and extending to a depth of 20 cm into the sediment (Matisoff et al., 1975). This study suggests that the upper part of the sediment column is responsive to changes in the composition of the overlying water (e.g. chlorinity) and that reactions within the sediment involving iron, manganese, phosphate, silica and other dissolved species respond seasonally to factors such as temperature changes and, perhaps, organic activity. Sampling of interstitial waters must be repeated in a time frame that can resolve these temporal changes.

Inhomogeneities in sediment composition, pathiness in benthic infauna distribution, gradients in salinity, and rapid changes in depth and flow regime over small horizontal distances in estuaries may lead to spatial variations in interstitial water chemistry. Preliminary work in Chesapeake Bay has disclosed that spatial variation in interstitial water chemistry is small around any one sampling station (152 / metres radius), but is significant between sampling stations at intervals of several miles (Matisoff et al., 1975). Spatial variation must be considered when designing an estuarine sampling programme.

Extracting Interstitial Waters

A variety of methods have been employed for separating interstitial waters from the sediment that contains them. The majority of these methods require that a core of the sediment be collected. Interstitial waters are then extracted from sections of the core by centrifugation (Powers, 1975), squeezing (Reeburgh, 1967), leaching (Emory and Rittenburg, 1952) or successive dilution (Murthy and Ferrell, 1973). Serious problems are associated with both the coring and extraction procedures. Cores cannot be collected without causing some disturbance of the sediment; the smaller the core diametre the greater the disturbance. This problem can be minimized but not completely eliminated by using a large diametre, box coring device (Goldberg, personal communication).

Further problems arise in recovery and handling of the core prior to extraction of interstitial water. Certain dissolved species are sensitive to changes in temperature, and their concentrations change if the core is allowed to warm up or cool down after collection (Fanning and Pilson, 1971; Bischoff et al., 1970; Mangelsdorf et al., 1969). This temperature effect is particularly troublesome in estuarine environments where there is often a temperature gradient of 10-15°C in the upper metre of the sediment column. This means that, in order to extract the interstitial waters at in situ temperatures, each sample from different depths in a single core would have to be maintained at a different temperature. Other dissolved species common in the interstitial waters of anoxic estuarine environments are sensitive to oxygen, so that sample handling must be done under an inert atmosphere to prevent changes in composition from occurring (Bray et al., 1973; Troup et al., 1974).

As a consequence of these effects, obtaining interstitial waters that accurately reflect their in situ chemistry is a slow and tedious process. Recently a method has been described for the in situ sampling of pore waters (Sayles et al., 1976). A probe with sampling ports at the desired intervals is lowered into the sediment. After emplacement, pore waters are drawn through filter covered sampling ports by a vacuum generated internally in the sampler. The probe is then recovered with the sediment-free pore waters. This method of sampling appears promising but there are serious problems associated with disturbance of the sediment during emplacement of the probe and channeling and flow of pore waters at the interface between the probe and the sediment. Other methods for collecting in situ samples using dialysis techniques or ion exchange media are being investigated. At the present time there does not seem to be a completely satisfactory method for the collection of interstitial waters; particularly for closely spaced sampling of these waters in the upper few centimetres of the sediment column.

## Analysis of Interstitial Waters

Assuming that it is possible to obtain representative samples of interstitial water, the next step is to perform analyses for the dissolved species of interest. In order to define the chemical environment of the interstitial water system, and decipher the reactions that are occurring, it is necessary to obtain data for a number of dissolved species. This necessitates using analytical methods that require only small amounts of sample because of the limited volume of interstitial water that can be collected in present sampling technqiues. Analytical procedures of high precision have been worked out for a number of dissolved species (Bray et al., 1973; Brooks et al., 1967; Goldhaber and Kaplan, 1975). Flameless atomic absorption appears to have adequate sensitivity for certain trace metals. Anodic stripping voltametry has been applied to the analysis of zinc in estuarine waters (Bradford, 1972) and may be adapted to interstitial water analysis. Neutron activation analysis has been used for the analysis of manganese and zinc in seawater (Rona et al., 1962) and has the sensitivity necessary for the direct determination of a number of elements in interstitial waters. No satisfactory analytical method has, however, been developed for aluminum in interstitial waters. This element is the key to interpreting clay mineral reactions in the sedimentary environment, efforts should be made to find a good analytical procedure for the levels expected in interstitial waters.

## Mineralogy and Composition of Solid Phases

The solid phases of most significance to interstitial water composition are those that are reactive in the sedimentary environment. These are often, on the one hand, poorly crystallized solids of variable composition derived from deeply weathered terrigeneous materials or, on the other, fine-grained, poorly crystallized authigenic phases. In either case, it is difficult to identify and characterize such materials. Interpretation of the chemistry of the interstitial water-sediment system requires knowledge of both the interstitial water chemistry and the composition and structure of the solid phases present.

Development in instrumentation over the past decade have provided a means to get at some of these problems. The electron probe and the scanning electron microscope are two promising tools for investigating the composition and identity of very small constituents of the sediment. These methods have draw-backs, however, in that many of the authigenic minerals in anoxic sediments are oxygen-sensitive and will not survive the preparation techniques necessary for the above analytical methods. More effort is necessary in the area of developing better methods for the identification and analysis of poorly crystallized and fine-grained sediment constituents.

## References

Bischoff, J.E.; Greer, R.E.; Luistro, A.O. 1970. Science, vol. 167, p. 1245-6.

Bradford, W.L. 1972. Chesapeake Bay Institute Tech. Report, vol. 76, 103 p.

Bray, J.T.; Bricker, O.P.; Troup, B.N. 1973. Science, vol. 180, p. 1362-4.

Bricker, O.P.; Troup, B.N. 1975. In : J.H. Carpenter, Proc. of the sound international estuarine conf., p. 1-27.

Brooks, R.R.; Presley, B.J.; Kaplan, I.R. 1967. Talanta, vol. 14, p. 809-16.

Emory, K.O.; Rittenberg, S.C. 1952. A.A.P.G., vol. 36, p. 735-806.

Fanning, K.A.; Pilson, M.E.Q. 1971. Science, vol. 173, p. 1228-31.

Goldhaber, M.G.; Kaplan, I.R. 1975. Soil Sci., vol. 119, p. 42-55.

Mangelsdorf, P.C.; Wilson, T.R.S.; Daniell, E. 1969. Science, vol. 165, p. 171-4.

Matisoff, G.; Bricker, O.P.; Holdren, G.R.; Kaerk, P. 1975. In : T.M. Church (ed.), Marine chemistry in the coastal environment, (ACS Symposium Series N° 18, p. 343-63.

Murthy, A.S.P.; Ferrell, R.E. 1973. Clays and clay minerals, vol. 21, p. 161-5.

Powers, M.C. 1975. J. Sed. Pet., vol. 27, p. 355-72.

Reeburgh, W.S. 1967. Ph.D. dissertion, 94 p. Baltimore, Maryland. The Johns Hopkins University, p. 67.

Rona, E.; Hood, D.W.; Muse, L.; Buglio, B. 1962. Limnol. Oceanogr., vol. 7, p. 201-6.

Sayles, F.L.; Mangelsdorf, P.C.; Wilson, T.R.S.; Hume, D.N. 1976. Deep Sea Res., vol. 23, p. 259-64.

Troup, B.N.; Bricker, O.P.; Bray, J.T. 1974. Nature, vol. 249, p. 237-9.

# Group 2
# Forms and species of particulates in estuarine systems

# Report

## 1. INTRODUCTION

The mass particulate balance in estuaries with contemporaneous sediment accumulation is of the greatest importance if chemical and biological processes within the sediment are to be evaluated as well as the fate of anthropogenic material. Mass particulate balance is established from the input ot the estuary from rivers, the atmosphere and other sources and the output of substances to the ocean and to the sediments. Natural factors which govern this balance are the subject of this discussion and concern processes involving sources and sinks of particulate material in estuary systems, the compositional characterization of particulate phases and the geologic time frame in which mass transfer processes take place.

These natural factors, however, govern the mass particulate balance only during times of low variability in sediment input when equilibrium may be attained; therefore evaluation of the transfer processes must always be seen with respect to the degree of variability found in each specific estuary. Such variability concerns composition and flux of material as a function of the physiography of the environment, the extent ot tidal range, the degree of isostatic disequilibrium and the prevailing meterological and climatic conditions. The residence time of water and its variability is an equally important factor in controlling biological and chemical processes.

It must be emphasized that these controlling factors change in frequency and intensity and are thereby responsible for the high variability of the estuarine environment. All events and processes must be evaluated within a time frame and the information contained in sedimentary particles on the history of sedimentation should be fully utilized.

## 2. PROCESSES INVOLVING PARTICULATE MATTER TRANSFER

2.1 Residence time of estuarine waters. Water in transit through an estuary provides the vehicle for material transfer. An important characteristic affecting biochemical and transfer processes is, therefore, the "residence time" of water. For fresh water the most commonly used measure is the "flushing time" which is defined as the total amount present in the estuary divided by the river inflow per unit time. Other time scales, both for the fresh and the seawater fraction, are the "mean age" and the "average transit time". For suspended matter the "mean residence time" is determined by the total amount in suspension and the total output per unit time. The residence time for suspended matter is much larger than that for water and dissolved substances. The relationship between these parametres determines the magnitude for material transfer in an estuarine system. Which of these time scales is most useful depends upon the phenomenon to be studied.

2.2. Settling and sorting of particles in estuarine waters. The fate of particles in an estuary is largely regulated by the settling velocity of the particles which is in turn determined by their size. In a well-stratified estuary, fine particles transported downstream will remain in the uppermost layer and are carried away into the ocean without entering the estuarine cycle proper. Coarser particles will sink through the boundary layer into the salt wedge and settle on the bottom or otherwise take part in mass transfer processes. This results in a separation of fine and coarse-grained particles of fluvial and atmospheric origin. Generally, in strongly stratified estuaries, little material is supplied to the estuary from the ocean since tidal currents tend to be weak.

In less well-stratified estuaries, the bottom currents are much stronger and carry particulate material upstream. This material may be of fluvial or marine origin, usually representing a mixture of both. In an unstratified estuary, fine-grained material, both from the sea and from the river, accumulates by other mechanisms. These mechanisms of accumulation of suspended matter favour grain sizes between 10 and 50 µm. The dimensions are sufficiently large to cause periodic settling but sufficiently small for easy transport and resuspension. A second factor in the settling of fine-grained material is flocculation of freshwater clay minerals upon entering the brackish water. This increases the effective grain size thereby increasing the settling velocity. Some dissolved substances, such as humic acids and certain inorganic compounds other than clays, may also flocculate. A third factor which results in the increase of particle size is the formation of fecal pellets. As a combined result of these factors, unstratified estuaries often have high turbidity and, in many cases, exhibit a turbidity maximum in the region where the freshwater and the salt water meet. As already noted, a second important feature resulting from accumulation mechanisms in estuaries is that the "residence time" of suspended material is much longer than that of water and dissolved substances.

A high flux rate of fine-grained suspended matter to an estuary does not always lead to a net sediment accumulation but is, however, a necessary condition for this process. Whether or not sediment accumulates depends on the "degree of saturation" of the estuarine waters with solid material. In shallow estuaries bordering broad shelves, the basic morphology is controlled, for the most part, by sand structures which are in hydrodynamic equilibrium with currents and waves, thus limiting the space for accumulation of fine-grained material. Permanent accumulation may occur only in sheltered places which may have a relatively short life span. Such areas are not suited for geochronological studies although accumulation when it does occur, may proceed very rapidly. The Rhine River, for example, has a "saturated" estuary, whereas fjords and deep water estuaries are "undersaturated" and provide better conditions for particle-by-particle sedimentation.

The implications resulting from these different mechanisms of deposition are discussed in this report in the section on the time frame of estuarine sedimentation.

2.3 Output of particulate matter from estuarine systems. In "undersaturated" estuaries the output of material will, by definition, be smaller than the input, and in "saturated" estuaries input in the long run will equal output. In both types the properties and composition of the material being exported may be quite different from those of the material imported. Processes of transformation and alteration of particulate matter will be more fully discussed at a later point in this report.

In principle, it appears that the output from estuaries is also influenced by particle size, particles smaller than those which are accumulating tending to escape from the estuary. Two main processes specifically decrease particle size to that range. The first of these is the microbial breakdown of organic detritus, ultimately producing either dissolved constituents which are exported or nutrients which may be re-incorporated into the living phytoplankton. The second is the pH and/or redox-dependent dissolution, precipitation and sorption reactions which can transfor large particles into smaller ones (or vice versa). These reactions occur dominantly at the water/sediment interface and the saltwater/freshwater boundary and often are controlled by the presence or absence of oxygen or hydrogen sulphide. Important constituents in these processes may be ferric hydroxides, manganese hydroxides, ferrous sulphides, iron-manganese phosphates and carbonates.

2.4 Research priorities. A central issue in determining mass transfer in estuarine systems appears to be the size distribution of particulate material. The sizes of suspended particles are not intrinsic properties; rather, they are subject to changes by various mechanisms such as flocculation, dissolution, sorption, fecal pellet formation, and microbial degradation. Monitoring changes in size distribution of particulate matter along the salt gradient will provide important information on where and how significant transformation of the particulate load occurs.

The size distribution of the particulate matter will control net accumulation, composition and grain-size distribution of estuarine sediments, in addition to the output of material to the ocean in response to changing hydrodynamic regimes.

The magnitude of particulate matter transfer may be gleaned from considerations of water and suspended matter budgets and from knowledge about internal cycling. Such investigations are always a prerequisite for any budgetary approach to understanding estuarine systems.

## 3. CHARACTERIZATION OF PARTICULATE MATTER

Characterizing particulate material serves two important purposes, providing (i) an indication of the source of particles and, (ii) information on the processes that have formed, transformed and altered the material.

3.1 Identification of source. Indicators of the source of particulate matter in estuaries are crucial to many natural and anthropogenic problems. These indications can be the simple identification of particles that are unique with regard to source, such as titanium paint chips and slag particles, or they can involve a complex investigation of properties, such as isotopic composition.

Particulate material in estuaries originates from numerous sources, a list of the most important sources and their specific materials follow :

Oceanic : The oceanic source of particulate matter is characterized by biogenic skeletons of diatoms, coccoliths and foraminifera and by clay minerals having specific properties that indicate materials from coastal sources due to marine erosion;

Fluvial : The fluvial source includes materials eroded from the drainage area (provenance), (Martin and Maybeck; Gibbs, this volume) hydroxyates from weathering reactions in the form of hydroxides, products of weathering reactions as clay minerals (Elderfield, this volume), organic remains from vascular plants or as humic matter (Hopner and Orliczek, this volume);

Atmospheric : Airborne anthropogenic particulate matter constitutes an important atmospheric source locally because of the close proximity of industries and urban settlements to most of the world's major estuaries (Suess; Bertine; Jedwab and Chesselet, this volume); and

Estuarine : Organic and inorganic flocculates in suspension are indicative of estuarine sources as are certain authigenic minerals such as manganese sulphides and iron-manganese phosphates.

While these source groupings are based on modes of transport of particulate matter, it is essential that material from anthropogenic sources be treated separately. This group includes metals, slag particles, tailings, fly-ash, inorganic pigments, building materials, radio nuclides, sewage and other waste matter. This material enters the estuary via surface runoff, atmospheric fall-out, and man-made effluents.

Of special importance appears to be the identification of the sources of organic matter and its behaviour. Organic matter is always part of the fine-grained particulate load and therefore behaves largely in the same manner. There are, however, some important differences. Input of organic matter into the sediment occurs not only by settling but also through incorporation of organisms. Ingestion by organisms supplies organic matter to coarse-grained estuarine deposits. As a result, anaerobic conditions prevail both in muddy and in sandy estuarine deposits. The other important difference in the behaviour of particulate organic matter compared that of fine-grained, inorganic particulate material is that its destruction by decomposition nearly always exceeds its input by production. Consequently there is a continuous net input of organic matter from other sources.

In "undersaturated" estuaries, most of the organic matter, whether imported or produced *in situ*, is buried and preserved (or, at least is only partly decomposed) whereas, in "saturated" estuaries much of the organic matter is largely decomposed within the water column. Very little is known about the relative significance of these two decomposition mechanisms.

The characteristic properties used to distinguish between materials from anthropogenic, fluviatile and oceanic, and estuarine sources are grouped here.

Anthropogenic Sources:

*Heavy metals*. Anthropogenic heavy metals have been identified by above-baseline concentrations, time-concentration profiles and coal-residues;

*Lead isotopes*. Lead isotope ratios can be used to characterize the source of lead from anthropogenic origin (Bertine, this volume).

*Carbon and sulphur isotopes*. Man's increased mobilization of fossil fuels has increased the concentrations of stable carbon isotopes relative to natural radiocarbon levels. Recognition of fossil carbon-bearing pollutants such as petrochemical substances and fossil fuel residues free of $^{14}C$, may be obtained from the dilution effects. Sewage sludges with large amounts of bomb-produced $^{14}C$ may considerably exceed the present $^{14}C$ level in the modern environment and may thus also be recognized. The isotope ratio of sulphur can also be used to identify the source of the material (Erlenkeuser, this volume).

*Particle morphology*. Smelting of ores and burning of fossil fuel produce airborne particles of characteristic shapes and sizes. Spheres of slag particles have been identified in estuarine sediments and marine coastal deposits.

Fluvial and Oceanic Sources :

*Clay minerals*. Where there are compositional differences between the fluvial and oceanic sources of clay minerals in estuaries, the source or proportions from each source can be identified;

*Sr isotope*. The isotope ratio of Sr can be used to determine sources of particles;

U and Th isotopes. The source of particles and some knowledge of the transport processes can be obtained by studying these isotopes (Martin et al., this volume); and

Organisms. The organisms that live in freshwater and in salt water are different and therefore provide good indicators of the source environment. The hard parts of these organisms are preserved in sediments and give paleoenvironmental information.

Estuarine Sources :

Carbonates. Microcrystalline iron carbonates and mixed iron-manganese carbonates appear to be indicative of estuarine sources; they precipitate as the direct result of microbial mineralization processes;

Phosphates. Iron and manganese phosphates, mixed phosphates and hydroxo-phosphato-complexes have been reported as authigenic constituents of estuarine sediments. Input of organic phosphorus compounds from anthropogenic sources and subsequent microbial mineralization may also affect these minerals; and

Carbon isotopes of organic matter. Organic matter in the water body of an estuary may be differentiated either as river or mine-derived matter based on the stable carbon isotope ratios, which, in general, are different for marine and river-borne particulates.

3.2 Processes of formation and alteration of particles. A second important aspect of characterizing the particles found in estuaries is the gaining of an understanding of the processes that have formed them and may have altered them. Particles in estuaries are complex in nature, having central cores and exteriors that have been altered many times and which may be coated with organic materials and/or metallic hydroxides. Particles might thus yield much information regarding their formation if each portion could be analyzed separately. This partitioning and analysis is essential to our understanding of the processes of particulate matter transfer.

The works of Burton, Förstner and Gibbs, in this volme discuss the partitioning of metals into the various phases and give the variety of techniques that have been employed. Although true partitioning to separate the metal coatings is desired, practical necessity dictates use of operational methods that give only approximate information of the phase composition. Progress in partitioning phases for substances other than the metals has not been extensively explored.

3.3 Research priorities. The characterization of particulate matter in estuaries may be improved by (a) systematic application of isotope ratio studies for the elements Pb, C, S, N and O; (b) identification of surface properties and changes in these properties along gradients in estuaries; and (c) special studies on submicronic particles.

Among the surface properties of particles, permeability, ion fixation capacity and selectivity and the chemical composition of surface coatings deserve special attention. Organic particles or organic coatings of mineral particles and their origin and stability appear to control the major transformation processes of particulate matter in estuaries.

Testing and standardization of partitioning techniques for metals is desirable to determine the chemical composition of surfaces and the mobility of metal ions and metal complexes. Development of techniques for similar partitioning of nutrients and organic constituents is also encouraged.

## 4. TIME FRAME OF ESTUARINE SEDIMENTATION

The mechanism of sedimentation must be ascertained before the historical time frame of a sediment profile can be established because estuaries have highly variable sedimentation conditions, as discussed in detail in Gibb's paper in this volume. A variety of sedimentation mechanisms can be recognized from profiles of $^{210}Pb$ with depth as illustrated in Fig. a-h.

Only after assuring that a sedimentation rate is obtainable can bulk accumulation rates be determined using $^{210}Pb$ distributions, $^{228}Th/^{232}Th$ ratios or bomb-produced nuclides such as $^{137}Cs$, $^{239+240}Pu$ and $^{14}C$, and ages be assigned to various depth levels for flux estimates of anthropogenic metals, pesticides, or PCB's.

It is also possible to utilize anthropogenic materials as time markers in recent sediments. It is necessary, however, to recognize that remobilization of materials occurs frequently in these deposits, especially through redox and diagenetic changes. Therefore parametres for normalization and differentiation of anthropogenic and natural sources are needed. Both Al and Mn are recommended as respective indicators for constant alumina-silicate input and diagenetic mobilization.

In summary, it appears that $^{210}Pb$ should be the initial parametre to be measured in core sample studies in order to determine whether or not a historical record is contained in the sediment profile and thus to obtain a rate estimate of bulk sedimentation. Subsequently, Al and Mn distributions should be used as criteria for possible variations in alumina-silicate input and diagenetic remobilization.

## 5. RECOMMENDATIONS

5.1 Variability of estuarine environments. Extreme variability of particulate matter input is a unique feature of estuaries. The temporal, spatial, and compositional variability of particulate input from rivers and oceans should be dealt with by rigorous statistical sampling and analytical designs to ascertain the validity of the data.

5.2 Composition and source of particulate matter. Particles which may indicate specific sources should be catalogued. All available techniques should be employed and new ones must be developed for the collection of information on morphological and compositional characteristics of particles. Knowledge of partitioning of the material into generic phases must be expanded to cover all significant phases including coatings and the altered exterior and interior portions of the particles. The methodology of partitioning should be perfected and standardized. The origin of phases in estuaries must be looked upon relative to the chemistry of the environment of formation.

5.3 Reactivity of particles. The reactivity of particles controls the transformation and alteration processes of matter in estuaries. Surface coatings, whether organic, metallic or otherwise, severely alter the intrinsic properties of mineral particles. It is these coatings which help determine the electrical properties and, hence, the degree of flocculation of particles. The coatings further control the type of reaction with surrounding solutions and the reaction rates involving the interior substance as well as the coating substance; hence these surface coating merit further study.

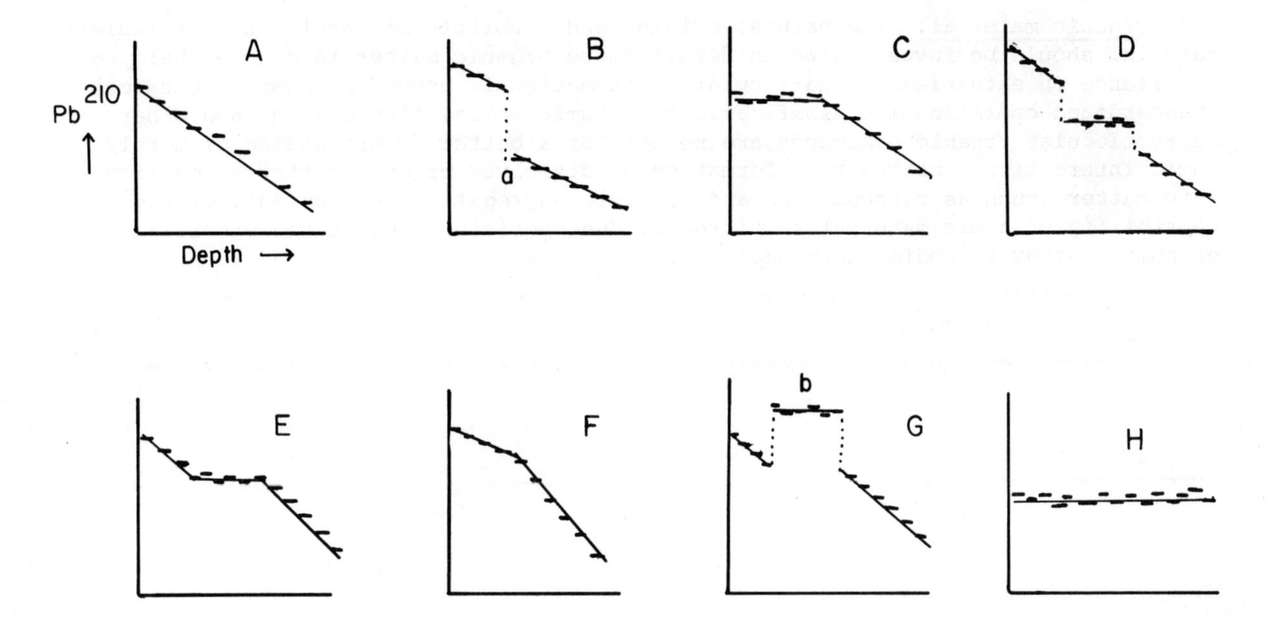


Fig. 1 The depositional history of areas as recorded in the $^{210}Pb$ content with depth.

A) Particle-by-particle deposition.

B) Erosional event (point a) in an area having particle-by-particle deposition.

C) Homogenization of the top sediment by mixing processes such as bioturbation or storms.

D) Homogenization of an older layer of sediment by storm mixing or bioturbation.

E) Rapid deposition of material by a major storm or slumping in an area normally having particle-by-particle deposition.

F) Change in sedimentation rate in an area experiencing particle-by-particle deposition.

G) Rapid deposition from a different source of material (b) in an area of particle-by-particle deposition.

H) Complete mixing of the sediment - possible only in theory.

5.4 Organic material. The nature, origin, and stability of non-living particulate material should be investigated in detail since organic matter is of overwhelming importance in estuaries. In particular, information is needed on humic substances. Standardized operational classifications of humic acids, fluvic acids and other macromolecular organic compounds are needed for a better understanding of metal/humic interactions, of the transformation of dissolved organic matter to particulate matter (such as flocculation and particle aggregation on bubbles), of the susceptibility of humic material to microbial decomposition, and of the stabilization of humic matter in sedimentary deposits.

# Paint pigments: an overlooked group of stable technological tracers

Jacques Jedwab* and Roger Chesselet[x]

The further away an industrially produced particle drifts from its source, both in time and space, the more difficult it becomes to rocognize it as such. Not only does the dilution by naturally produced particles increase exponentially, but also the possible presence of natural particles very similar to the man-made ones frequently introduces an insuperable difficulty.

The case of black magnetic spherules has been well documented for a long time (see Jedwab, 1970). These particles are extremely abundant in the atmosphere above urban and industrial areas, where they are introduced by arc welding, metal cutting and foundries , and by power plants. They are found worldwide in air and seawater, but their origin becomes obscure, since other possible sources of magnetic spherules include volcanoes or more exotically, meteorite ablation; this latter source, which can be dismissed in most cases as quantitatively negligible, becomes noteworthy in oceanic manganese nodules which may form at places where the sedimentation rate is extremely low. In spite of these difficulties, the need for reliable tracers of a technological influence in various environments is apparent and will certainly grow in the future.

Considered formally, the properties of the ideal particulate technological tracer may be described as follows :

1. It should be stable in time and space and, above all, it should be able to bear geological dissolution and diagenesis phenomena;
2. It should be easily and quickly recognizable amongst millions or billions of other particles;
3. It should present clear-cut, specific properties which allow no doubt as to its technological origin;
4. It should be easily distinguishable from similar, nature-born particles; and
5. It should present physical properties which allows it to follow the main dispersion and concentration patterns of the other fine-grained minerals.

In order to discover man-made particulates which potentially meet the requrements, a systematic investigation of all products released by technological operations could be mounted. Unfortunately, this investigation - which somebody should eventually undertake - would be a very long one. A valuable step in this direction has already been made by McCrone and Delly (1973). Another more empirical approach

---

* Laboratorie de Géochimie, Université de Bruxelles, 50, ave. Roosvelt, 1050, Brussels, Belgium.

x Centre des Faibles Radioactivités du CNRS-CEA, Gif-sur-Yvette, France.

is presented below whose results are based on the observation of man-made particles which were selected by the natural processes themselves.

An extensive study of all kinds of coloured and opaque particles suspended in ocean waters has been carried out over several years by the authors within the framework of the GEOSECS program. A novel method of detection and determination of micron-sized particles was devised for this study (see Jedwab, 1975), a short description of which will be useful for the understanding of the results which will follow.

NUCLEPORE membrane filters, on which the particles are to be collected, are fixed semi-permanently on glass slides in such a way that the latter can be successively scanned under the light microscope, the scanning electron microscope and the electron microprobe (SEM-EMP). The light microscope is equipped with both transmitted and reflected light sources and polarizing filters, so that all transparent, semi-opaque and opaque particles can be conveniently observed. The glass slide is scanned visually under oil immersion at various magnifications, and the noteworthy particles are precisely marked and documented with polaroid pictures. The filter is then brought under the SEM-EMP with the particles still adhered. They are relocated, and their morphology and chemistry studied. In favourable cases, several particles belonging to the same optical/chemical category are collected and a characteristic pattern determined with the help of a micro-X-ray diffraction camera. It is thus possible to collect a set of optical, morphological, chemical and structural data for each kind of particle, which not only leads to a definitive diagnosis, but which also allows for the recognition of subtle individual features.

It is worthwhile to pinpoint a few particularities of the visual approach.

1. The immersion oil, through which the observations are made, obliterates all colourless, low-refracting species, such as silica, quartz, felspars and phyllosilicates. This enhances the opaque, highly refracting and coloured particles, even when they are present in very low concentrations.

2. The petrographical, polarizing microscope employed here is extremely useful in detecting particles which present a high optical dispersion. Under the strong, polarized light source used in the reflective mode, such particles seem to light up against the dark background. This property is especially useful for the detection of the paint pigments.

3. The eye/brain couple is an extremely efficient device to sort out an object with some predetermined property, e.g. "red colour", or "sperical shape". This allows an practiced observer to scan a slide of $10^6$ particles in a couple of hours and detect a few individual particles possessing the sought-after properties.

Through the use of this method about 40 specific kinds of particles have been found in the Atlantic and Pacific Oceans. One of the first aims of this study was of course to determine the origin(s) of each kind of particle. But due to the exploratory nature of this study and the relative novelty of the field, these origins are generally determined by inference. There are particles whose origin is rather clear. For example, the small crystals of ilmenite $FeTiO_3$, which are emitted in abundance by volcanoes and are never made by man, or the steel and aluminum flakes, which have no known natural source, but are extensively produced by man either on purpose or as byproducts of the metal industries.

A recurrent group of particles which posed a considerable problem of source-identification was comprised of the following particles :

(A) Agglomerates of micron-sized, calibrated crystals of colourless $TiO_2$, sometimes intermixed with calibrated crystals of $Fe_2O_3$;

(B) Agglomerates of micron-sized, calibrated crystals of green $Cr_2O_3$, sometimes intermixed with $TiO_2$;

(C) Loose fragments or agglomerates of a deep blue, transparent compound containing Co and Al; and

(D) Loose fragments of a sky blue, isotropic compound made up of the peculiar assemblage of Na, Al, S and Si.

The first, titanium oxide, is by far the most frequently encountered on the filters, and does not immediately indicate a specific source, but B, C and D are much more meaningful. Chromium trioxide has been manufactured and used since 1800 as a green paint pigment, but exists in nature only as the very rare mineral eskolaite; cobalt aluminate has been manufactured since 1777 and is used as a blue paint pigment, but is entirely unknown as a natural compound; and sodium sulfo-alumino-silicate has been manufactured since 1828 under the name of ultramarine, and is chemically equivalent, although not identical, to the very rare mineral lazurite.

These three compounds thus belong to the same class of man-made materials, i.e. inorganic paint pigments. The titanium oxide agglomerates also belong to the paint pigments; in fact, the rutile form, which was introduced in 1942, is the most widely manufactured pigment (the 1974 world production capacity was estimated at about $2x10^6$ metric tons). It must, however, be said that other forms of titanium oxide were found on the filters both as prismatic needles and rounded monocrystals. These may come from the decay of almost any type of natural rock.

The discovery of man-made pigments in the deep waters of two of the largest oceans immediately raises the question of the utility of the whole class as technological tracers in marine and estuarine sediments. A more general approach is thus timely, but this can obviously not be undertaken here, and below are briefly outlined the important facts. The basic source of data pertaining to the vast subject of pigments will be found in the reference text by Patton (1973).

1. Man makes or has made about 50 different inorganic pigments. Each industrial epoch is characterized by the introduction of new kinds of pigments.

2. The dates of discovery, commencement of massive industrial production and of possible disappearence are rather well known, at least for the purpose of short term geological dating.

3. The annual production (or capacity of production) of each specific pigment, will also be known so that quantities ultimately injected in the environment can be estimated.

4. Pigments generally have high refractive indices, high optical dispersions, and each one is specifically manufactured with a fixed morphology and grain size. This makes them comparatively easy to detect visually in complex mixtures of particles using the described microscopic methods. It should be noted that bright green or blue minerals are extremely rare in naturally produced sediments.

5. Pigments generally contain one or more heavy elements in their formulae, or have compositions which are unknown among natural compounds; this greatly facilitates their microscopic discrimination as man-made products.

6. Pigments are generally highly inert both in sea water and in air.

## CONCLUSIONS

The discovery of man-made inorganic paint pigments among suspended particles collected from the oceans points to the possible practical use of these materials as tracers of technological influxes. They fulfil most of the requirements of ideal tracers, and especially those of tracers of the industrial activity of man during the last two centuries.

## References

Jedwab, J. 1970. Les sphérules cosmiques dans les nodules de manganèse. Geoch. Cosmochim. Acta, vol. 34, p. 447.

Jedwab, J., 1975. A method of extraction and analysis of possible cosmic particles from manganese nodules. Meteoritics, vol. 10, p. 233.

McCrone, W.C.; Delly, J.G. 1973. The Particle atlas (second edition), Ann Arbor, Ann Arbor Science Publ., 4 vols.

Patton, T.C. (ed.) 1973. Pigment handbook. New York, Wiley, 3 vol.

# The content of major elements in the dissolved and particulate load of rivers

J.-M. Martin and M. Meybeck*

## INTRODUCTION

Rivers are by far the most important supply of material to the ocean system. The total flux of dissolved and particulate matter is estimated at $2.0x10^6$ g $yr^{-1}$ ($1.55x10^{10}$ g $yr^{-1}$ for the solid load and $4.0x10^{15}$ g $yr^{-1}$ for the dissolved load) which is 6 times the glacial scour - Antarctica included - and roughly 200 times the atmospheric fall-out (Livingston, 1963; Holemna, 1968; Meybeck, 1977; Goldberg, 1971). Many studies have been devoted to the determination of river load since the pioneering work of Forel on the Rhône river and the Lake of Geneva in the late 1800's. If the major dissolved element composition is now well documented (Alekin and Brazhnikova, 1960; Livingstone, 1963; Meybeck, 1977), the chemical composition of suspended sediment is still poorly known except for some elements (Clarke, 1924; Moore, 1967; Konovalov and Ivanova, 1970; Martin, *et al.*, 1976) in a few rivers (Gould and Howard, 1960; Turekian and Scott, 1967; Georgescu *et al.*, 1973; Brunskill *et al.*, 1975). Chemical speciation of elements in rivers is of major concern to chemists and hydrologists (Gibbs, 1973; Förstner and Müller, 1975); however there is still no comprehensive data on the average composition of suspended sediments nor on the overall relationship between dissolved and solid transport of elements, except for some local studies in the U.S.S.R. (Strakhov, 1967; Morozov, 1969).

This type of information is needed for a better understanding of weathering processes, geochemical mass balance and characterization of the upstream boundary conditions of estuaries.

## SAMPLING AND ANALYTICAL PROCEDURES

The data presented in this paper are partly derived from literature for the Mississippi and Nile (Clarke, 1924), Colorado (Gould and Howard, 1960), Mac Kenzie (Brunskill *et al.*, 1975), Danube (Georgescu *et al.*, 1973 , sample B 15), while new analyses have been performed on the rivers listed in Table 1. These new analyses include some of the major world rivers (Amazon, Congo, Ganges, Orinoco and Parana) and various European rivers.

Whenever possible, large river water samples up to 300 litres have been collected. The Ems and Ganges samples reflect newly-deposited and fine-grained sediment taken on the river bank. Suspended sediment has been usually recovered by filtration with 0.45µ Millipore filters except for the Congo and Rhône river samples for which continuous centrifugation and settling respectively have been used.

* Laboratoire de Géologie, Ecole normale supérieure, 46, rue d'Ulm - 75230 Paris Cedex 05.

The samples were ashed at 600°C and dissolved for flame atomic absorption (Al, Fe, Mn, Ca, Mg, Na, K) and spectrophotometric analysis (Si, Ti, P).

## RESULTS

Analyses of suspended matter from the western European rivers are given in Table 2, those from the large rivers are given in Table 3.

The average composition of the world's rocks exposed to weathering is given for purposes of comparison in Table 3 and has been computed on the basis of 56% shales, 16% sandstones, 8% limestones and 20% of composite igneous rocks (Gould and Howard, 1960). The elemental composition taken for this computation is that given by Fairbrige (1972) and Wedepohl (1968). The average composition of suspended matter has been weighted by the sediment discharge of each river according to Holeman (1968) data.

## REPRESENTATIVITY OF DATA

As seen in Table 1, many rivers were sampled only once and a large uncertainty of the representativity of the results could be expected. However, a five-year survey of the Rhône river reveals that the suspended load composition is not highly variable for major elements. For 17 samples the coefficient of variation, $s/\bar{x}$, (Table 2) ranges from 0.1 (Si) to 0.32 (Fe); the same computation has been performed with four individual values in the Mekong river and the coefficient of variation is generally less than 0.2. In both cases the only exception is phosphorus. Therefore as a first assessment, a single analysis may be considered as reasonably representative of the suspended load composition of the river, with errors usually less than 30 per cent especially for Si, Al, Fe, Ca, Mg.

The similarity of the suspended river matter from analogous basins (e.g. Amazon, Congo, Orinoco) obtained by different sampling methods, suggests that the different procedures used in this study do not greatly affect the major element composition of the samples.

The selected rivers represent a wide spectrum of morphoclimatic features ranging from the subarctic to the equatorial environment and from mountain rivers to plain rivers. This sample makes up about 25 per cent of the world's total drainage areas into the ocean and about 15 per cent of the world's rivers solid transport even allowing for the Chinese rivers, still unknown to western scientists and which carry about 35 per cent of the world solid load, this sample may be considered as an initial assessment of the solid load carried by the rivers of the world.

Finally, it must be pointed out that the average Na and K contents presented here are very similar to the average found in Soviet rivers (Morozov, 1969) (see Table 3).

## DISCUSSION

### DISSOLVED AND SOLID TRANSPORT

The relation between dissolved and solid transport is a much debated subject: while American authors, following Judson and Ritter (1964) used to consider that the total dissolved transport $T_d$ is inversely related to the total solid transport $T_s$, Soviet scientists (Strakhov, 1967), following Alekin and Brazhnikova (1962) assume that $T_d$ is directly related to $T_s$. In a recent paper based on 30 of the world's major rivers, the Soviet findings have been clearly validated on a world basis (Meybeck, 1976).

Due to the predominance of the relief factor, the $T_s/T_d$ ratio ranges from 0.1 in plains basins to more than 10 in semi-arid mountainous basins but the climate may affect this general pattern. If similar relief types are considered under variable climatic conditions (mainly characterized by the river runoff), $T_d$ and $T_s$ may appear to be inversely related such as it was first reported in the U.S.A. (Judson and Ritter, 1964).

Except for some Soviet rivers, the river elemental transport relationships are still practically unknown (Strakhov, 1967; Morozov, 1969).

Elemental transport for some selected rivers are shown in Table 4. We have considered the major rivers and some additional smaller rivers such as the Po (Turekian and Scott, 1967), the Dranse and alpine Rhône, and two tributaries of the Lake of Geneva (Gibbs, 1973), the Chari (Carmouze, 1976) and the Ouham (Gac and Pinta, 1973). The computation has been done on the basis of Table 3 for solid load, and using Meybeck (1976) review for dissolved load. In the case of the Blue Nile, it is assumed that all the suspended matter measured by Clarke (1964)

originates from this tributary, since there is a known deposition of White Nile suspended load in the Bahr el Gazal swamps. Only the most abundant elements in both dissolved and suspended matter have been considered. Aluminium and iron were discarded because of lack of reliable data on the dissolved load of a great majority of the rivers.

The variations in elemental transport rates are high : more than two orders of magnitude for the dissolved transport and more than three orders of magnitude for the solid. Figures 1-4 represent the relationships between the two kinds of transport for Ca, Mg , K-Na, and $SiO_2$, respectively, numbers in the Figure refering to Table 4. There is a definite direct relationship between Ca, Mg, K and Na. The figure for Na is somewhat different from the others and the dissolved transport rate always exceeds 0.6 ton $km^{-2}$ $yr^{-1}$. This amount can be easily accounted for by the contribution of marine recycled salts which have been estimated to be that same order of magnitude in various rainwater chemical composition studies (Junge, 1963; Caroll, 1970; Zverev, 1971).

The silica variation is much more random and Figure 4 does not show any significant trend. However, if the rivers which belong to similar climatic conditions are grouped, i.e. as tropical rivers (Ouham, Chari, Congo, Amazon and Orinoco) or as temperate and/or mountainous zone rivers (Danube, Rhône, Mac Kenzie, Ganges) two direct relationships between dissolved and solid transport appear. Even so, there is no sharp limit between these two sets of rivers, the Parana (# 18) and Mekong (#12), which would be considered as tropical rivers on the basis of their suspended load chemical composition, but are obviously linked to the temperate zone on this figure. Finally, it must be pointed out that the Colorado lies far below the general trend and further data is needed to improve the silica relationship, especially in the semi-arid environment.

Strakhov (1967) has previously stressed the major importance of the geological and geomorphological features, which are even more important for the understanding of element migration than for their specific physicochemical properties. The reason put forward is that the suspended load composition (see Table 3) and the proportion of elements in the dissolved load of rivers are much less variable than the total dissolved and solid-transport rates (Meybeck, 1976) which are primarily governed by the relief and climate.

The slope of the various elemental graphs gives a basic idea of the relative importance of the dissolved transport versus solid transport. However, it has been found more suitable to compute for each given chemical element the so-called "soluble transport index" (S.T.I.) which is the dissolved transport rate divided by the total transport rate. Fig. 5 represents the world average and range of S.T.I. for Si, Ca, Mg, Na and K. The individual S.T.I. values have been computed from Tables 3 and 4 and from dissolved transport rates in rivers (Morozov, 1969). The rivers have been registered into three different sets according to their morphoclimatic features. As a whole, the dissolved transport is dominant for the warm and/or low relief basins while the solid transport clearly dominates in the mountainous regions.

The world S.T.I. averages (Table 5) have been computed on the basis of the average composition of river suspended matter (Table 3) considering a total solid discharge of $15.5 \times 10^{15}$ g $yr^{-1}$ of which 90 per cent is assumed to be mineral (Meybeck 1977) and a yearly discharge of dissolved matter corrected for pollution (Alekin and Brazhnikova, 1960; Livingstone, 1963; Meybeck, 1977). A first assessment of S.T.I. values for other major elements such as S, Cl, Al, Fe, Ti, P and Mn has been performed on the basis of the following river dissolved concentrations : Al = 400 $\mu g\ l^{-1}$ (Durum, et al., 1960) Fe = 40 $\mu g\ l^{-1}$ (Gibbs, 1975), Ti = 10 $\mu g\ l^{-1}$ (Durum, et al., 1960) and Mn = 8.2 $\mu g\ l^{-1}$ (Konovalov, 1973). An approximation of S and Cl suspended matter content has been made assuming their concentration levels close to the average shale concentration i.e. $2.0 \times 10^{-3}$ g $g^{-1}$ and $5.0 \times 10^{-3}$ g $g^{-1}$ respectively (Wedepohl, 1968). This figure may be higher than the real one and is thus likely to be an underestimate of the S.T.I. for these elements. The world water discharge used in the computation is 37,400 $km^3\ yr^{-1}$ (Baumgartner and Reichel, 1975).

As expected, Cl, S, Ca, Na and Mg are mostly carried by rivers to the estuarine system in dissolved form, while Al, Ti and Fe are almost entirely bound to particulates. A significant percentage of K, P and Si is carried in the dissolved form, especially in plains regions. Therefore, if for any given element the forms of transport are mainly related to the geological and morphoclimatic characteristics of the basins, the different behaviour observed between the respective major elements is clearly derived from their geochemical properties.

## FLUXES OF MAJOR RIVERINE ELEMENTS TO THE ESTUARINE SYSTEM

The elemental composition of river suspended load is highly variable (Table 3). However, as the drainage areas are very extended ($> 5 \times 10^5$ $km^2$), they integrate various lithological features and the difference between their climates and orography are most probably prevailing. Obviously, there are definitely two sets of data concerning the suspended load composition according to the morphoclimatic features of the basins. The suspended load of tropical rivers (Amazon, Congo, Mekong, Orinoco and Parana) is richer in Al, Fe, Ti than the average surficial rocks, and depleted with respect to Ca, Na and possibly Mg. In contrast, the suspended loads of mountain, temperate and arctic rivers is similar to the surficial rocks, except for Na which is significantly depleted. In both cases, the mean silica content is much closer to the average surficial rocks. As was noted by Alekin and Brazhnikova (1962), solid material carried by mountain rivers consists mainly of rock debris or poorly weathered soil particles. On the contrary, tropical rivers usually carry soil particles that are relatively enriched in the less soluble elements (Al, Fe, Ti) and depleted in the more soluble.

These depleted elements are, to a large extent, carried by rivers in the dissolved form. However, if the dissolved load - corrected for the atmospheric and pollution inputs - originates mainly from rock weathering, the particulates carried by the river may also derive from other processes (wind abrasion, glacial scour, mass wasting, etc.) that do not cause any release of dissolved elements. Moreover, even if all dissolved material is generally exported from the basin, all particulates (weathered particles and/or rock debris) are not always completely transported by the river out of the basin since they may be retained in alluvial cones, flood plains, etc. ( Martin et al, this volume). It is thus misleading to compare the total annual river discharge (particulate and dissolved load) to the theoretical erosion supply.

A first assessment of the flux of riverine material entering the estuarine zone is given here. As new physico-chemical conditions prevail in the estuaries the concentration and/or the speciation of some elements can become altered (Bien, et al, 1958; Martin et al., 1973; Wollast et al., 1973) so that these data may have to be corrected to set up the actual geochemical mass balance between the continent and the ocean.

Acknowledgement

We thank J.P. Carbonnel, P.J. De Petris, A. Deleglise, S. Krishnaswamy, H. Sioli for providing us with some precious samples.

Table 1. Sampling procedures of river suspended load.

| River | Date of collection | Location | Volume sampled | Suspended matter separation | Sampled by |
|---|---|---|---|---|---|
| Amazon | Sept. 1968 | Manaus | 30 litre | millipore filter 0.45μ | Sioli |
| Congo | July 1969 | Brazzaville | 30 litre | millipore filter 0.45μ | Authors |
| Congo | July 1970 | Brazzaville | 180 litre | continuous centrifugation | Authors |
| Ganges | March 1973 | Allahabad | river bank | freshly deposited sediment | Krishnaswamy |
| Mekong | Sept. 1968 to June 1969 | Pnom Penh | 4 x 30 litre | millipore filter 0.45μ | Carbonnel |
| Orinoco | July 1971 | Mouth | 10 x 2 litre | millipore filter 0.45μ | Authors |
| Parana | 1. Dec. 1975 | Santo Dome | 65 litre | continuous centrifugation | De Petris |
| Ems | 1962-1967 | Diele | river bank | freshly deposited sediment | De Groot |
| Garonne | Nov. 1968 | La Réole | 30 litre | millipore filter 0.45μ | Authors |
| Meuse (Maas) | July 1968 | Zalthommen | 10 x 30 litre | millipore filter 0.45μ | Authors |
| Loire | May 1969 | Le Pellerin | 5 litre | millipore filter 0.45μ | Authors |
| Rhône | Dec. 1962 to June 1967 | Montélimar | 17 x 300 litre | settling | C.N.R. |

Table 2. Chemical composition of western European rivers (expressed in % of dry weight at 600°C)

| | Ems | Garonne | Maas (Meuse) | Loire | Rhône conc. | Rhône $s/\sqrt{x}$ (1) | Rhine (2) | Average (3) |
|---|---|---|---|---|---|---|---|---|
| $SiO_2$ | 64.6 | 57.7 | 60.6 | 60.1 | 46.5 | (0.1) | 57.2 | 57.8 (267) |
| $Al_2O_3$ | 9.6 | 22.4 | 14.2 | 18.4 | 10.0 | (0.15) | 14.0 | 14.8 (78.3) |
| $Fe_2O_3$ | 15.9 | 8.3 | 9.65 | 7.76 | 5.74 | (0.32) | 6.2 | 6.9 (48.2) |
| $TiO_2$ | 0.63 | 0.83 | 1.16 | 0.88 | 0.48 | (0.15) | (0.9) | 0.78 (4.7) |
| MnO | 0.36 | 0.22 | 0.91 | 0.21 | 0.12 | (0.23) | ? | 0.36 (2.8) |
| $P_2O_5$ | 1.31 | 0.31 | (2.7) | 0.33 | 0.55 | (0.42) | (0.65) | 0.63 (2.7) |
| CaO | 1.42 | 2.73 | 4.8 | 6.59 | 18.3 | (0.2) | 10.2 | 7.34 (52.4) |
| MgO | 1.05 | 2.89 | 1.7 | 2.49 | 1.94 | (0.14) | 2.95 | 2.17 (13.0) |
| $Na_2O$ | 0.59 | 0.76 | 0.88 | 1.03 | 1.02 | (0.29) | 0.13 | 0.74 (5.5) |
| $K_2O$ | 1.43 | 4.09 | 2.38 | 3.76 | 2.12 | (0.18) | 1.3 | 2.52 (21 ) |
| Drainage area ($10^3$ $km^2$) | 2.5 | 32 | 30 | 110 | 96 | | 159 | 450 |
| Solid discharge $10^6$ $t.y^{-1}$ | / | 2 | 0.075 | 1.1 | 30 | | 4.5 | 40 |

(1) $s/\sqrt{x}$ = coefficient of variation.

(2) recomputed from Hellman and Bruns, 1968.

(3) arithmetic average. Number in parenthesis is elemental content expressed in $10^{-3}$ g $g^{-1}$ of dry weight at 600°C.

Table 3. Chemical composition of river suspended load (expressed in $10^{-3}$ g $g^{-1}$ of dry weight at 600°C).

| | Al | Ca | Fe | K | Mg | Mn | Na | P | Si | Ti |
|---|---|---|---|---|---|---|---|---|---|---|
| Amazon | 115 | 16 | 53.8 | 18 | 11.2 | 0.8 | 7.85 | 1.65 | 267 | 6.95 |
| Congo | 112 | 7.1 | 58 | 11.8 | 5.8 | 1.4 | 2.1 | 1.5 | 259 | 8.4 |
| Ganges | 77 | 26.5 | 37 | 21.3 | 12.4 | 1.06 | 10.7 | 0.56 | 285 | 5.3 |
| Mekong | 112 | 5.9 | 55.7 | 24 | 13.5 | 0.94 | 5.2 | 2.0 | 275 | 3.6 |
| Orinoco | 113 | 3.0 | 57.5 | 27 | 5.8 | 0.74 | / | / | 292 | 8.6 |
| Parana | 106 | 5.9 | 48.1 | 17.8 | 10.9 | 0.27 | 6.7 | 1.3 | 289 | 9.4 |
| Western European rivers | 78.3 | 52.4 | 48.2 | 21 | 13 | 2.8 | 5.5 | 2.7 | 267 | 4.7 |
| Nile | 98.4 | 40.3 | 108 | 19.0 | 18.4 | / | 7.35 | / | 244 | / |
| Mississipi | 60 | 16.7 | 26.2 | 20.7 | 9.0 | 0.51 | 12.0 | 0.83 | 349 | 3.9 |
| Danube | 63.2 | 45 | 55 | 20.6 | 21 | 0.60 | 17.4 | 0.68 | 299 | 4.2 |
| Mackenzie | 83.6 | 71 | 40.5 | 24.8 | 3.9 | 0.65 | 2.24 | / | 295 | 4.7 |
| Colorado | 43.2 | 34 | 22.7 | 14.9 | 10.4 | 0.43 | 4.8 | 0.54 | 362 | 3.0 |
| River suspended sediment | 89 | 22.5 | 46.5 | 20.3 | 11.4 | 0.9 | 8.2 | 1.15 | 293 | 5.3 |
| Soviet river suspended sediment | | | | 18.7 | | | 9.5 | | | |
| Surficial rocks | 69.3 | 45 | 35.9 | 24.4 | 16.4 | 0.8 | 14.2 | 0.61 | 175 | 3.8 |
| Tropical rivers suspended sediment | 111.6 | 7.6 | 54.6 | 19.7 | 9.4 | 0.75 | 5.5 | 1.6 | 276 | 7.4 |
| Temperate and cold rivers suspended sediment | 67.5 | 40.9 | 38.2 | 20.5 | 11.6 | 1.0 | 8.8 | 1.1 | 309 | 4.3 |

Table 4. Elemental rates of dissolved ($T_d$) and suspended ($T_s$) transport by rivers in t. $km^{-2}$ $yr^{-1}$
( ) approximate values.

| River | Si | | Ca | | Mg | | Na | | K | |
|---|---|---|---|---|---|---|---|---|---|---|
| | $T_d$ | $T_s$ | $T_d$ | $T_s$ | $T_d$ | $T_s$ | $T_d$ | $T_s$ | $T_d$ | $T_s$ |
| 1 Danube | 0.93 | 22.1 | 12.85 | 3.33 | 2.77 | 1.55 | (2.66) | 1.3 | (0.33) | 1.5 |
| 2 Mac Kenzie | 0.93 | 15.8 | 5.6 | 3.8 | 1.75 | 0.21 | 1.18 | 0.12 | 0.18 | 1.33 |
| 3 Colorado | 0.14 | 110 | 3.0 | 10.4 | 0.88 | 3.15 | 3.45 | 1.45 | 0.16 | 4.6 |
| 4 Blue Nile | 1.78 | 108 | 4.6 | 18 | 1.16 | 8.5 | 1.24 | 3.3 | 0.44 | 8.5 |
| 5 Mississipi | 0.96 | 30.3 | 6.0 | 1.45 | 1.58 | 0.79 | 1.95 | 1.06 | 0.5 | 1.8 |
| 6 Alpine Rhône | 1.5 | 278 | 44 | 60 | 5.4 | 36.5 | 3.3 | 15.5 | 1.35 | 32 |
| 7 Dranse | 2.3 | 175 | 100 | 125 | 12 | 19 | 2.3 | 3.9 | 1.6 | 19 |
| 8 Po | 1.8 | 30.8 | 33 | 18.9 | 6.8 | 5.0 | 8.3 | | 1.7 | |
| 9 Ganges | 1.75 | 148 | 9.4 | 13.8 | 3.0 | 6.45 | 4.1 | 5.56 | 1.3 | 11.1 |
| 10 USSR | | | | | | | 1.25 | 0.15 | 0.455 | 0.30 |
| 11 Rhône (Beaucaire) | 1.0 | 60 | 40.8 | 36.9 | 4.5 | 3.23 | 4.6 | 2.1 | 0.92 | 5.0 |
| 12 Mekong | 2.9 | 120 | 10.1 | 2.6 | 2.2 | 5.9 | 2.6 | 2.3 | 1.4 | 10.5 |
| 13 Chari | 0.77 | 1.0 | 0.26 | 0.01 | 0.11 | 0.01 | 0.19 | 0.02 | 0.11 | 0.02 |
| 14 Ouham | 2.27 | 1.9 | 0.55 | 0.01 | 0.33 | 0.02 | 0.64 | 0.01 | 0.46 | 0.06 |
| 15 Amazon | 4.56 | 21.2 | 5.7 | 1.27 | 2.7 | 0.89 | 0.87 | 0.62 | 0.87 | 1.43 |
| 16 Congo | 1.67 | 3.43 | 0.9 | 0.09 | 0.6 | 0.08 | 1.05 | 0.03 | 0.57 | 0.16 |
| 17 Orinoco | 5.3 | 26.4 | 3.3 | 0.27 | 1.0 | 0.53 | (1.15) | / | (1.15) | 2.44 |
| 18 Parana | 1.18 | 11.6 | 1.0 | 0.24 | 0.42 | 0.44 | 1.0 | 0.27 | 0.35 | 0.71 |
| World average | 2.0 | 4.1 | 5.5 | 3.15 | 1.4 | 1.6 | 1.9 | 1.15 | 0.5 | 2.85 |

Table 5. Flux of major river elements to the estuarine system $10^{12}$ g $yr^{-1}$ ( ) approximate values.

| | Dissolved | Particulate | Dissolved / Total % (S.T.I.) |
|---|---|---|---|
| Si | 201 | 4100 | 4.7 |
| Al | (15) | 1250 | (1.2) |
| Fe | (1.5) | 651 | ( .2) |
| Mg | 143 | 160 | 47 |
| Ca | 547 | 315 | 64 |
| Na | 190 | 115 | 62 |
| K | 51 | 284 | 15 |
| Ti | ( .38) | 74 | ( .5) |
| P | (1.5) | 16.1 | (8.5) |
| Mn | ( .3) | (12.6) | (2.3) |
| S | 105 | (28) | (79) |
| Cl | 197 | (70) | (74) |

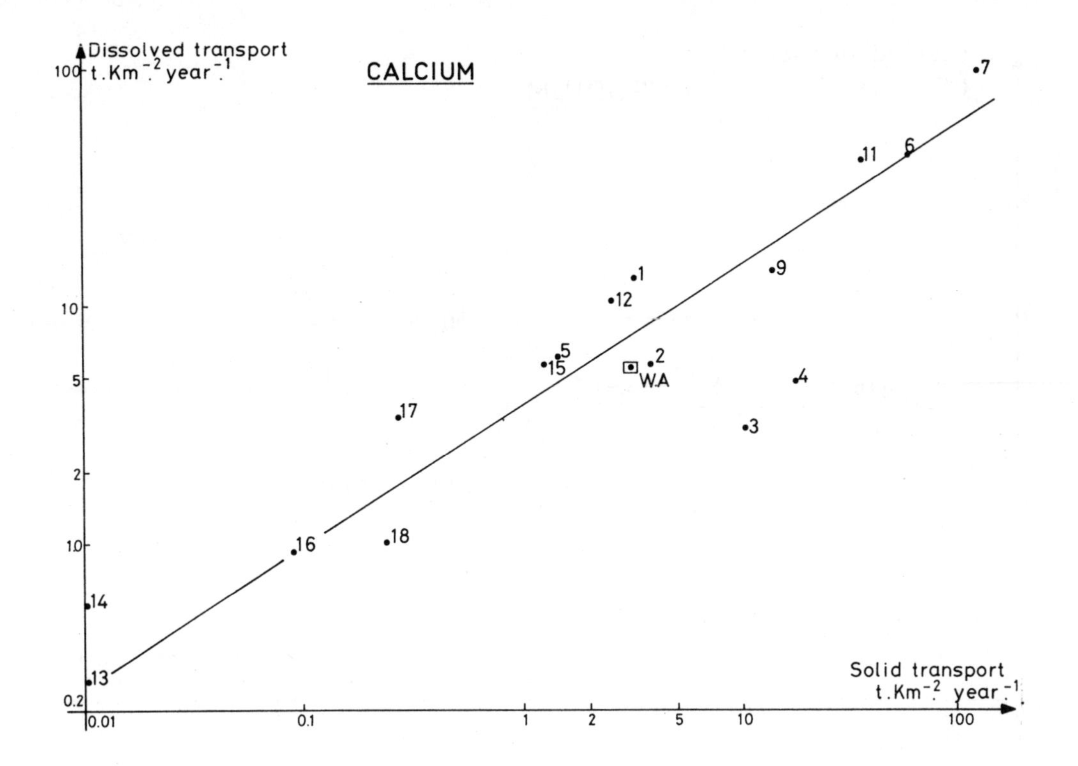


Fig. 1 Relationship between dissolved and solid transport of calcium.

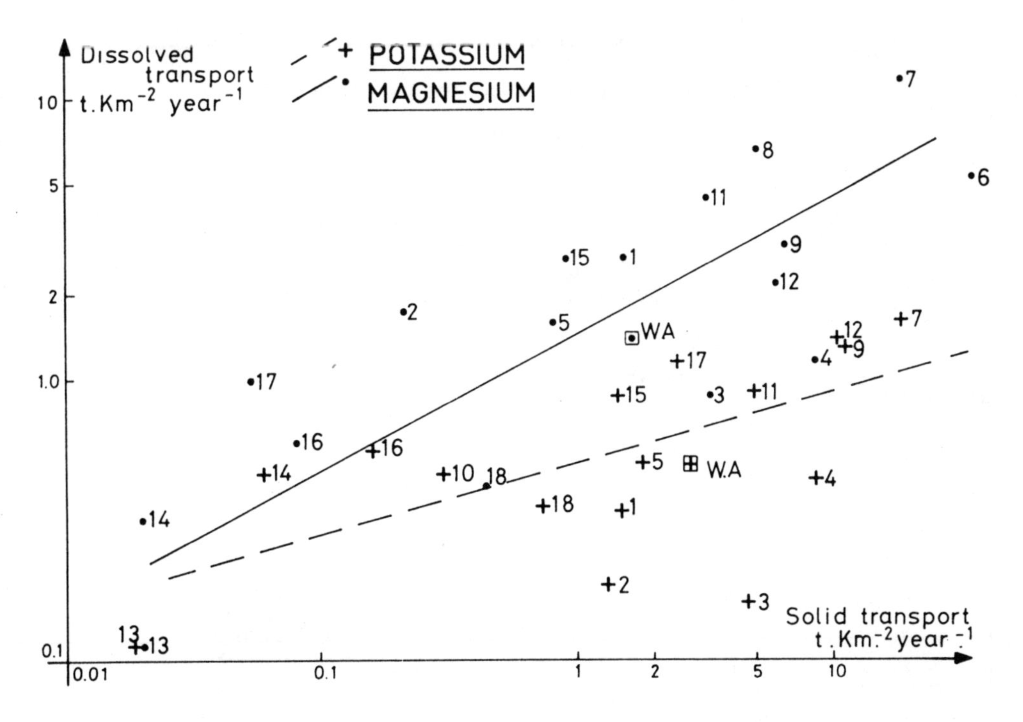


Fig. 2 Relationship between dissolved and solid transport of magnesium and potassium.

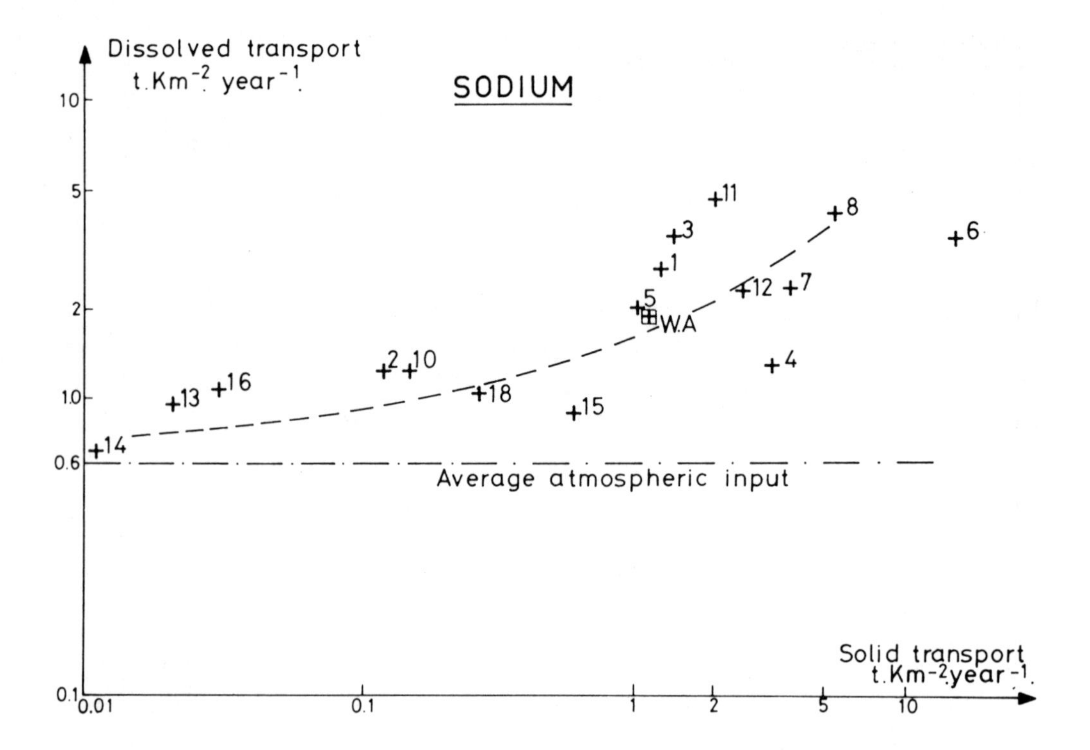


Fig. 3 Relationship between dissolved and solid transport of calcium.

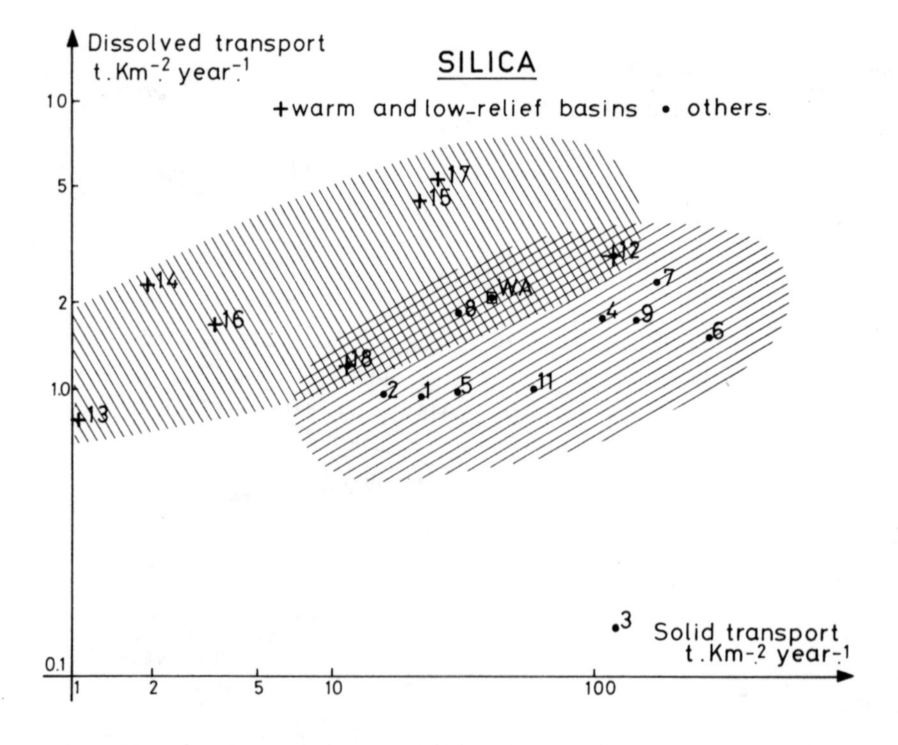


Fig. 4 Relationship between dissolved and solid transport of silica.

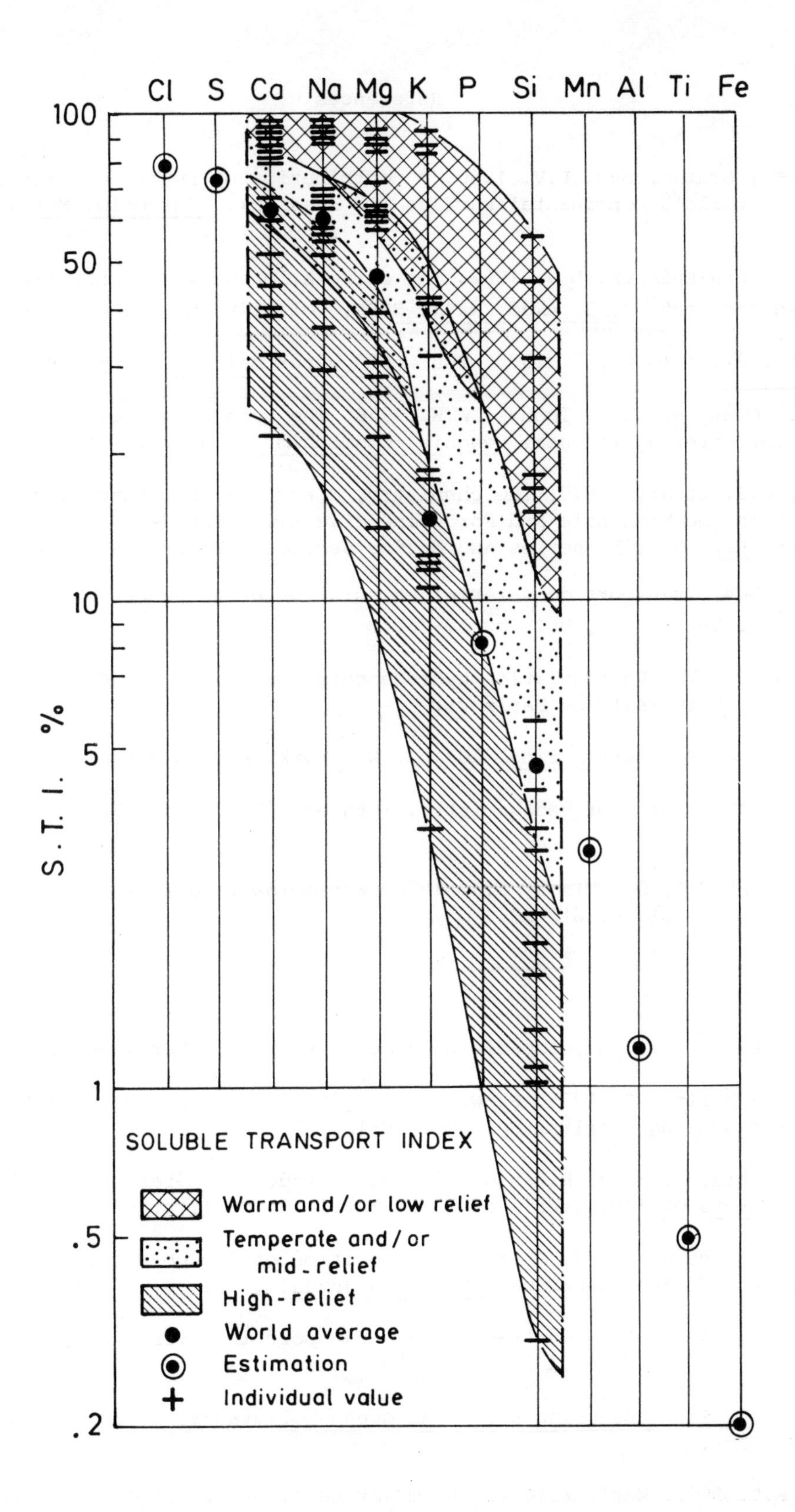


Fig. 5 Rivers soluble transport index for the major elements (world range and average). STI = dissolved transport over total transport in per cent.

## References

Alekin, O.A.; Brazhnikova, L.V. 1960. A contribution on runoff os dissolved substances on the world's continental surface (in Russian). Gidrochim. Mat., vol.32, p.12-34.

Alekin, O.A; Brazhnikova, L.V. 1962. The correlation between ionic transport and suspended sediment. Dokl. Akad. Nauk. SSSR, vol. 146, p. 203-6 (in Russian).

Baumgartner, A.; Reichel, E. 1975. The world water balance, 179p., Elsevier.

Bien, G.S.; Contois, D.E.; Thomas, W.H. 1958. The removal of soluble silica from fresh water entering teh sea. Geochim. Cosmochim. Acta, vol. 14, p. 35-54.

Brunskill, C.J. et al., 1975. The chemistry, mineralogy and rates of transport of sediments in the Mackenzie and Porcupine rivers watersheds, N.W.T. and Yukon 1971-73. Techn. Rep. 566, Fisheries and Marine Services, Environment Canada.

Canali, L. 1964. Transport de matériel en solution et en suspension du Pô. Ass. Int. Hydrol. Sci. Bull., vol, 9 1, p. 17-26.

Carmouze, J-P. 1976. La régulation hydrogéochimique du lac Tchad. Thèse de Doctorat ès Sciences. Université de Paris VI.

Caroll, D. 1970. Rock weathering, 203 p., New York, Plenum Press.

Clarke, F.W. 1924. Data of geochemistry, (5th ed.) U.S. Geol, Survey Bull., vol. 770, 841 p.

Durum, W.H.; Heidel, G.; Tison, L.J. 1960. Worldwide runoff of dissolved solids. Ass. Int. Sci., publ. 51, p. 618-28.

Fairbridge, R.W. 1972. Encyclopedia of geochemistry. New York, Rheinhold Book Corp.

Förstner, U.; Müller, G. 1975. Schwermetalle in Flüssen und Seen als Ausdruck der Umweltverschmutzung. Berlin, Springer-Verlag.

Gac, D.Y.; Pinta, M. 1973. Bilan de l'érosion et de l'altération en climat tropical humide. Cahiers ORSTOM, (série géologie,) vol. 5, N° 1, p. 83-96.

Georgescu, I.I. et al. 1973. Chemical composition of Danube water, sea water, algae and sediments of the Black Sea. Thalassia Jugosl., vol. 9, N° 1/2, p. 87-99.

Gibbs, R. 1973. Mechanisms of trace metal transport in rivers, Science, vol. 180, p. 71-3.

Gibbs, R. 1975. In : Assessing potential ocean pollutants, p. 305. Washington D.C., Nat. Acad. Sci.

Goldberg, E.D. 1971. Man's role in the major sedimentary cycle. In : Dyrssen and Jagner (eds.) "Nobel Symposium", p. 267-83.

Gould, H.R.; Howard, C.S. 1960. Comprehensive study of sedimentation in lake Mead, 1948-1949. U.S. Geol. Survey, Prof. Paper 295.

Hellmann, H.; Bruns, F.J. 1968. Die chemische Zusammensetzung der Ungelöstenstoffe des Rheins. Deutsche Gewasserkindl. Mitt., vol. 12, N° 6, p. 162-6.

Holeman, J.N. 1968. The sediment yield of major rivers of the world. Water Resources Res., vol. 4, N° 4, p. 737-47.

Judson, S.; Ritter, D.F. 1964. Rates of regional denudation in the United States. Jaurnal Geophys. Res., vol. 69, N° 16, p. 3395-401.

Junge, C.E. 1963. Air chemistry and radioactivity, 382 p., New York, Academic Press.

Konovalov, G.S. 1973. The dynamic of rare and trace element content and their drainage from the USSR territory into sea basins. Proc. Ist Symp. Hydrogeochem. Biogeochem., Tokyo Seot. 70, Washington, D.C. Clarke Co., p. 606-12.

Konovalov, G.S.; Ivanova, A.A. 1970. River discharge of microelements from the territory of the USSR to the sea basins. Okeanologiia SSSR., vol. 10, N° 4, p. 482-88 and 628-36.

Leopold, J.B.; Wolman, M.G.; Miller, J.P. 1964. Fluvial processes in geomorphology. 522 p. San Francisco, London, W.H. Freeman Ed.

Livingstone, D.A. 1963. Chemical composition of rivers and lakes. Data of geochemistry (chapter G) U.S. Geol. Survey Prof. Paper 440 G, G1-G64.

Martin, J.M.; Høgdahl, O.; Philippot, J.C. 1976. Rare earth element supply to the ocean., J. Geoph. Res., vol. 81, N° 18, p. 3119-24.

Martin, J-M.; Kulbicki, G.; DeGroot, A.J. 1973. Terrigeneous supply of radioactive and trace elements to the ocean. Proc. Ist Symp. Hydrogeochem. Biogeochem. (Tokyo, Sept. 70) p. 463-83., Washington D.C., Clarke Co.

Meybeck, M. 1972. Bilan hydrochimique et géochimique du lac Léman., Verh. int. Ver. Limnol., vol. 18, p. 442-53.

Meybeck, M. 1976. Total dissolved transport by world major rivers. Hydrol. Sci. Bull. vol. 21, N° 2, p. 265-84.

Meybeck, M. 1977. Dissolved and suspended matter carried by rivers : composition, time and space variations, and world balance. SIL-UNESCO. Symposium on interaction between sediments and freshwater. Amsterdam, Sept. 1976.

Moore, W.S. 1967. Amazon and Mississippi river concentrations of Uranium, Thorium and Radium isotopes. Earth Planet. Sci. Lett., N° 2, p. 231-4.

Morozov, N.P. 1969. Geochemistry of the alkali metals in rivers. Geokhimiya, vol. 6, p. 729-39.

Strakhov, N.M. 1967. Principle of lithogenesis, vol. 1, 245 p., London, Oliver and Boyd.

Stumm, W.; Morgan, J.J. 1970. Aquatic chemistry, 583 p. New York, Wiley Intersciences.

Turekian, K.K.; Scott, M. 1967. Concentrations of Cr, Ag, Mo, Ni, Co and Mn in suspended material in streams. Envir. Sci. Techn., vol. 1, p. 940-2.

Wedepohl, L.K.H. 1968. Chemical fractionation in the sedimentary environment. In : L.H. Ahrens (ed.): Origin and distribution of the elements., Oxford Pergamon Press, p. 999-1015.

Wollast, R. et al. 1973. Origine et mécanisme de l'envasement de l'estuaire de l'Escaut. Rapport de synthèse. Lab. Rech. Hydrauliques, Barger hunt, Min. Travaux Publics, Bruxelles, Déc. 1973.

Zverev, V.P. 1971. Hydrochemical balance of the USSR territory. Dokl. Akad. SSSR., vol. 198, p. 161-3 (in Russian).

# Uranium and thorium isotope behaviour in estuarine systems

J.-M. Martin, V. Nijampurkar and F. Salvadori *

INTRODUCTION

Radioactive disequilibrium within the uranium-thorium isotope series in the aquatic environment is of major importance in terms of geochronological studies and geochemical mass balances of these radionuclides in the oceanic system. A great deal of work has been done in this field during the last decades (Broecker, 1963; Ku, 1966; Turekian and Chan, 1971; Cherdytse , 1971; Goldberg and Bruland, 1974), but until now few studies (Moore, 1967; Bhat, 1970; Scott, 1967; Miyake, et al., 1973) of these elements have been made in rivers or estuaries. This paper deals mainly with uranium and thorium concentrations, and the activity ratios of these radionuclides during estuarine mixing of river and seawaters.

In estuarine environments, dissolved and particulate elements are subjected to new physico-chemical, biological, and hydrodynamic processes which can modify the speciation, the concentration and finally the budget of these elements to the ocean. Indeed, the geochemical behaviour of particulate and dissolved elements in estuaries is still poorly understood.

As far as uranium is concerned, it has been recognized that its removal rate by authigenic material is much lower than its input rate from rivers (Koczy et al., 1957; Veeh, 1967). Furthermore, the observed $^{234}U$ excess, as well as the constant $^{234}U/^{238}U$ activity ratio in the ocean is not yet clearly understood (Ku, 1966; Bhat, 1970) and it is therefore important to examine whether estuarine processes are able to give some insight into these problems.

SAMPLING AND ANALYTICAL PROCEDURES

Samples have been collected from the Charente and Gironde estuaries in south-western France (Fig. 1) both of which are well known from dynamic, sedimentational and chemical standpoints (Martin et al., 1970, 1971, 1973, 1976; Allen, 1973; Allen et al., 1975).

The Gironde estuary is formed by the junction of the Garonne and Dordogne rivers and represents at high tide a system of more than 600 km$^2$. The mean annual freshwater inflow is $3.0 \times 10^{10}$ m$^3$ while the tidal intrusion at the mouth reaches $9.0 \times 10^{11}$ m$^3$ yr$^{-1}$. The sediment discharge is approximately $2 \times 10^6$ ton yr$^{-1}$ and the drainage basin 80,000 km$^2$.

* Physical Research Laboratory, Navrangpura, Ahmedabad, India.

Compared to the Gironde the Charente estuary is much smaller. The mean annual freshwater flow is $1.6x10^9$ $m^3$ while the suspended sediment discharge is $9.6x10^4$ ton $yr^{-1}$ and the drainage basin 9,300 $km^2$.

Both estuaries can be classified on the average as partly mixed following Pritchard's classification (1955). Drainage areas of each consist mainly of sedimentary rock with a higher percentage of carbonate rock in the Charente watershed. In both these estuaries we have measured U and Th isotopes, turbidity, pH, oxygen, nutrients (including phosphates and silicates) fluoride, the trace and major elements of water, suspended and bottom deposits, all as a function of chlorinity ranging from zero to 18.1%. Oxygen, pH and salinity have been measured by instruments placed in situ. Small volume samples have been collected for nutrient, trace and major elements and fluoride analyses. For uranium measurement, 30-50 water samples were collected in pastic arboys after centrifuging in a continuous flow centrifgue (11,000 rpm) in the field, and the suspended sediment was recovered for analysis. The water samples, acidified to pH 1 with 50% nitric acid and "spiked" with $^{232}U$ were also transported to the laboratory for measurement of uranium isotopes. Uranium has been recovered from the water samples using the activated charcoal procedure of Kosjlakov and Ezova (1966) and used in river and seawater by Huu Van and Lalou (1969).

In this procedure an acidified water sample was stirred for a few hours to bring about complete homogeneity and then partially neutralised with hexamethylenetetramine to pH 4.5 -5. The activated charcoal (2-3 g/10 l of water) was added and stirred for 4-5 hours. The water was then filtered by filter paper blue No 111 and was then refiltered (through the same filter paper) to increase the efficiency of extraction. The charcoal was ashed at 600°C for 24 hours. The ashes were processes for uranium by the classical radiochemical method (slightly modified) of Ku (1966).

Intercomparison of the preconcentration method with the classical technique using coprecipitation with $Fe(OH)_3$ has been previously performed for river and sea waters. It did not show any significant differences either for $^{238}U$ concentration or $^{234}U/^{238}U$ activity ratios. The suspended and bottom sediments were processed for uranium and thorium by the methods given by Ku (1966). The planchets were counted in a vacuum chamber with a surface barrier detector coupled with a 400 channel analyser.

## RESULTS AND DISCUSSION

### 1. The environmental background

The major physico-chemical and sedimentational parametres, i.e. turbidity, chlorinity, $O_2$, pH, phosphate and fluorides are summarized in Fif. 2 - 7. These data are discussed in detail elsewhere (Salvadori, 1976) and only the major relevant parametres which control the behaviour of U-Th isotopes along the Charente and Gironde estuaries are briefly described here. Chlorinity, which is the most critical parametre, is represented on a logarithmic scale against distance from the river mouth for both the Gironde and Charente estuaries (Fig.2). The variations indicated are roughly the observed ranges of chlorinity due to seasonal tide and microlocational variations and vertical distribution. As can be seen, the degree of chlorinity varies considerably with time and space but at least three different areas are easily distinguishable : the upper estuary, where chlorinity slowly increases in the downstream direction, the middle estuary, where the gradient is a maximum; and the lower estuary where chlorinity slowly attains the same level as the seawater.

Equally important are the sedimentational characteristics of the estuaries. From the various turbidity measurements taken in the Gironde estuary, we can describe the general distribution of turbidity (Fig. 3). The most strkinig phenomenon is the so-called "turbidity maximum" (or "bouchon vaseaux") which has been observed in many estuaries of northern Europe and the United States (Martin et al., 1970; Allen, 1973; Allen et al., 1975). Its main feature is that turbidity is maximal in the middle part of the estuary and diminishes both landward and seaward. Also the area of turbidity maximum moves with the tide: landwards with flood, seawards with ebb.

In the case of dredged estuaries, alteration of the natural profile of the bottom and possibly incorrect disposal of dredge spoils may contribute to high sediment concentrations, as in the Gironde estuary where levels as high as 100 g/l ("fluid mud") have been measured (Martin et al., 1970.

No overall picture is presently available for the Charente estuary, but preliminary surveys show the same kind of sedimentational pattern with turbidities as high as 15 g/l in the upper estuary.

In any case, increasing concentration and residence time of sedimentary particles will ease solid-liquid chemical interactions. Moreover, these sediments are rich in organic matter which is the most important source of chemical energy. This organic material is far from being at equilibrium and undergoes rapid chemical changes which may influence in turn many physico-chemical reactions. This organic matter degradation is partly responsible for the dissolved oxygen and pH distributions which have been observed in both Gironde and Charente estuaries. River water pH is slightly alkaline (7.5 - 8.0 in Gironde, 7.2 - 8.1 in Charente), decreasing in areas of low salinity but never reaching acidic values. Where it meets the sea it again increases to attain sea water values (Fig. 4.). The same trend is observed for oxygen concentrations with minimum values in the low salinity areas (Fig. 5). This minimum is especially important in the Charente estuary where the naturally occurring phenomenon is accentuated by heavy pollution.

The major feature relevant to uranium study is the large excess of orthophosphate (Fig. 6) occurring in the Charente estuary, below 2‰ chlorinity. This is primarily due to calcium fluorophosphate processing which is involved in the manufacturing of phosphoric acid and fertilizers and which have greatly polluted the Charente estuary with phosphate and fluoride, the conservative behaviour of fluoride under these conditions being particularly evident (Fig. 7). On the contrary, if we compare the theoretical dilution curve of water enriched in phosphate by sea water, the experimental curve lies below that of the theoretical curve, showing the non-conservative behaviour of phosphate in this situation.

If we now consider the Gironde estuary (Fig. 6 and 7), both fluoride and phosphate behave conservatively and show that this estuary is virtually unpolluted with respect to these elements.

## 2. Isotopes of uranium and thorium

The $^{238}U$ concentration and $^{234}U/^{238}U$ activity ratios in the estuarine waters along with chlorinity are shown in Table 1; results on the suspended matter and bottom sediments are presented in Tables 2 and 3. The variation of $^{238}U$ concentration in water (Fig. 8), suspended matter and bottom sediments (Fig. 9), $^{234}U/^{238}U$ activity ratios in water (Fig. 10) and $^{230}Th/^{232}Th$ activity ratios in suspended sediment (Fig. 11), are also shown and presented as a function of chlorinity.

### 2.1 The river system

As far as rivers are concerned, our attention has been primarily focussed on the determination of dissolved uranium - 238 against total uranium-$^{238}$ discharge in the Charente and Garonne rivers. Because the turbidity varied widely, this measure has been calculated using the mean annual water and solid discharge multiplied by elemental concentrations.

The dissolved uranium percentages for the Charente and the Garonne rivers are 69 and 71% respectively. Both these figures clearly pinpoint the major importance of dissolved transport of uranium in these rivers. This agrees well with the known ability of uranyl ions to form highly soluble complexes with carboxyl ions and humic acids. Moreover, this percentage is obviously higher than the world average due to the lower solid discharge existing in these two rivers.

If we consider the riverine suspended material, it shows a drastic depletion in uranium compared to ionium ($^{230}Th/^{238}U$ are respectively 1.5 and 3.4 in the Garonne and Charente). This clearly emphasizes that thorium is poorly solubilized compared to uranium. With regards to $^{234}U/^{238}U$ activity ratios, the main feature is the excess of $^{234}U$ found in Charente and Garonne river water. These activity ratios are in close agreement with available data from river water (Moore, 1967; Scott, 1967; Bhat, 1970).

It is of interest to compare the excess of $^{234}U$ in water with its deficiency in suspended sediment.

If we assume a steady state model for land erosion processes (where input of fresh uranium from rocks is assumed to be balanced by an output in suspended matter and water) and a short residence time of uranium in the weathered layer (the age of the soils in the considered area are probably less than 20,000 years) compared to the $^{234}U$ half-life, the $^{234}U$ excess in the dissolved phase should be equivalent to its deficiency in the solid phase. Likewise, forcomputing the ratio between the dissolved and total $^{238}U$ we use the average annual water and solid discharge. This computation shows that the $^{234}U$ excess in water is not balanced by the deficiency in suspended sediment and reinforces our previous computation in the Garonne river (Martin, 1971). These observations show either that our primary assumptions do not hold or that dissolved uranium does not originate from the same kind of material as that in the suspended phase, or most probably that the excess in the dissolved phase comes from leached particulates which are not all transported to the river. The validity of the latter conclusion would allow a first estimate of the weathered material compared to that effectively transported in rivers.

### 2.2 The estuarine environment

Considering how the fate of dissolved $^{238}U$ entering the estuarine system, it is useful to compare its observed distribution with that predicted from the theoretical dilution curve of river water by sea water. The index of mixing employed being chlorinity (Fig. 8).

In the Gironde estuary, which can be considered as a natural estuarine environment, we observed a conservative behaviour of $^{238}U$, in good agreement with the conservative nature of uranium in the ocean system. Similar observations have been made in the Godavari estuary by Borole *et al.* (in press).

In the Charente estuary, the release into the environment of phosphate processing plant effluents, which are naturally enriched in uranium, increased the soluble uranium concentration from 0.4 to 2 ppb while the chlorinity remained nearly constant. If we assume complete homogeneity of the effluent in river water and consider the theoretical dilution curve of water enriched in uranium, in sea water, then all experimental points lay below this line showing that uranium is taken up from the solution during estuarine mixing.

As mentioned previously, phosphate acts non-conservatively in the Charente estuary; uranium, which can associate strongly with orthophosphate, is most likely removed from the solution in concert with phosphates. Flocculation of humic material (Gardner and Menzel, 1974; Sholkovitz, 1976) iron hydroxydes (Aston and Chester, 1973), and colloidal particulates, which is an important process in estuaries, may enhance uranium deposition.

If we compare the $^{238}U$ concentration in the waters, sediments and suspended matter of the Charente, we observe in the sediments two maxima which could correspond to the two minima existing in the water ; on the contrary, the suspended matter does not appear to be affected by the uranium removal (Fig. 9).

Our observations show that this estuary behaves as an additional sink for soluble uranium. Contrary to other sinks, such as fjords or the Black Sea (Kolodny and Kaplan, 1973; Presley et al., 1972) the Charente estuary is an oxidizing environment. Similar observations have been made in Tampa Bay, Florida by Kaufman (in press).

The different uranium behaviour observed in the Gironde and Charente estuaries emphasizes the importance of specific forms of trace metals and the difficulty of predicting their fate in pollution studies.

The $^{234}U/^{238}U$ activity ratio in water shows a very surprising evolution. While we were expecting a preferential leaching of $^{234}U$ during estuarine mixing (owing to the increase of complexing ions such as $Cl^-$), we observed in both estuaries a rapid decrease from approximately 1.3 to 1.08 in the low salinity areas. This suggests a preferential removal of $U^{234}$ from the solution as the river water mixes with sea water (Fig. 10).

In the case of the Charente estuary, we could suppose that the artificially introduced uranium due to pollution is responsible for the decrease of the activity ratio; but this is in conflict with the 1.3 activity ratio extant at the maximum soluble uranium concentration. In fact, the same phenomenon was observed in the Gironde estuary which is assumed to be "unpolluted", so we are forced to consider it to be natural process. A possible explanation would be that total or partial removal of the riverine uranium is followed by its substitution by uranium with a lower activity ratio leached from the sedimentary particles. This would agree with laboratory experiments performed by Moreira and Lalou (1972) who observe that uranium is easily leached from various rocks during water percolation but that the $^{234}U/^{238}U$ activity ratio is comparatively low whenever sedimentary rocks are examined. If such an interpretation is true, it would show that both the removal and mobilization of uranium could take place simultaneously and this might explain some of the contradictory results obtained when the fate of trace metals in estuaries has been studied. However, it must be borne in mind that this interpretation implies that the charcoal preconcentration procedure does not discriminate $^{234}U$ from $^{238}U$. As mentioned, this has been checked for river and open ocean waters, but we cannot exclude a possible fractionation mechanism due to a differentiation of uranium isotopic specific forms occurring in the estuarine system, especially in the low salinity areas, where $^{234}U$ should be less adsorbed by the charcoal powder than $^{238}U$ species. This has been observed in nearshore waters (Goldberg, personal communication), and would explain the different pattern observed in the Godavari estuary (Borole et al., in press) where the ferric hydroxide procedure has been used. In this estuary, the $^{234}U/^{238}U$ activity ratio appears to be much more conservative and even shows a small excess compared to the theoretical dilution curve (Fig. 10). Preliminary observations performed in the Zaire estuary are more similar to the Godavari data and additional experiments are being made to explain these discrepancies and to evaluate their geochemical consequences.

A major difficulty in the understanding of estuarine goechemistry is related to an inadequate knowledge of the sedimentational balance, especially the relative importance of marine and terrestrial particles within the estuarine system. As there is a strong association between isotopes and suspended sediment, we attempted to use them as sedimentation tracers. In fact, if the $^{230}Th/^{232}Th$ activity ratio in Gironde suspended matter is plotted against chlorinity, a gradual decrease is seen (Fig. 11). The same trend occurs in the Charente estuary when we distinguish between the two sets of samples collected in July 1974 and July 1975. The important feature here is that while the upstream ratios are different, the various curves converge towards a unique marine value. These observed decreases could then be easily explained by a constant mixing of riverborne detrital material with suspended matter originating from the sea. This hypothesis, which supposes that the estuarine suspended matter is partly "generated" from the ocean, would be in good agreement with the data on C and rare earth elements in suspended sediment (Martin et al., 1976; Letolle and Martin, 1970). However, clay mineralogy data for the Gironde estuary (Martin, 1971) show that differential settling of terrestrial suspended matter due to smaller size and settling velocity of montmorillonite particles occurs along the estuary, so that the different concentrations and activity ratios encountered in estuarine suspended sediment can also reflect different mineral compositions.

Finally, it must be pointed out that $^{228}Th/^{232}Th$ activity ratios have been determined in a few Charente estuarine samples of sediment and suspended matter. These ratios are always less than one (except for river sediment), this being due to the well-known diffusion of $^{228}Ra$ from the sediment (Koczy et al., 1957; Moore, 1969) which subsequently contributed to the high $^{228}Th/^{232}Th$ activity ratio observed in the superficial and deep waters of the oceans (Nikolayev et al., 1962; Somayajulu and Goldberg, 1966).

| Sample | $Cl^-$ (‰) | $^{238}U$ (µg $l^{-1}$) | $^{234}U/^{238}U$ (A.R.) |
|---|---|---|---|
| CH 04-35 | 0.030 | 0.39± 0.10 | 1.10 ± 0.05 |
| CH 04-39 | 0.028 | 0.63± 0.18 | 1.27 ± 0.09 |
| CH 04-33 | 0.025 | 2.01 ±0.31 | 1.22 ± 0.06 |
| CH 04-23 | 0.47 | 1.25± 0.05 | 1.08 ± 0.02 |
| CH 04-18 | 2.43 | 1.73± 0.18 | 1.11 ± 0.08 |
| CH 04-07 | 6.57 | 1.48± 0.20 | 1.11 ± 0.04 |
| CH 04-05 | 10.76 | 1.92± 0.20 | 1.03 ± 0.03 |
| CH 04-02 | 18.20 | 2.65± 0.16 | 1.15 ± 0.02 |
| G 76-01 | 0.011 | 0.65± 0.09 | 1.33 ± 0.08 |
| G 76-20 | 0.57 | 0.67± 0.05 | 1.08 ± 0.03 |
| G 76-04 | 2.52 | 1.01± 0.08 | 1.15 ± 0.03 |
| G 76-10 | 13.49 | 2.14± 0.12 | 1.16 ± 0.05 |
| G 76-26 | 18.10 | 2.70± 0.25 | 1.14 ± 0.08 |

Table 1. $^{238}U$ and $^{234}U/^{238}U$ activity ratios in Charente (CH) and Gironde (G) estuaries.

(A.R.) = activity ratio; errors quoted are due to counting statistics ($1\sigma$) only.

| Sample | $Cl^-$ (‰) | $^{238}U$ ppm | $^{234}U/^{238}U$ (A.R.) | $^{230}Th$ dpm $g^{-1}$ | $^{232}Th$ ppm | $^{230}Th/^{232}Th$ (A.R.) | $^{228}Th/^{232}Th$ (A.R.) | $^{230}Th/^{238}Th$ (A.R.) |
|---|---|---|---|---|---|---|---|---|
| CH. TOC.5 | 0.134 | 2.9 ± 0.2 | - | 7.3 ± 0.1 | 20.4 ± 0.3 | 1.32 ± 0.03 | 1.13 ± 0.04 | 3.4 ± 0.10 |
| CH. SOU.2 | 1.56 | 2.9 ± 0.1 | 0.95 ± 0.03 | 6.2 ± 0.1 | 18.0 ± 0.3 | 1.37 ± 0.03 | - | 2.9 ± 0.06 |
| CH.04.18 | 2.43 | - | - | - | - | 1.22 ± 0.03 | 0.81 ± 0.04 | - |
| CH.SOU.5 | 4.43 | 2.4 ± 0.1 | 0.93 ± 0.05 | 5.2 ± 0.2 | 15.8 ± 0.6 | 1.31 ± 0.08 | - | 2.9 ± 0.06 |
| CH.04.07 | 6.57 | - | - | - | - | 1.16 ± 0.04 | 0.90 ± 0.04 | - |
| CH.POR.1 | 7.15 | 2.5 ± 0.1 | 1.01 ± 0.02 | 4.9 ± 0.1 | 14.9 ± 0.3 | 1.30 ± 0.04 | - | 2.7 ± 0.06 |
| CH.POR.4.02 | 11.0 | 2.2 ± 0.15 | 0.91 ± 0.09 | 5.1 ± 0.1 | 15.9 ± 0.3 | 1.29 ± 0.03 | - | 3.2 ± 0.2 |
| CH.POR.4.03 | 13.6 | 2.3 ± 0.15 | 0.96 ± 0.04 | 4.5 ± 0.2 | 16.3 ± 0.7 | 1.2 ± 0.06 | 0.91 ± 0.05 | 2.7 ± 0.10 |
| CH.04.01 | 16.33 | - | - | 4.5 + 0.8 | 20.1 + 1.0 | 0.90 ± 0.06 | - | - |
| G 76-01 | 0.011 | 3.6 ± 0.1 | 0.99 ± 0.03 | 4.0 ± 0.1 | 13.0 ± 0.4 | 1.23 ± 0.05 | - | 1.51 ± 0.05 |
| G 76-20 | 0.57 | 3.0 ± 0.1 | 0.90 ± 0.10 | 5.0 ± 0.1 | 17.4 ± 0.2 | 1.16 ± 0.03 | - | 2.29 ± 0.06 |
| G 76-04 | 2.52 | 2.8 ± 0.05 | 0.88 ± 0.04 | 5.3 ± 0.2 | 19.15 ± 0.6 | 1.12 ± 0.05 | - | 2.61 ± 0.18 |
| G 76-16 | 8.17 | 2.6 ± 0.2 | 0.87 ± 0.03 | 4.3 ± 0.1 | 15.6 ± 0.2 | 1.09 ± 0.03 | - | 2.24 ± 0.05 |
| G 76-10 | 13.49 | 3.1 ± 0.2 | 0.94 ± 0.03 | 4.3 ± 0.1 | 16.7 ± 0.3 | 1.03 ± 0.03 | - | 1.90 ± 0.05 |
| G 76-26 | 18.10 | 3.0 + 0.05 | 0.97 + 0.09 | 4.1 + 0.1 | 18.8 + 0.6 | 0.87 + 0.06 | - | 1.86 + 0.12 |

Table 2. U - Th isotopes in suspended matter from Charente (CH) and Gironde (G) estuaries - (Carbonate-free basis)
leg - (A.R.) = Activity ratio
= errors quoted are due to counting statistics (1σ) only

| Sample | $Cl^-$ ‰ (average) | $^{238}U$ ppm * ($dpm\ g^{-1}$) | $^{234}U/^{238}U$ | $^{228}Th$ * $dpm\ g^{-1}$ | $^{230}Th$ * $dpm\ g^{-1}$ | $^{232}Th$ * ppm ($dpm\ g^{-1}$) | $^{230}Th/^{232}Th$ | $^{228}Th/^{232}Th$ | $^{230}Th/^{238}Th$ |
|---|---|---|---|---|---|---|---|---|---|
| CH.1 | 0.02 | 3.1 ± 0.03 (2.26) | 0.84 ± 0.08 | 2.02 ± 0.03 | 2.93 ± 0.04 | 9.81 ± 0.13 (2.45) | 1.20 ± 0.04 | 0.82 | 1.29 |
| CH.2 | 0.02 | 3.0 ± 0.02 (2.19) | 1.06 ± 0.06 | 1.81 ± 0.09 | 2.57 ± 0.11 | 9.68 ± 0.40 (2.42) | 1.06 ± 0.09 | | 1.17 |
| CH.6 | 0.7 | 3.2 ± 0.02 (2.34) | 0.97 ± 0.04 | 1.24 ± 0.05 | 2.21 ± 0.15 | 6.06 ± 0.30 (1.51) | 1.46 ± 0.15 | | 0.95 |
| CH.PO | 1.4 | 3.5 ± 0.01 (2.56) | 0.99 ± 0.03 | 2.78 ± 0.05 | 3.86 ± 0.06 | 13.47 ± 0.24 (3.37) | 1.15 ± 0.08 | | 1.51 |
| CH.9 | 3.9 | 2.15 ± 0.02 (1.57) | 0.91 ± 0.04 | - | - | - | - | | - |
| 04 - 11 | 5.5 | 2.7 ± 0.04 (1.95) | 0.96 ± 0.05 | 3.2 ± 0.12 | 5.94 ± 0.16 | 15.49 ± 0.51 (3.88) | 1.55 ± 0.08 | 0.87 | 3.04 |
| 04 - 07 | 11.6 | 3.7 ± 0.08 (2.70) | 0.92 ± 0.10 | 3.65 ± 0.04 | 4.82 ± 0.05 | 17.68 ± 0.19 (4.42) | 1.10 ± 0.02 | 0.80 | 1.96 |
| 04 - 04 | 14.5 | 2.8 ± 0.04 (2.05) | 0.90 ± 0.05 | 2.89 ± 0.05 | 3.8 ± 0.05 | 14. 08 ± 0.21 (3.52) | 1.08 ± 0.03 | | 1.85 |
| CH.11 | 17.0 | 2.5 ± 0.01 (1.83) | 0.99 ± 0.03 | 2.62 ± 0.07 | 4.43 ± 0.09 | 17.76 ± 0.31 (4.45) | 1.00 ± 0.05 | 0.79 | 2.42 |

Table 3 U - Th isotopes in Charente estuarine sediment.

(* carbonate free basis)

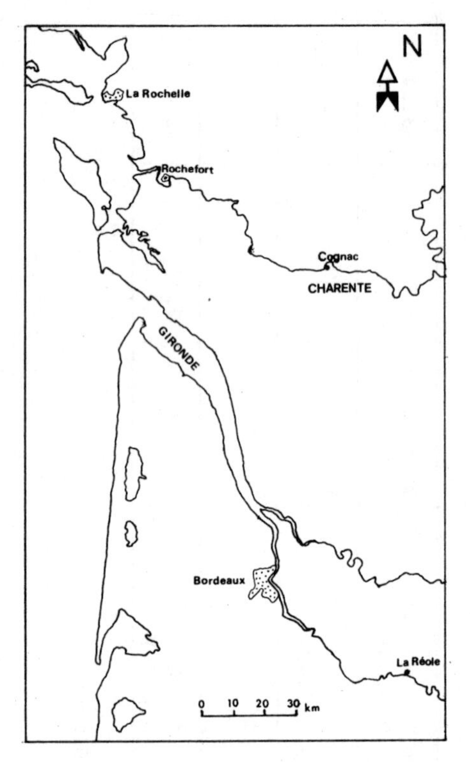


Fig. 1. General map of Gironde and Charente estuaries.

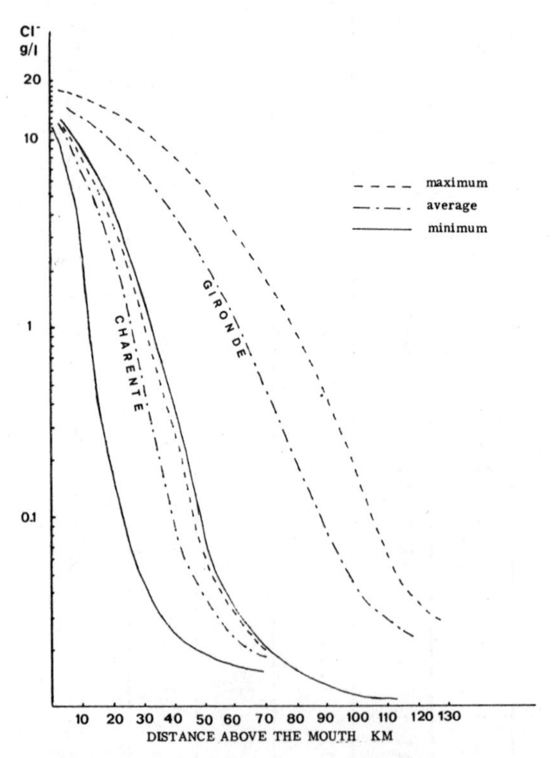


Fig. 2. Chlorinity variation in Charente and Gironde estuaries: average and extreme values.

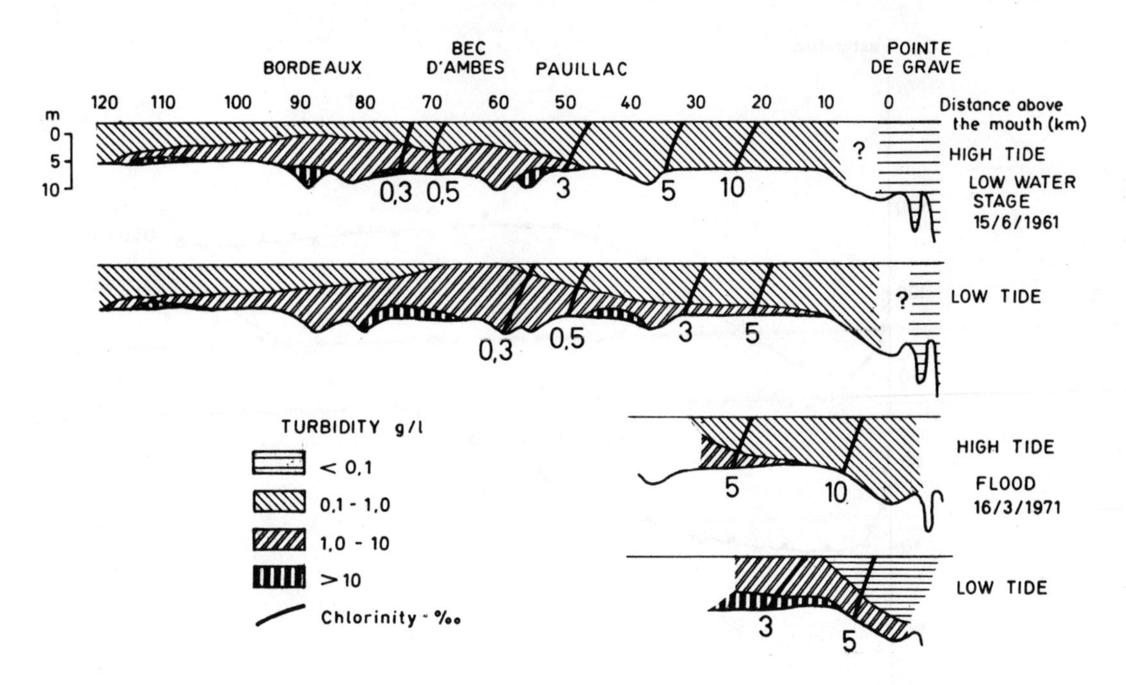


Fig. 3. General distribution of turbidity in Gironde estuary (data from Port autonome of Bordeaux and Laboratoire National d'Hydraulique of Chatou France).

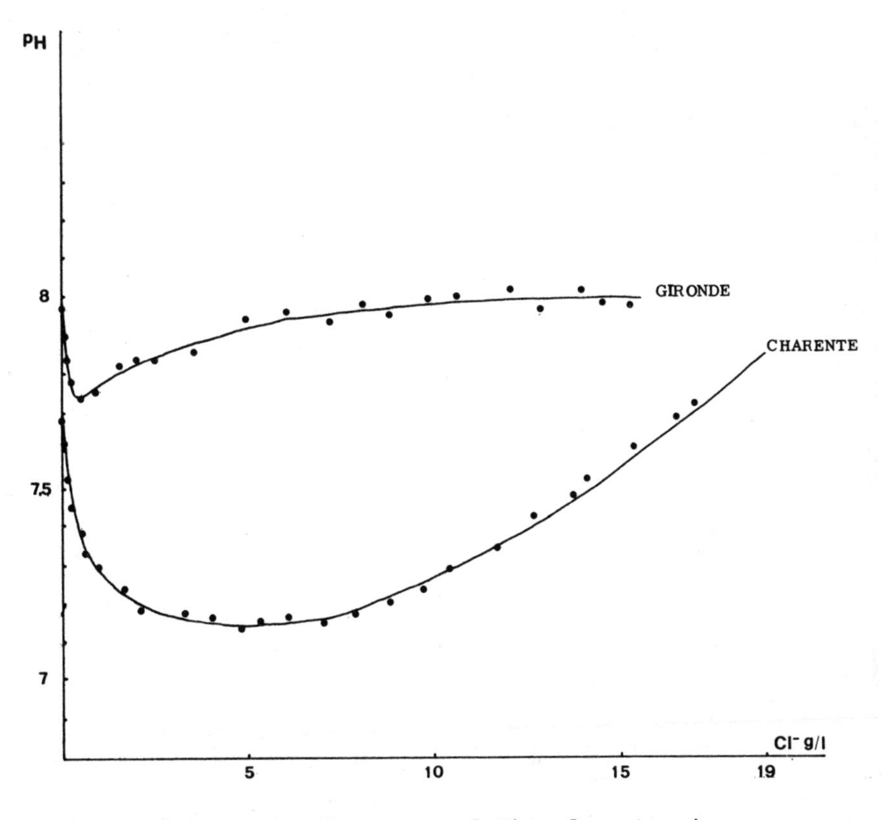


Fig. 4. pH vs chlorinity in Charente and Gironde estuaries.

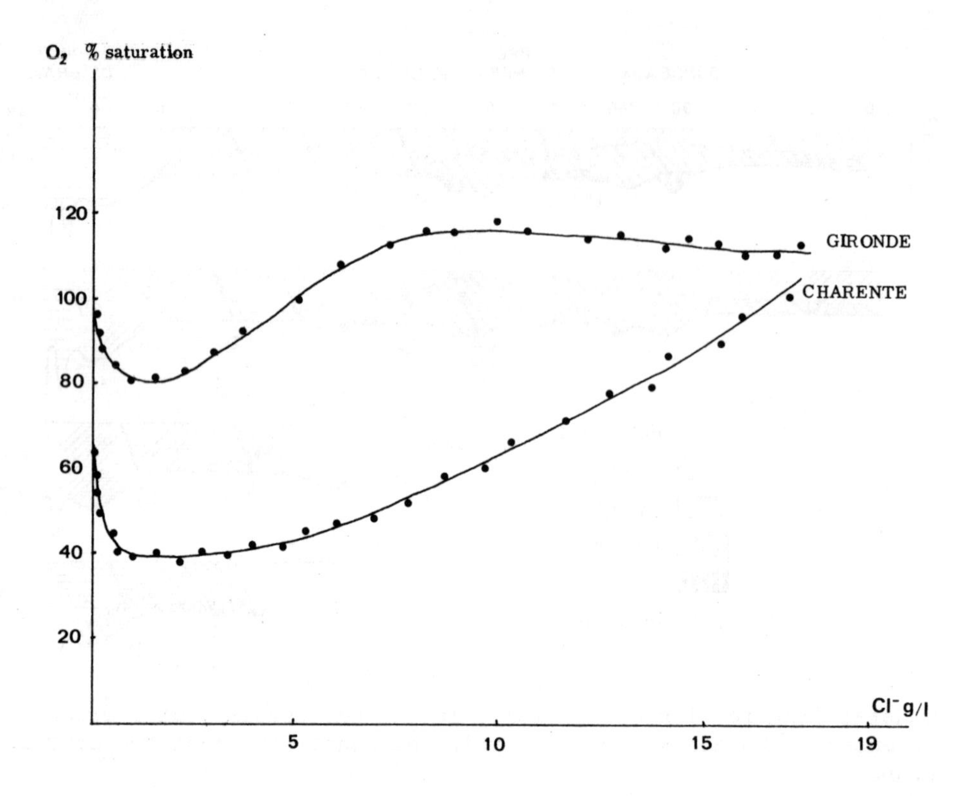


Fig. 5. Oxygen content vs chlorinity in Charente and Gironde estuaries.

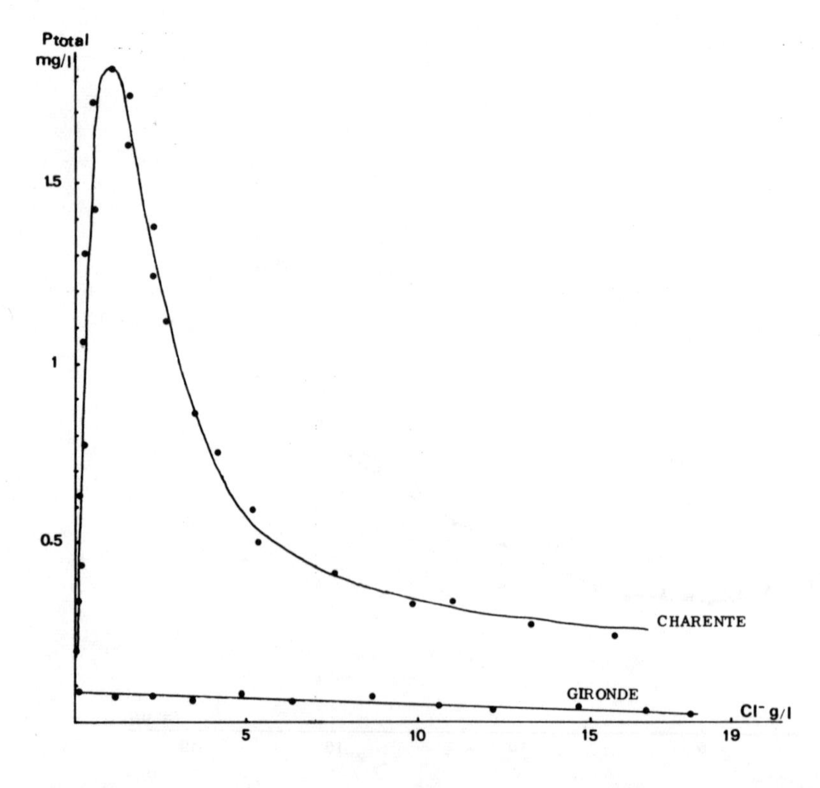


Fig. 6. Phosphate concentration vs chlorinity in Charente and Gironde estuaries.

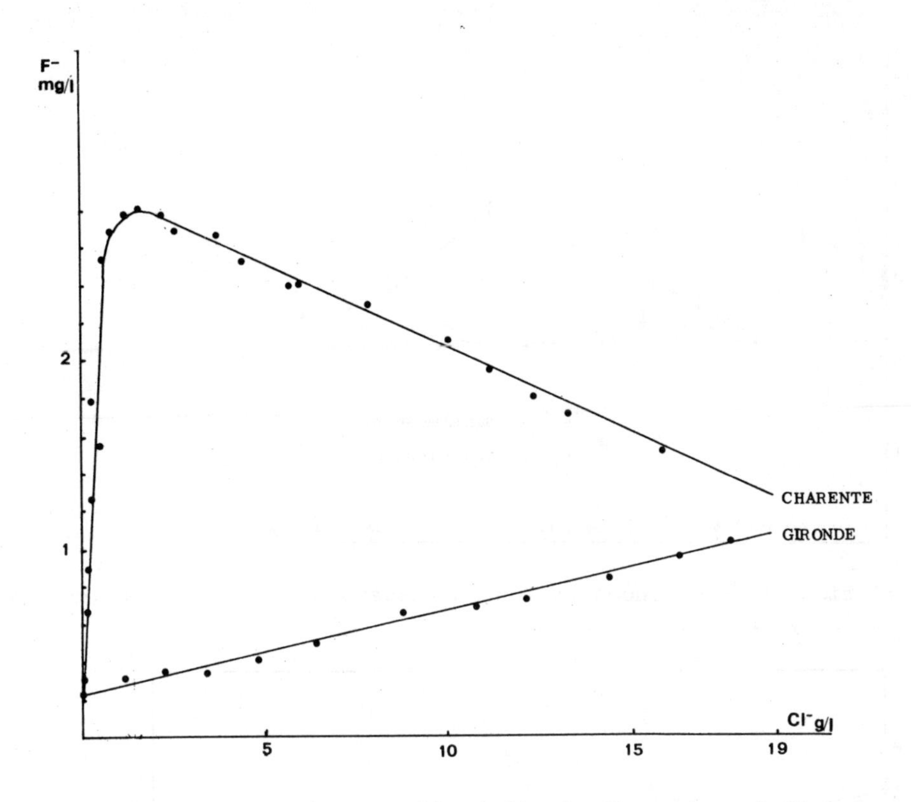


Fig. 7. Fluoride concentration vs chlorinity in Charente and Gironde estuaries.

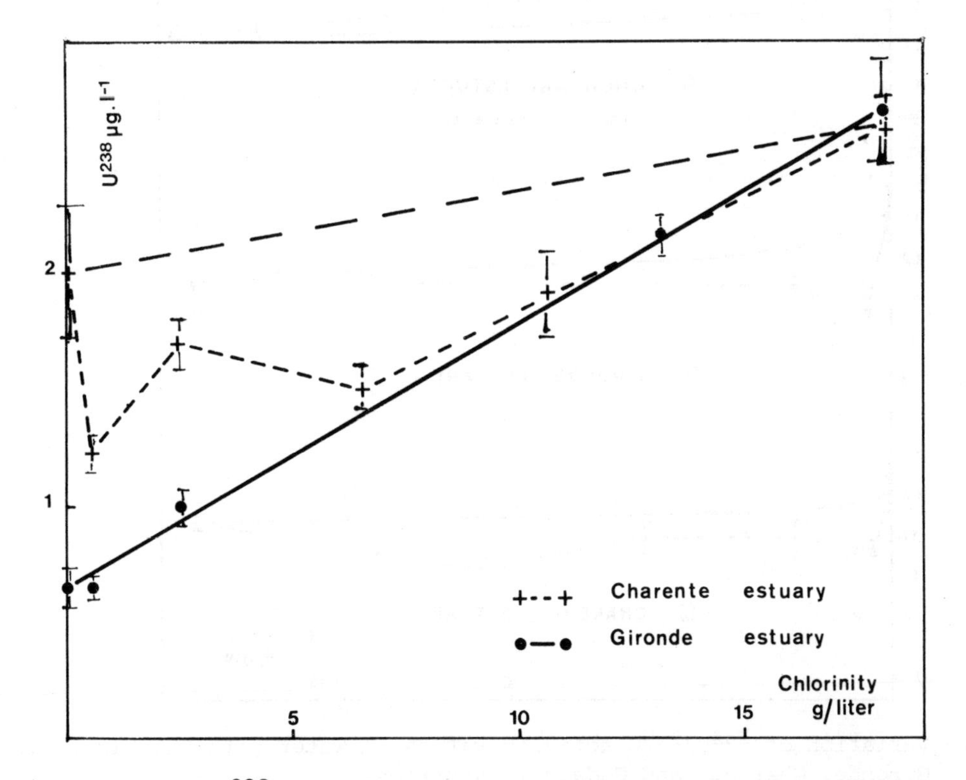


Fig. 8. Variation of $^{238}U$ concentration in water with chlorinity in the Gironde and Charente estuaries.

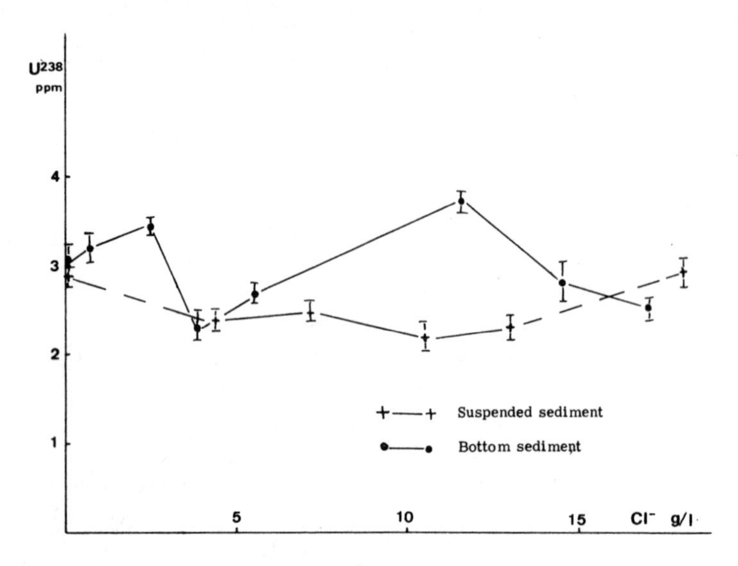


Fig. 9. Variation of $^{238}U$ concentration in suspended and bottom sediment with chlorinity in Charente estuary.

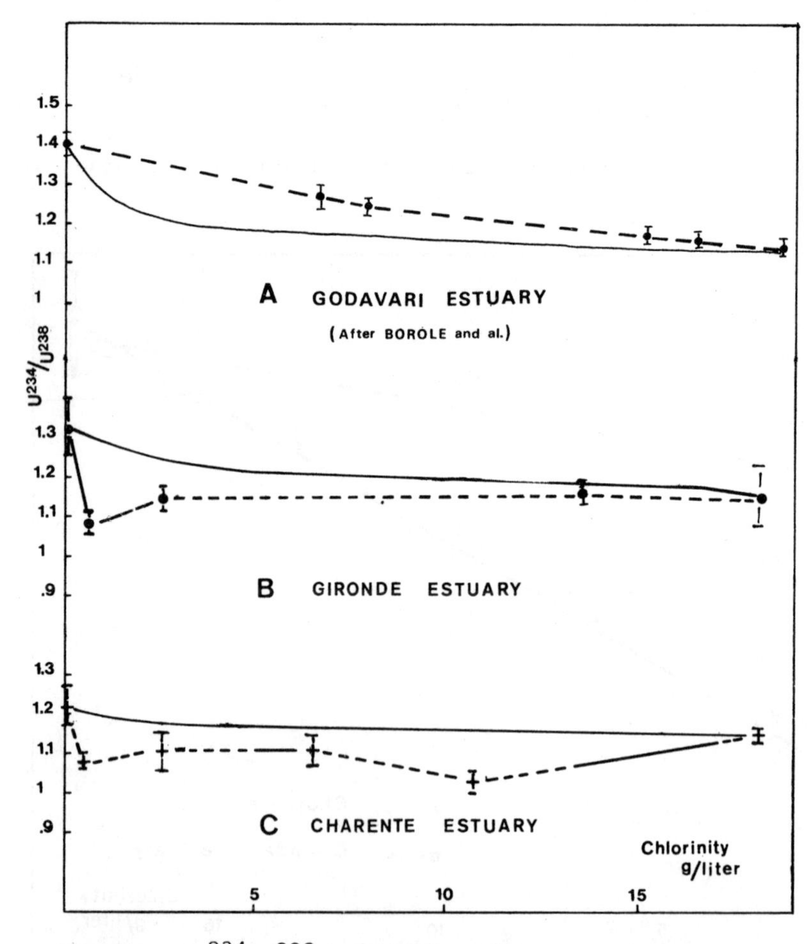


Fig. 10. Variation of $^{234}U/^{238}U$ activity ratios in water with chlorinity in the Gironde, Charente and Godavari estuaries.
(Full lines: theoretical dilution curves
dot lines: experimental measurements).

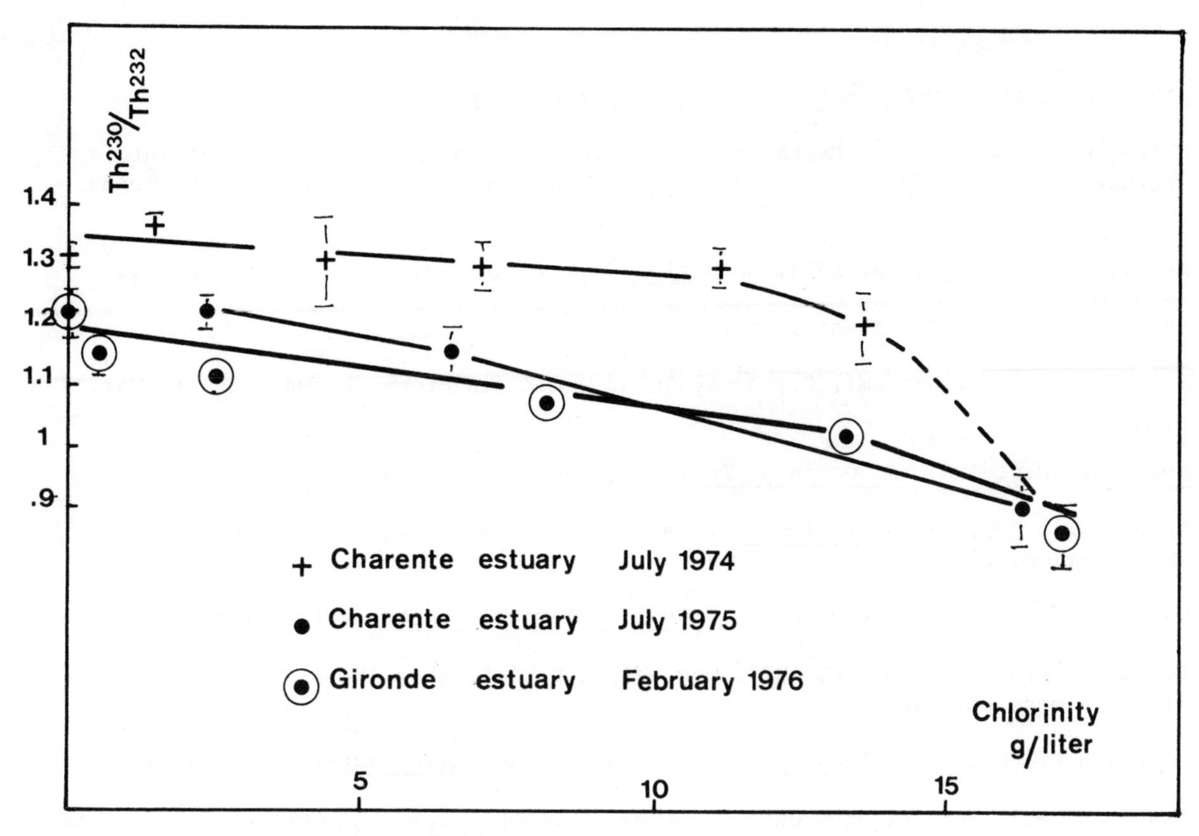


Fig. 11. Variation of $^{230}Th/^{232}Th$ activity ratios in suspended sediment with chlorinity in the Charente and Gironde estuaries.

Allen, G.P. 1973. Mem. Inst. Geol. Aquit., vol. 5, p. 314.

Allen, G.P.; Sauzay, C.; Castaing, P.; Jouanneau, J.M. 1975. 3rd Biennal Int. Estuarine Res. Conf. : Recent Advances in Estuarine Research. Galveston, Texas, Oct. 6-9.

Aston, S.R.; Chester, R. 1973. Estuarine Coast. Mar. Sci., vol. 1, p. 225-31.

Bhat, S. 1970. M. Sc. Thesis, Univ. of Bombay.

Borole, D.V.; Krisnaswami, S.; Somayajulu, B.L.K. in press, Estuar. Coast. Mar. Sci.

Broecker, W.S. 1963. J. Geophys. Res., vol. 68, p. 2817-34.

Cherdytsev, C.C. 1971. Uranium-234. Jerusalem, Israel Program for Scientific Translation, 234 p.

Gardner, W.S.; Menzel, D.W. 1974. Geochim. Cosmochim. Acta, vol. 38, p. 813-22.

Goldberg, E.D.; Bruland, K. 1974. The sea, E.D. Goldberg, (Ed.), New York, Wiley-Interscience, p. 451-89.

Huu Van N.; Lalou, C. 1969. Sonderdruck aus Radiochimica Acta, vol. 12, p. 156-60.

Kaufman, M.I. (unpublished data). quoted in Osmond, J.K. and Cowart, J.B.(in press)

Koczy, F.F.; Picciotto, E.; Poulaert, G.; Wilgain, S. 1957. Geochim. Cosmochim. Acta, vol. 11, p. 103-29.

Koczy, F.F.; Tomic, F.; Hecht, F. 1957. Geochim. Cosmochim. Acta, vol. 1, p. 86-102.

Kolodny, Y.; Kaplan, I.R. 1973. In Proc. symp. hydrogeochim. biogeochim., vol. 1, p. 418-42, Washington, D.C., Clarke Company.

Kosjlakov, V.X.; Ezova, M.P. 1966. Radiochim., vol. 8, n°. 3.

Ku, T.L. 1966. Ph.D. thesis, New York, Columbia University.

Letolle, R.; Martin, J.M. 1970. Modern. Geol., vol. 1, p. 275-8.

Martin, J.M. Doct. ès Sciences, Univ. of Paris, juin 1971.

Martin, J.M.; Hogdahl, O.; Philippot, J.C. 1976. J. Geophys. Res., vol. 81, N° 18, p. 3119-24.

Martin, J.M.; Jednacak, J.; Pravdic, V. 1971. Thalassia Jugosl., vol. 7, N° 2, p. 619-37.

Martin, J.M.; Kulbicki, G.; Groot, A.J. (de) 1973. Proc. symp. Hydrogeochim. biogeochim., vol. 1, p. 463-83. Washington, D.C. Clarke Company.

Martin, J.M.; Meybeck, M.; Heuzel, M. 1970. Sedimentology, vol. 14, p. 27-37.

Miyake, Y.; Sugimura, Y.; Yasujima, T. 1973. Papers in Meteo. and Geophys., vol. 24, N°. 1, p. 67-73.

Moore, W.S. 1967. Earth Planet Sci. Lett., vol. 2, p. 231-34.

Moore, W.S. 1969. Earth Planet Sci. Lett., vol. 6, p. 437-66.

Moreira, L.; Lalou, C. 1972. Ann. Acad. Brazil. Cienc., vol. 44, N°. 1, p. 13-8.

Nikolayev, P.S.; Lazarev, K.F.; Grashchenko, S.M. 1962. Dokl. Akad. Nauk. SSR., vol. 138, p. 489-90.

Presley, B.J.; Kolodny, Y.; Nissenbaum, A.; Kaplan, I.R. 1972. Geochim. Cosmochim. Acta, vol. 36, p. 1073-90.

Pritchard, D.W. 1955. Proc. Am. Soc. Civil Eng., vol. 81, p. 717/1 - 717/11.

Salvadori, F. Doct. 3è cycle, Univ. of Paris, Juin 1976.

Scott, M.R. 1967. Earth Planet Sci. Letter, vol. 4, p. 245-52.

Sholkovitz, E.R. 1976. Geochim. Cosmochim. Acta, vol. 40, N° 7, p. 831-45.

Somayajulu, B.L.K.; Goldberg, E.D. 1966. Earth Planet Sci. Lett., vol. 1, p. 102-6.

Turekian, K.K.; Chan, L.H. 1971. In : A.O. Brunfelt, E. Steiness, ed., Activation analysis in geochemistry and cosmochemistry, p. 311-20, Oslo, Universitetforlaget.

Veeh, H.M. 1967. Earth Planet. Sci. Lett., vol. 3, p. 145-50.

# The pertinent physical, chemical and biological characteristics of the input of sedimentary source materials entering the estuarine zone

Sanit Aksornkoae*

Sedimentation in estuarine or mangrove environments is a significant characteristic of these biotopes. Sediment is an important source of nutrient materials for plants and of food for suspension feeders living therein. The potential source of sedimentary materials entering the estuarine or mangrove zones can be classified into three groups : litter, water, and animals.

## Litter

Two means of the input of sedimentary source materials into the estuarine zone by litter are biomass input and bio-degradation. The biomass input mostly derives from producers (leaves, branches, trunks and roots), consumer decay and organic material and decay of living organisms from the river, estuary and ocean. Biodegradation is caused by bacteria and fungi (directly and at different stages of recycling of the organic material through consumers), invertebrates and fish.

Only a few investigations have been made on this subject. From the study of Florida mangrove forest, the total litter fall was approximately 650 g $m^{-2}$ $yr^{-1}$ (Snedaker and Lugo, 1973). There was no definite amount of litter decomposition rate recorded but the rate was very high during the first 20 days and then gradually decreased. The nutrient elements contained in litter have not yet been studied. Information on bio-degradation or decomposition is still lacking.

Regarding the upper ground mangrove biomass, the biomass of a red mangrove has been reported to surpass 11,000 g $m^{-2}$ of dry material for the forest and 6.4 g $m^{-2}$ of dry weight for animal biomass (Golley _et al._, 1962). In the survey of a Thai mangrove forest, it was found that the dry weight is about 52 metric tons/ha, for mangrove plantations of ages about 3, 6, 9, 11, 12, 13 and 14 years, the drey weight was approximately 20, 50, 93, 116, 149, 167 and 188 metric tons/ha respectively (Aksornkoae, 1975). An increase of 14 g $m^{-2}$ $day^{-1}$ for the yield of timber was reported by Noakes (1955) for Malaysian mangrove forest. Therefore, it is not so surprising that there is a gross production of more than 15 g $m^{-2}$ $day^{-1}$ of organic matter in mangrove forest (Golley _et al._ 1962). Due to high productivity, exportation of dead organic matter which originates from forest turnover and enters the estuarine zone is considerable.

---

* Faculty of Forestry, Kasetsart University, Bangkok, Thailand

Water

The input of sedimentary source materials entering the estuarine zone via a water route depends on : (a) river input volume, currently velocity and seasonal variations; (b) sea water input including the tidal cycle and ranges, and tidal and coastal current; and (c) water turnover, rate of mixing and gradients of freshwater to salt water.

This input can be indicated by the interpretation of ERTS - 1 images along the southern coast of Thailand (Sukhato, 1975). The study indicated the areas of heavy sedimentation occurring near to shore along the Indian Ocean western coast. These sedimentary materials comprised sand, silt and mud. Due to the strong current bordering the Indian Ocean, all these sediment particles were widely distributed and extended over the slope of the estuarine region or mangrove area. The study of mangrove soils has shown that they are very dark bluish grey, muddy, silty (silty clay), massive, structureless, plastic and cheesy. These soils may have originated from a variety of source rock materials which include sandstones and shales, metamorphosed slate, quartzite, mica schist and some kinds of igneous rocks.

The soil analyses indicated a range in textures of soils from sands and silts in the lower tide zone, silts in the normal tide zone to clays in the high tide zone. The bulk densities of the soils ranged from 0.1 to 1.5 (weight per unit volume). The chemical compositions of the soils ranged as follows : carbon from 0.5 to 38% and nitrogen from 0.05 to 0.8%, and their cation exchange capacity approximately from 5 to 68 m.e./100 grams (Zinke, 1974).

Analyses of mangrove water showed that the content of phosphate ranged from 0.56 to 1.08 µg litre$^{-1}$ while the nitrate and nitrite contents were approximately 0.62 µg litre$^{-1}$ and 0.23 µg litre$^{-1}$ respectively (Vuthisathirapinyo, 1975). It was found that at spring tides phosphate concentration was higher and the nitrate and nitrite concentrations were lower than those a neap tides.

Macro-detritus generally floats (leaves, wood debris, and roots), and it can escape from the labyrinth of trees, reaching the mud-flat or the estuarine zone through currents. However, small particles (usually up to 1 cm), remain in real suspension as a part of the leptotel (total suspended particles) and slowly settle down in mangrove channels, on mud-flats or in estuarine regions. The settling, induced by the decrease of tidal current speed, constitutes the first sediment layer. Analysis of tropical estuarine suspended matter and sediments has already proved that the composition of the leptotel is almost the same as the composition of the sapropel.

Animals

The input of sedimentary source material into the estuarine zone from animal sources is very diverse, being derived from both benthic and pelagic sources and often including shell and skeletal fragments of foraminifera, hydroids and corals, polyzoa, crustacea, echinoderms and molluscs. Minute animals may be seen in samples from the surface layer of the sediment, for example, flagellates, ciliates, foraminifera, nematodes and copepods. These animals enter the estuarine zone mostly by wave action and complement those already existing in the area. The decomposition of these organisms produces the organic matter and nutrient elements which accumulate in the sediment. Information regarding the quantity of the organic matter or nutrient elements derived from these materials in the deposits is not yet available.

Such macro consumers as birds or mammals existing in an estuarine zone or mangrove area or migrating from other habitats also provide sedimentary source materials in the estuarine zone. Certain kinds of birds, bats, primates, rodents and carnivores are commonly found in these regions, although thereis no information about the quantity of material that is incorporated from these sources.

## References

Alsornkoae, S. 1975. Structure, regeneration and productivity of mangroves in Thailand. Ph.D. Thesis (Microfilmed), Dept. of Botany and Plant Pathology, Michigan State University, 109 p.

Golley, F.; Odum, H.T.; Wilson, R.F. 1962. The structure and metabolism of Puerto Rican red mangrove in May. Ecology, vol. 43, p. 9-19.

Noakes, P.S.P. 1955. Methods of incre-sing growth and obtaining natural regeneration of the mangrove type in Malaya. Malayan Forester., vol. 18, p. 23-30.

Snedaker, S.C.; Lugo, A.E. 1973. The role of mang-ove ecosystems in the maintenance of environmental quality and a high productivity of desirable fisheries. Centre for Aquatic Science, Univ. of Florida, 404 p.

Sukhato, P. 1975. Status report of physics, chemistry, geology, microbiology of mangroves. Paper presented in national workshop/seminar on mangrove ecosystems of Thailand, 10-16 January, 1976. 13 p.

Vuthisathirapinyo, P. 1975. A comparative study of chemical and physical parametres in mangroves and open - sea waters at Phuket Island, Thailand. Paper presented in national workshop / seminar on mangroves ecosystems of Thailand, 10-16 January, 1976. 35 p.

Zinke, P.J. 1974. Effect of herbicides on soils of South Vietnam. National Academy of Science - National Research Council., Washington, D.C. 39 p.

# Group 3
# Transfer processes between the water and the sediment

# Report

## 1. INTRODUCTION

The estuaries are usually classified with respect to their hydrodynamical properties as well mixed, partially stratified or highly stratified systems. The mixing properties of the fresh and sea water masses reflect fundamental physical characteristics of the estuary : river discharge, topography, tidal effect, air-water interactions, etc.

These parametres are also of primary interest for the understanding of the chemical and biological processes. They control not only the salt intrusion into the estuary but also define residence times of the water masses in the various estuarine regions, the zone of maximum turbidity and accumulation of sediments, the exchange coefficients between river, water and air, and the relative benthic and planktonic plant activity. They thus affect directly or indirectly the behaviour of chemical species and the activity of organisms.

The knowledge of the hydrodynamical properties of an estuary is thus an indispensable preliminary step to any biogeochemical study of such an environement. In most cases these properties are poorly known due to the complexity of such systems but, despite this complexity, a large effort should be made in order to improve the understanding of physical, chemical and biological processes. We need more basic information about the time scales related to processes, such as the residence times of the water masses in various regions of the estuary, the residence times of deposited sediments, their reworking and episodic transport over large distances by storms or by high tide waters, etc.

Though the hydrodynamical parametres are of fundametnal importance, they are, however, insufficient to characterize an estuary from a chemical or biological point of view. The nature and the amounts of dissolved and suspended matter and organic constituents transported by streams are, for example, useful parametres which could be used to classify further an estuary in the context of our approach.

## 2. INPUTS

Estuaries receive inputs from multiple sources. It is necessary from a geochemical standpoint to identify and quantify these inputs. Three major sources contribute material to estuaries are (1) rivers, (2) coastal marine waters and (3) shoreline erosion. A fourth source, biogenic production of organic material, calcium carbonate, and opaline silica within the estuary, may be of great significance to the accumulation of sediment in many estuaries. In the northern hemisphere, inputs from man's activities such as domestic and industrial sewage outfalls, dredge spoil disposal, dumping and agricultural runoff may now have reached proportions equal to or exceeding natural inputs. Atmospheric inputs of some elements such as lead may be significant. Materials transported to estuaries via river systems consist of inorganic and organic terrigeneous debris, particulate and dissolved. The highly degraded products of rock weathering, often coated with trace metal-rich iron and manganese films, are reactive in the estuarine environment. Dissolved organic material may play an important role in fluvial transport of heavy metals. Bacterially mediated degradation of particulate and dissolved organic material in the sediment is the driving force for redox and pH sensitive reactions that take place beneath the sediment - water interface. Coastal marine waters transport inorganic and organic particulate and dissolved materials into the estuary. The sulphate component of sea water is of major importance as an electron acceptor and determines the amount of sulphide mineral that can precipitate in anoxic estuarine sediments. Suspended matter transported into the estuary from the coastal marine environment may contain trace metals and other materials such as hydrocarbons and chlorinated hydrocarbon compounds.

Materials contributed by shoreline erosion are highly variable and will depend upon the local geology. The flux of material from an eroded shoreline is poorly understood and is in need of further study. In coastal plain estuaries of the eastern United States, shoreline erosion produces large amounts of sand and clay from unconsolidated sediments of Cretaceous and younger formations. Commonly this material is different in chemical composition from riverborne sediments. For example, in the northern Chesapeake Bay the benthic environment is dominated by sediment derived from input. In the southern part sediment derived from shoreline erosion dominates. Striking differences in pore water chemistry result. The location and the biological productivity of an estuary can have a profound effect on the cycling and accumulation of materials on the bottom. Estuaries in temperate climates commonly contain sizeable amounts of siliceous diatom tests whereas sediments in some tropical estuaries may consist predominantly of biogenic calcium carbonate debris. These differences will be reflected in the chemistry of the sedimentary environment.

For the last several hundred years the activities of man have played an increasingly important role in changing the estuarine environment. The land areas surrounding temperate climate estuaries have been favoured by man for habitation. The use of land areas for agricultural and industrial purposes, and of the estuaries themselves for waste disposal, has had profound environmental effects. One of the most direct effects has been a change in sedimentation rates. In general, man's land use practices have drastically increased erosion and, as a consequence, increased the rate of deposition of sediment in estuaries perhaps the most significant being the clearing of land for agriculture and man's continuing agricultural practices. Construction and mining activities have large but shorter term effects on sedimentation rates. The results of these activities have not as yet been sufficiently quantified. Accompanying these sediment inputs are large quantities of anthropogenic materials such as organic matter, heavy metals, fertilizers, biocides and other toxic susbstances. The amounts, transport mechanisms and behaviour of these materials at the present time in estuarine sediments is largely unknown.

## 3. MASS BALANCE INVESTIGATIONS

Any attempts at a mass balance for elements and sediments in an estuary must involve, at the least, a knowledge of the fluvial dissolved and particulate fluxes to sediments within the estuary, and the net flux from the estuary to the ocean. Additionally, airborne fluxes are important in some cases. The fluvial flux, although variable with season, can be ascertained by detailed survey. The determination of the remaining parametres, however, may pose serious problems. The following are two examples of contributions which must be assessed in both space and time frames: a desorption of elements from estuarine sediments; and transport of sediments into, within and out of an estuary.

"Book-keeping" exercises for trace elements within the estuaries are generally expected to provide a completely different picture depending of the interval of time over which the balance is made. Nor can it be judged, either a priori or absolutely, as to what time scales should best be used. Estuarine time scales are usually much shorter than those encountered in the open ocean and range from days to years. Because of these changes in the parametres of an estuary, however, long time series measurements are called for, along the length and breadth of the estuary, both in waters and in sediments.

The entire problem is possibly not as difficult as it appears; some experimental simplifications can presumably be made. These rest on the fact that short term averages over periods of the order of weeks to a few years may be recorded in the hard or soft parts of organisms. Temporal variations in dissolved concentrations of the estuarine waters (surface and sediment-water interface) may be studied in this manner. These studies, coupled with direct determination of elemental concentrations in waters, may provide meaningful time averages for book-keeping purposes.

The isotope $^{210}Pb$ is a very convenient radionuclide for studying chronology of sediments for the last century. Considering the well documented episodic disturbancas that occur in estuarine sediments due to storms, it seems extremely important to determine the character of sediment chronology over longer intervals of time. The elements $^{32}S_i$ and $^{226}Ra$, whose half-lives are about 350 years and 1600 years respectively, seem to be useful radionuclides for this purpose and should be particularly useful in regions of high productivity.

Studies in coastal waters should provide useful additional information. Of particular use here would be studies of the concentrations of elements in coastal sediments as a function of distance from the coast.

## 4. SORPTION, FLOCCULATION AND SETTLING

Dominant mechanisms of transfer of chemical species to and from the sediment are :

*(a) Sorption of dissolved species to suspended matter (alive and dead); and (b) flocculation of dissolved and settling of suspended matter.*

(a) Sorption of dissolved species to suspended matter. This process may be split up into adsorption and absorption, the first being through ion exchange, precipitation or accumulation through bacteria or a combination of the three. Most of these adsorption processes are of a rather rapid character (minutes, hours, days). Additional incorporation into the suspended matter occurs in living material for those chemical species which permeate cell walls and/or membranes. Much slower is the incorporation (or absorption) of chemical species into dead suspended matter. Absorption may continue for certain substances or elements for months and years.

As long as these processes are of a reversible character, desorption occurs and when desorption balances sorption an equilibrium will be attained. This probably occurs between adsorption and desorption, obviously less with absorption.

Thus a redistribution of adsorbed substances among suspended particles is possible. It takes place inside the seabed where diffusion in the interstitial water causes the vertical transfer in the seabed. In the water column the sorption/desorption processes (uptake and loss for living material) cause in particular the turnover of nutrients.

Sorption to sediments from the supernatant water is more intensive when sediments are regularly brought into suspension. The flux (as also the loss from) to the seabed may increase by several order of magnitudes, as compared with conditions with a stable undisturbed seabed.

(b) Flocculation of dissolved and settling of suspended matter. It is, in particular, the increase in ionic strength and pH of the water which may cause flocculation of river organic and inorganic matter. The reactions are caused by changes of surface charge, but precipitation reactions may also occur due to changes in solubility. The settling of flocculated and suspended matter in the estuary will depend on the hydrophysical conditions such as currents and waves. Salt water counter-currents from river outflows frequently cause their return and long residence in the estuary.

All these processes in their reversed form may be responsible for the mechanisms of loss of chemical species from the sediments. Again the flux from the sediments is more rapid for regularly disturbed sediments. For the undisturbed layers below, the movement upwards of species can occur only when there exists a gradient in concentration and/or an upward water movement. Such gradients in concentration occur for substances which are brought into solution by complexation reactions or through production caused by mainly reducing conditions deep in the sediment.

## 5. BIOFILTRATION

Estuaries with substantial population of filtering shellfish have an additional mechanism of chemical species transfer to the sediments. Filter feeding organisms can collect small particulates that would not normally settle and produce aggregates that are deposited. The total water mass in the upper layers of many estuaries may be subjected to biofiltration in a few days, and this can cause a major flux of microparticulate aggregates to the sediments. The chemical composition of these aggregates can be very different from that of the original sources (e.g., enriched in transition metals) and the resulting deposits will be enriched in those elements primarily associated with the organic and inorganic microparticulates. This process leads to increased retention of small particulates in the estuaries. Where it is relevant the quantitative significance of filter feeders in settling processes must be established.

## 6. ESTUARIES AS SOURCES FOR GASES ENTERING THE ATMOSPHERE; MAN'S IMPACT ON THESE PROCESSES.

In the estuaries large amounts of organic matter are produced by the biota and contributed by river runoff. This is subsequently degraded by microbiological processes. The organic matter or inorganic electron acceptors are the limiting source for many of these processes depending upon the specific estuary. The metabolic products from some of these bacterial processes are gases such as

$CO_2$, $CH_4$, $H_2S$, $N_2$, $N_2O$ (?), $NO_2$ (?), $NH_3$.

In many estuaries man has increased the supply of inorganic nutrients and decomposable organic matter and hence stimulated the production of such gases. These gases may contribute significantly to their atmospheric burdens and to the global cycling of the elements sulphur, nitrogen and carbon. The factors governing these processes and man's impact upon these factors should, therefore, be investigated to a much greater extent in the future.

Sulphate analysis of European river samples between 1848 and 1906 showed "excess" sulphate, that is, all sulphate could not be accounted for by weathering, volcanic sulphur or sea-spray. In contrast, analyses of the river runoff of the non-industrialized continents, Asia, South America and Africa, conducted during this century show no excess sulphate. The excess sulphate is thus a consequence of the pollution in Europe. Since it is unlikely that all excess sulphate could be accounted for by anthropogenic emissions into the atmosphere and by chemical fertilizers, the inference is that organic and/or nutrient pollution may have caused an increasing

$H_2S$ production in the coastal areas. The release of sulphur is a consequence of coal and metal mining.

The main conversion by which molecular nitrogen can be released as an end product is its denitrification by bacteria. The bacteria reduce nitrate to molecular nitrogen under oxygen - limiting conditions. This process takes place in the water column when anoxic conditions are established but occurs more frequently in the upper layer of organic rich deposits. It is now suggested that the estuaries are some of the privileged regions where the nitrogen fixed by man in fertilizers may be restored to the atmosphere by the denitrification process. In the case of the Scheldt estuary, the amount of nitrogen released to the atmosphere has been estimated at 25% of the input of total nitrogen in this estuarine system. The quantification of the reduced nitrogen compoonds in estuaries is urgently needed with respect to the potential impact upon climate and the ozone layer.

Estuarine sediments act as a source of reduced and methylated species of some elements to the overlying matter as a consequence of their production in the anoxic zones, i.e. high electron acceptor availability. The organic matter produced in the upper layers of the water column and in the benthos through primary production, as well as its consequential degradation products, provides an energy source for microbial activity. As a result, the sedimentary zone can become anoxic leading to the production amongst others of hydrogen sulphide, nitrogen oxide, nitrogen gas and iodide. In addition, methylation reactions can produce methane, methyl iodide, dimethyl mercury, methyl arsines, methyl selenides, methyl tins, etc.

Through advective and convective processes and molecular diffusion these products can enter the overlying estuarine water and hence the open ocean or atmosphere. Thermodynamically unstable, but oftimes persistent, reduced species of elements can thus come into the open ocean system or atmosphere.

The fluxes of these substances from estuarine sediments are yet to be determined but may be a significant part of the cycles of some elements in the marine environment and in atmosphere.

## 7. TIME FRAMES IN ESTUARINE SEDIMENTS

The solid phase and interstitial waters of estuarine sediments may have entirely different time frames with respect to their accumulation and retention. In some estuaries, such as Chesapeake Bay, there appears to be particle-by-particle accumulation at rates of the order of millimetres per year on the bases of $^{210}Pb$ and artificial radioactive geochronologies.

On the other hand, the interstitial waters in the upper twenty centimetres approximately may be replaced in time periods of a month or even less (see papers by Bricker and Billen in this volume). Further, redox reaction such as the microbial reduction of sulphate, can reach environmental steady states in periods of a month or less.

The determination of the time frames involving the flow of dissolved constituents into and out of the interstitial waters is a problem of prime significance. The factors governing the movements of these waters in the sediments are yet to be identified and pose challenges to estuarine scientists.

The rate of mass transfer across the benthic boundary layer and the migration of dissolved substances in the interstitial water of the sediment column is controlled by the mass transfer coefficient and the concentration gradient of the dissolved substances in the pore water. The layer close to the sediment-water interface may be subjected to intensive bioturbation or physical perturbations which markedly increase the transfer of both dissolved and particulate matter in that layer.

The physical perturbations are of particular importance in shallow waters where high flow velocities occur above the water-sediment interface which is characteristic of many estuarine systems. There are several examples suggesting that the mass transfer coefficient may be several orders of magnitude higher than the molecular diffusion coefficient. A similar effect has been attributed to pressure gradients created by wave motion. The fundamental mechanisms and the effects are in need of quantitative investigation.

The fluxes across the benthic boundary layer are often calculated by multiplying the mass transfer coefficient by the vertical concentration gradient estimated graphically from the analytical data. The uncertainties in the concentration gradients resulting from sampling difficulties in the zone near the sediment-water interface may lead to erroneous flux values.

Mathematical modelling of the diagenetic processes may provide more representative approximations of these fluxes if sufficient reaction rates for appropriate species are known. The determination of the reaction rates and their variation with depth in the sediments constitute a first order problem.

## 8. SUMMARY OF PRIORITY RESEARCH AREAS

1. The knowledge of the hydrodynamical properties of an estuary is an indispensable preliminary step to any biogeochemical study of such an environment.

2. A large effort should be made in order to improve the understanding of physical, chemical and biological processes, including the time scales related to these processes, such as the residence times of the water masses in various regions of the estuary, the residence times of deposited sediments, and their reworking and episodic transport over large distances by storms or by high tide waters.

3. It is necessary from a geochemical standpoint to identify and quantify the inputs to estuaries of inorganic and organic, particulate and dissolved material from its sources: (i) rivers, (ii) coastal marine waters, (iii) shoreline erosion, (iv) biogenic production within the estuary, (v) atmosphere and (vi) from man's activities.

4. Man's land use practices over the last several hundred years (such as sewage disposal, dumping from mining, and agricultural practices resulting in extensive soil erosion) have greatly increased the introduction of sediment into estuaries, the results of which have not been sufficiently quantified. Accompanying these sediment inputs are large quantities of anthropogenic materials such as organic matter, heavy metals, fertilizer, biocides and other toxic substances. The amounts, transport mechanisms and behaviour of these materials in estuarine sediments is largely unknown at the present time and needs to be established.

5. The mass balance of elements and sediments in the estuaries should be estimated. This will require a knowledge of fluvial dissolved and particulate fluxes to sediments within the estuary, net flux from estuaries to oceans and airborne fluxes.

6. Because of the temporal changes in the parametres of an estuary, which range from days to years, long time series measurements are called for along the length and breadth of the estuary, both in waters and in sediments.

7. It is important to determine the character of sediment chronology over long intervals of time. Radionuclides are useful for this purpose, particularly in regions of high productivity.

8. The quantitative significance of filter feeding organisms in settling processes needs to be established. Since they can collect small particulates that would not

normally settle and produce aggregates that are deposited, they can cause a major flux to the sediments of microparticulate aggregates whose chemical composition may be very different from that of its sources.

9. The fluxes of certain substances from estuarine sediments need to be investigated because they may constitute a significant part of the cycles of some elements in the marine environment and in atmosphere. Estuarine sediments act as a source of reduced and methylated species of some elements due to their production in the anoxic zones; species such as hydrogen sulphide, nitrogen oxide, nitrogen gas, iodide, methane, methyliodide, dimethyl mercury, methyl arsines, methyl selenides, methyl tins, etc., are involved. Through advective and convective processes, and molecular diffusion, these products can enter the overlying estuarine water and hence the open ocean or atmosphere.

10. In many estuaries man has increased the supply of inorganic nutrients and decomposable organic matter and hence stiumulated the production of gases. These gases may contribute significantly to their atmospheric burdens and to the global cycling of the elements sulphur, nitrogen and carbon. The factors governing these processes and man's impact upon these factors should, therefore, be investigated to a much larger extent in the future.

11. The determination of the time frames, which vary from weeks to centuries, involving the flow of dissolved constituents into and out of the interstitial waters is a problem of prime significance. The factors governing the movements of these waters in the sediments are yet to be identified and pose challenges to estuarine scientists.

12. There is need for the quantitative investigation of the fundamental mechanisms and the effects involving the mass transfer across the benthic boundary layer and the migration of dissolved substances in the interstitial water of the sediment column. These are controlled by the mass transfer coefficient and the concentration gradient of the dissolved substances in the pore water, which in turn are affected in the layer close to the sediment-water interface by bioturbation or physical perturbations.

13. Mathematical modelling of the diagenetic processes may provide representative approximation of fluxes across the benthic layer - if reaction rates for appropriate species are known. The determination of the reaction rates and their variation with depth in the sediments constitute a first order problem and therefore need to be studied.

# The use of radiocarbon in estuarine research

Helmut Erlenkeuser*

## 1. Introduction

Much impressive work has been carried out both into the present state and, in some cases, the history of pollution in estuarine and coastal marine sediments. A major, yet poorly investigated problem concerns the long-term fate of these pollutants. With regard to the physical processes involved in the accumulation of pollutants, much more knowledge is needed of the time scales of the particle transport through the estuary to the ocean, of the characteristic times of short-term accumulative and erosion processes, and of the degree of sediment mixing due to these effects. All these processes strongly influence the level of pollution that may be built up in a given environment and its variation with time. An understanding of transient behaviour with respect to the increasing discharge of pollutants is of primary importance for any prediction of the possible endangerment of the estuarine and the adjacent marine environment.

The study of these problems may be facilitated through the use of radioactive isotopes. Different time scales can be examined by choice of natural radionuclides of different half-lives, e.g., the radioisotopes of polonium, thorium, lead and carbon. Man-made radionuclides, such as plutonium, may also provide chronological information.

The radio-carbon isotope ($^{14}C$ half-life: 5,700 years) provides information on the long-term depositional behaviour of biogenic carbon. Sedimentation rate studies are the most common application of $^{14}C$ in sediment research. Moreover, isotope balance calculations, which can be performed for $^{14}C$ much more easily than for most of the other radionuclides, are particularly applicable to the study of sediment mixing and redistributional processes. This application is not restricted to long-term phenomena only. Due to the entry of bomb-produced $^{14}C$ into the atmosphere, and subsequently into the other exchangeable carbon reservoirs, even short-term phenomena with time scales in the order of 10 years may be investigated by radiocarbon methods. This method has long been in use in oceanography to study the carbon dioxide exchange between the atmosphere and the ocean and also in biological systems to study widely ranging problems. The relative concentrations of modern organic matter and certain classes of pollutants in the sediments and the re-introduction of these components into the food chain are further fields of interest where the radiocarbon isotope may be successfully used.

---

* Institut für Reine und Angewandte Kernphysik, University of Kiel, - 23, Kiel, Federal Republic of Germany.

## 2. The radiocarbon age

The experimental methods in natural radiocarbon measurements have been treated in detail in the literature and will not be discussed here (see e.g., the various proceedings of the international radiocarbon conferences, or volumes of RADIOCARBON).

Occasionally confusion arises when presenting $^{14}C$ activity figures, which are the quantities actually measured, so the radiocarbon activity of a sample is usually measured against a reference activity which is, according to common agreement, 95% of the $^{14}C$ activity of the NBS oxalic acid standard. This reference activity (=0.95 NBS oxalic acid activity) is adjusted to the activity of wood from A.D. 1850 with $\delta^{13}C = -25$ ‰ and age corrected to A.D. 1950. It thus represents the activity of a terrestrial biosphere unaffected by dead $CO_2$ from fossil fuel burning and artificial $^{14}C$ from the fusion bomb tests in the atmosphere. If based on this reference activity, the radiocarbon chronology of samples from the terrestrial biosphere starts at A.D. 1950.

Other reservoirs, however, such as the ocean or fresh waters may have lower activities than the atmosphere, and the chronology of such samples, if expressed on the conventional $^{14}C$ age scale, starts with a non-vanishing "zero-age".

The so-called conventional radiocarbon age is calculated from the ratio of sample to reference activity by means of the law of radioactive decay on the basis of a $^{14}C$ half-life of 5568 years (note that the best known physical half-life is 5730 ± 40 years). The so-defined $^{14}C$ ages are denoted as years BP (i.e. years before present = years before A.D. 1950) and should not have been corrected in any way, neither for isotope fractionation, nor for dendrochronology, nor for other effects, in order to avoid confusion with the BP notation. Corrected ages should be termed "years BP", with a comment added on how these corrections were made.

## 3. Radiocarbon in organic sedimentary source materials

The processing of oil and coal introduces considerable amounts of industrial waste products and petrochemical pollutants into natural water systems, particularly rivers. The burning of fossil fuels releases aerosols and ash particles which have a noticeable content of residual carbon and are highly enriched in certain heavy metals (Erlenkeuser *et al.*, 1974; Suess and Erlenkeuser, 1975). Large quantities of these materials are transported to the hydrosphere either directly - by dry fallout or washout from the atmosphere - or by runoff from land.

These pollutants have a pronounced effect on the radiocarbon levels of the natural environment. Since petrochemicals and fossil fuel residues contain carbon which is of infinite age compared to the $^{14}C$ half-life, they are essentially free of radiocarbon activity and hence dilute the natural $^{14}C$ level. For example, Kolle *et al.* (1972) reported $^{14}C$ activities for the suspended load and the sediments of the Rhine river which were much lower than the estimated natural level. In a similar way, Rubin and coworkers (cf. Rubin and Spiker, 1972; Spiker and Rubin, 1975) studied pollution effects in groundwater, lakes, and rivers by $^{14}C$ measurements. Because of the large difference in $^{14}C$ activity between fossil carbon and recently grown organic matter, the $^{14}C$ level is a very sensitive indicator for this type of pollution.

Stable carbon isotopes can yield additional information on the possible sources of the materials investigated, but the difference in levels of $\delta^{13}C$ between pollutants and natural matter is often too small to allow a conclusive distinction. For example, fossil fuels or domestic wastes and natural organic matter in river deposits have about the same stable carbon isotope composition.

Difficulties may first arise from the problem of providing samples large enough to allow a $^{14}C$ determination. Conventional low level counters as used for radiocarbon dating need about 1 to 2 g carbon. Lower carbon amounts and smaller counters result in less statistical accuracy, but may be appropriate when sampling is difficult and $^{14}C$ effects are large. Counters for extremely small samples down to 10 mg of carbon have been discussed for tracing petrochemical pollutants in the air (Currie et al., 1976; Oeschger, 1976).

Sediment cores or particle traps usually provide large enough samples and organic fractions may be concentrated by floatation techniques. Rubin and Spiker (1972) used ultraviolet radiation and oxygen stripping gas in order to oxidize the dissolved organic matter at its natural concentration in the water.

The second problem concerns the determination, by measurement or estimation, of the natural $^{14}C$ level in the particular system under investigation. This level must be known for reference if isotope balances are to be calculated.

The initial concentration of $^{14}C$ in natural fresh waters has been extensively discussed in connexion with ground water dating (cf. IAEA, 1970). Disregarding any local or regional pecularities, the general ideas are that the dissolution of $^{14}C$ free fossil carbonates by modern soil-$CO_2$ dissolved in the seepage water and the further exchange with excess soil-$CO_2$ results in $^{14}C$ activities which are about 80 to 90% of the reference activity for temperate, non-arid climates and well developed soil profiles. For open waters then, the $^{14}C$ level gradually increases, by $CO_2$ exchange, towards that of the atmosphere. The isotope fractionation effects must also be taken into account. These occur between terrestrial plants, atmospheric $CO_2$, dissolved molecular $CO_2$, bicarbonate, and fresh water organic matter.

Allochthonous carbonaceous materials from different sources (each of which is presumably at a different activity level, in a different form and either dissolved or particulate) may further alter the $^{14}C$ level of a river. No general rule can be given as to how to estimate the relative contributions of the different effects, and a good background knowledge on the hydrogeology, climatology, and geography of the catchment areas of the river and its tributaries must be available - including the history of human settlement, agriculture, and industrial development. Such data may provide a more conclusive interpretation of the sedimentary record, which must not be spoken of exclusively in terms of pollution as it may reflect other, natural or man-induced changes of environmental conditions.

The natural $^{14}C$ level in a given environment can hardly be measured today because it is masked by various effects of pollution. However, undisturbed sediment sequences may be found which allow the reconstruction of $^{14}C$ levels in the past. The situation has become even more complex today because of the dramatic increase of the $^{14}C$ level in the atmosphere caused by nuclear weapons tests. The behaviour of the exchangeable carbon in nature may be described as a system of interacting reservoirs of different sizes and exchange rates, almost all of which are unknown. It is thus very difficult to predict accurately the response of the radiocarbon isotope in a river based on increased atmospheric $^{14}C$ level. Sampling along the river and its tributaries may help to elucidate the natural effects, and equally will help to identify the major sources of pollution.

Sewage sludges and domestic wastes which are major sources of pollution show relatively high $^{14}C$ activities at present since the organic constituents are mainly of terrestrial origin. They are at the same high level as today's atmosphere and percolate through to the natural water system without major delay.

Sedimentary particulates in the estuary may originate in part from the adjacent sea. In the past, the radiocarbon activity of marine organic matter was not far removed from that of the atmosphere - before large-scale fossil fuel burning and atomic weapons tests caused the steady state distribution of radiocarbon in the different reservoirs to be dramatically disturbed. Due to the exchange between the oceanic surface layer and the deep sea, the $^{14}C$ level of the surface water was slightly lower than in the atmosphere, at about 96% when corrected for isotopic fractionation to $\delta^{13}C = -25$ ‰ (Mangerud and Gulliksen, 1975; Gillespie and Polach, 1976). Marginal seas such as the Baltic, where the water residence time is large, might have had higher $^{14}C$ levels due to $CO_2$ exchange with the atmosphere.

The release of inactive $CO_2$ by the burning of fossil fuel caused a gradual decrease of the atmospheric $^{14}C$ level by about 2% up to A.D. 1950. The corresponding effect in the open sea probably would have been much smaller. Drastic changes, however, have taken place since the fusion bomb tests in the atmosphere. Models have been developed which described the uptake of bomb-produced $^{14}C$ by the ocean (Gulliksen and Nydal, 1972). The present $^{14}C$ in the ocean surface water of the northern hemisphere is about 110% of the $^{14}C$ reference activity, while it is about 140% in the present-day atmosphere. Separate models including hydrographic factors must be developed for marginal seas - and in some cases, for coastal waters as well - to reconstruct the variation of the $^{14}C$ level of the past decades as a basis for an appropriate understanding of the radiocarbon stratigraphy in recent sediments.

## 4. The redistribition of sediments

It has long been recognized (Emery and Bray, 1962) that the $^{14}C$ activity of near-surface marine sediments is significantly lower than one would expect from the $^{14}C$ level of the overlying water. The radiocarbon ages of these sediments are generally too old by 1000 to 2000 years for the organic fraction and even by more for the carbonate carbon. At least for coastal deposition zones, the high surface ages of the sediment appear to reflect redistribution and mixing of differently-aged deposits by the action of waves and currents. These processes seem to affect the sediments on time scales typically of the order of $10^3$ years. We do not know how effective these processes are with respect to the redistribution of newer particles, to what degree modern pollutants might become diluted in the long term by these effects, and what might be the long-term behaviour of the coastal sediments which have been accumulating particle-by-particle throughout the recent period.

We have studied sediment cores from the North Sea and the Baltic to gain a more detailed insight into the depositional behaviour of the radiocarbon isotope and the carbonaceous particulates in the off shore environment (Erlenkeuser, 1976 a). The cores were collected from different locations: from the Fladenground area in the northern North Sea; from the tidal flats off the western coast of Schleswig-Holstein, Federal Republic of Germany; from the Kiel Bight; from the western Baltic Sea; and from the Bornholm Basin, southern Baltic.

The radiocarbon distribution with depth for the organic fraction of muddy sediments from Kiel Bight is presented in Fig. 1. The cores were taken at depths greater than 20 m where bottom currents and wave action are usually small and fine-grain size classes prevail. Sediment accumulation proceeds approximately linearily with time at a rate of 1.4 mm/year for core KI-620 (from a local basin at 28 m depth), and at 0.45 mm/year for the upper part of core KI-1105, which was taken from a submarine channel at a depth of 27 m where meteorologically induced bottom currents modified the accumulation of sedimentary particles.

Disregarding the $^{14}C$ age deviation found immediately below the depositional interface, a sediment surface age of about 800 years BP was obtained for both cores by extrapolation. This age is in contrast with the apparent age of Kiel Bight water (in the pre-industrial era), which was lower than that in the open ocean (cf. Mangerud and Gulliksen, 1975) and amounted to 100 - 150 years only. Even then, the corrected surface age was too old by far to represent the true depositional age. These errors of the $^{14}C$ ages from the extrapolated sedimentation line in the upper part of the core (as well as the coinciding increase of certain heavy metals as found by E. Suess in the same depth ranges) must be ascribed to pollution effects and thus prove the young depositional age of this sequence. The stable carbon isotope ratios are about -20‰ to -22‰ on the PDB scale and are well within the range of marine plankton. The high (extrapolated) sediment surface age thus demonstrated the presence of old, natural organic carbon, probably of marine origin.

Another excellent example of the low radiocarbon level in modern deposits is shown in Fig. 2. The sediments originate from the tidal flats of the North Sea off the coast of Schleswig-Holstein. The cores were taken from large, man-made pits which were excavated in 1952 and 1962, respectively, each in turn silting up very rapidly (Wagner, 1974). Winter layers are laminated due to tidal effects while the stratification of summer layers is destroyed by burrowing organisms. X-ray photography thus allowed a precise determination of the true depositional age (Unsöld, 1974).

Marine deposits have been accumulating in this area since the late Atlantic time when the sea entered this shallow coastal zone. Wave action and tidal currents caused the deposits to be thoroughly mixed and frequently redistributed; the resulting $^{14}C$ age of these sediments was comparatively high, about 2000 to 3000 years. In Fig. 2, the $^{14}C$ dates of the organic fraction are arranged by depositional age instead of depth. They are about the same for both cores and are fairly uniform although the grain size composition indicates widely differing weather conditions and varying strengths of tidal effects during the depositional history of these sediments.

A residual systematic variation of the radiocarbon activity between deep and superficial layers may be ascribed to the input of bomb-produced radiocarbon to the marine environment, mainly after 1962/63. The analysis of the activity differences between samples taken at different depths and the water indicated that the modern organic matter makes up only 10% of the total organic content. In the upper layer of core KI-973, the modern carbon fraction was increased by rootlets of pioneer plants now growing at the surface of this pit.

The independence of the $^{14}C$ activities from grain size composition, a feature also found for Kiel Bight sediments, demonstrated that the high $^{14}C$ age of these young deposits was not an effect of, say, local erosion of varying intensity, but rather resulted from redistribution processes in a more extended area. The results suggested that the modern particulates supplied to the sediments were considerably diluted by the older deposits accumulated in the past.

As a further example, two sediment cores from the Fladenground area in the northern North Sea were studied (Fig. 3). According to the high sand content of these deposits, relatively strong bottom currents appear to prevail, and low accumulation rates were found. Sediment surface ages were about 1700 years BP for the organic fraction and about 4000 to 5000 years BP for the carbonate fraction. Further results recently obtained confirmed the general shape of the sedimentation curves as shown here. The internal age composition was quite different for the organic fraction and the carbonate carbon. The $^{14}C$ age distributions have nevertheless developed quite similarly and gave about the same sedimentation rates.

The above results show that the radiocarbon age is reproducible with respect to the physical processes affecting the $^{14}C$ composition of the sediments. It appears that the radiocarbon content of the sediments is influenced by a great number of small-scale interactions between the sea and the bottom; these produce a statistically well-defined mixing effect which may remain constant over longer periods of the depositional history.

More detailed insight into the sediment redistribution phenomenon was obtained from a sediment core from the Bornholm Basin, southern Baltic Sea, taken at 71 m depth (Fig. 4). This core could be dated independently from $^{14}C$ (Suess and Erlenkeuser, 1975). The sediments from about A.D. 1900 have a $^{14}C$ age of 1200 years BP. At less than 14 cm into the sediment an increasing amount of dead carbon from fossil fuel residues and/or petrochemical pollutants diluted the natural $^{14}C$ level and resulted in higher $^{14}C$ ages ("ash-effect"). Similarly, certain heavy metals were significantly enriched above their natural background - as represented by the deeper samples. Finally, the $^{14}C$ age decreases in the uppermost 3 cm layer due to the input of man-made radiocarbon arising from the fusion bomb tests in the atmosphere.

The $^{14}C$ age distribution in this core was interpreted on the basis of a model which considered the sedimentary organic matter to be a three-component mixture of a redistributed or eroded fraction ($c_{er}$), a recent organic fraction ($c_{rec}$), and a pollutant fraction ($c_{foss}$) of fossil carbon. The details are discussed in a following section. It turned out that the eroded organic fraction amounted to more than 70% of the total organic carbon and had a $^{14}C$ age of 1500 years. Since the stable carbon isotope ratios do not present evidence that land-derived organic matter is the possible source of this old carbon, redistribution processes affecting young marine deposits must have been continuous throughout the depositional history of this core. The mixed sediments are thought to have been subsequently transported to and re-deposited in the basin studied, along with the autochthonous organic matter settling down from the water column.

## 5. Differentiation between natural and anthropogenic organic carbon in the sediments

The muddy deposits in the deeper parts of the estuarine or near-shore marine environment, at locations where accumulation is undisturbed, offer a good opportunity for studying the effects and the history of pollution, by comparing the deeper, unaffected layers with the modern deposits. However, the complex interference of natural conditions and pollution effects can only be explored by cooperation of a broad variety of different scientific disciplines. The carbon isotopes, in this context, may be helpful to differentiate natural materials and certain classes of pollutants insofar as these are different with respect to their isotopic compositions. In particular, the important class of petrochemical pollutants and fossil fuel residues - which are essentially free from radiocarbon - may be sensitively detected by the dilution effect they have on the radiocarbon levels of the sediments. In addition, waste products discharged from sewage plants may often be identified since due to the high level of bomb-produced $^{14}C$ in the terrestrial biosphere, their content will significantly exceed the present $^{14}C$ level in either the marine or most fresh water environments.

The effect of fossil carbon bearing pollutants is seen in Fig. 1 in samples from core KI-620 from the Kiel Bight. The sedimentation rate was fairly constant at about 1.4 mm/year for depths greater than 20 cm. In the upper strata, dead carbon from fossil fuel residues gradually increased towards the depositional interface and diluted the natural $^{14}C$ level, resulting in higher $^{14}C$ ages. Similar results were reported by Baxter and Harkness (1975) for a Scottish estuary. It is easily calculated from an activity balance that a contribution of 1% of fossil carbon reduces the $^{14}C$ age by about 80 years. The fossil carbon content in Kiel Bight sediments was thus about 12% of the total sedimentary organic carbon. As has been already mentioned, this ash-effect was paralleled by an increase of certain heavy metals such as cadmium, lead, zinc and copper.

The age distribution for a core from the Bornholm Basin, is presented in Fig. 4 (Suess and Erlenkeuser, 1975). The radiocarbon ages demonstrate first the effect of redistributed, older marine organic matter which caused a $^{14}C$ age of about 1200 years BP for the deposits of the pre-industrial era; secondly, the contribution of fossil carbon pollutants which have entered the sediment since about 1910; and thirdly, the increase of the $^{14}C$ level in the marine environment due to the entry of man-made $^{14}C$ arising from the nuclear weapons tests in the atmosphere.

For purposes of interpretation, the sedimentary organic carbon is assumed to represent a three-component mixture of a recent autochthonous fraction, $c_{rec}$, with the respective activity of the water body, a fossil fraction, $c_{foss}$, free of $^{14}C$, and a so-called eroded or re-distributed fraction, $c_{er}$. The activity of the eroded fraction is not known a priori, but we may assume that this activity and the rate of supply of the eroded organic carbon have been constant throughout the core. The $^{14}C$ level of the eroded fraction generally appears to be fairly reproducible in near-shore sediments, and the uniform lithofacies in the present core suggests similar sedimentation conditions during this short accumulation sequence.

Based on these assumptions, we can compare the $^{14}C$ activities in the pre-bomb era sediments with the pre-industrial sediments from about A.D. 1900. The corresponding variations of the carbon concentration and $^{14}C$ activity can be analyzed then for (a) the fossil carbon content, $c_{foss}$, and (b) the excess concentration, $\Delta c_{rec}$, of the recent carbon fraction over that at the reference depth. No information is obtained through this model on the amount and the $^{14}C$ age of the eroded fraction. The results are shown in the right hand column of Fig. 4. The fossil carbon content amounts to about 4 or 5% of the total organic carbon and is essentially constant at less than 10 cm of sediment depth. Equally, the model suggests considerable variations of recent carbon supply.

A check on this model is provided by the bomb effect seen in the uppermost layers of the core. The response of the sediment to the altered water activity during the nuclear era can be analysed for the relative concentrations of the recent and the eroded carbon fractions and also gives the $^{14}C$ age of eroded matter. The procedure is again to set up the activity balance equations for the two different depths under the above assumptions of constant supply and $^{14}C$ activity of the eroded matter over the depositional period considered.

The initial $^{14}C$ activity of the recent organic carbon in the respective sediment layers was obtained from an activity model of the Baltic Sea water, which could be derived from $^{14}C$ measurements of modern water samples and older shells from the pre-bomb era. The response time of the water against atmospheric $^{14}C$ variations was found to be 10 years. As a consequence, the $^{14}C$ level of the Baltic during the last few years when the uppermost sediment layer was deposited should have been essentially the same as today.

Comparing the surface layer with a depth just sufficient to insulate the material from the bomb effect, the recent carbon fraction was found to make up only one quarter of the total sedimentary organic carbon while the bulk of organic matter appeared to be allochthonous (Fig. 5). The eroded fraction was assumed to be the same at the two depths compared. Its $^{14}C$ age, as calculated from the model was 1500 years.

Secondly, the surface layer could be compared with the depth below the industrial era sediments. The recent carbon fraction, $c_{rec}$, at that depth was found to be only 10% of the total organic carbon, while the eroded carbon, $c_{er}$, and its $^{14}C$ age were about the same as before. This strongly supported the hypothesis of a constant content of eroded matter in the core, which could be calculated at about 3.6% on a dry weight basis (heavy line in Fig. 5). These results also confirmed the previous model by which the fossil carbon pollution was estimated.

The present study indicated that the recent organic fraction may be but a very small part of the total sedimentary organic matter and may be subject to large variations even over short periods of depositional history: the recent carbon amounts to 0.5% on dry weight basis or 10% of total organic C at 17 cm of sediment depth, and 1.3% of dry matter or 27% of total organic C at the surface. What are the implications for the microbiological processes in these deposits?

The variation of the recent carbon supply to Baltic Sea sediments is probably related to hydrographic events. Occasional inflows of large water masses with high salinity from the Kattegat area into the deeper basins of the central Baltic must cause the older bottom water layer to be shifted towards the surface. The high nutrient content of these older waters will strongly stimulate primary production in the water column. The increasing amount of the recent organic matter as shown by our model parallels the strong increase of the fish yield in the Baltic over the same period of time (Dietrich and Kalle, 1965).

## 6. Biological application of natural radiocarbon in sedimentary habitats

Radiocarbon measurements on near-shore marine sediments have demonstrated that most of the sedimentary organic carbon is allochthonous and probably has been frequently eroded and redistributed. The time scales of these processes seem relatively long, and the radiocarbon activity has significantly decreased, by radioactive decay, as compared to the $^{14}C$ level in the modern or even the pre-bomb marine water and biosphere. In modern deposits of the North Sea and the Baltic Sea sediments $^{14}C$ ages of up to 2000 years have been found and the autochthonous fraction of modern organic carbon was only about 10 or 20% of the total organic matter. If microbial activities in the sediment attack this bulk of old organic matter to an appreciable extent, then the bacterial biomass should have a correspondingly low radiocarbon level - whereas a high level would be expected if modern carbon were the dominant fraction degraded. Similarly, the microbial degradation of $^{14}C$ free petrochemical hydrocarbons - particularly from oil - in heavily polluted estuaries and marine sediments might be investigated by radiocarbon.

One major difficulty will arise from the problem of isolating a large enough quantity of bacteria to allow $^{14}C$ determination. The amount of carbon should ideally be about 1 g. However, microcounters for samples down to 10 mg carbon may be used. The accuracy obtained with these counters is rather poor (in the order of 10% of the NBS reference activity), and much of the sensitivity of the method is then lost.

If bacteria cannot be analysed directly, larger organisms feeding on them may be collected, such as worms and polychaetes and possibly certain bottom-dwelling fishes. As a first approach we studied mussels from sedimentary habitats (Erlenkeuser, 1976 b). These bivalves take their food from the suspension layer at the immediate sediment-water interface. In the tidal area of the North Sea the activity of the water, and hence of modern plankton, was about 120% STD (STD = NBS reference activity) after correction for isotopic effects (i.e. at $\delta^{13}C = -25‰$), while the sedimentary organic matter had a $^{14}C$ level of 75 to 80% STD. The mussel's tissues were at the same level as the modern water, and no uptake of old - and hence probably degraded - organic constituents could be detected. Selective digestion of sedimentary organic matter has been also reported by George (1964) who for a polychaete found that only about 8% of the organic material available for feeding in the muddy sediments was digested by the animal. This may be also indicated by the low uptake of certain sediment-bound radionuclides in sediment feeders (Beasley and Fowler, 1976).

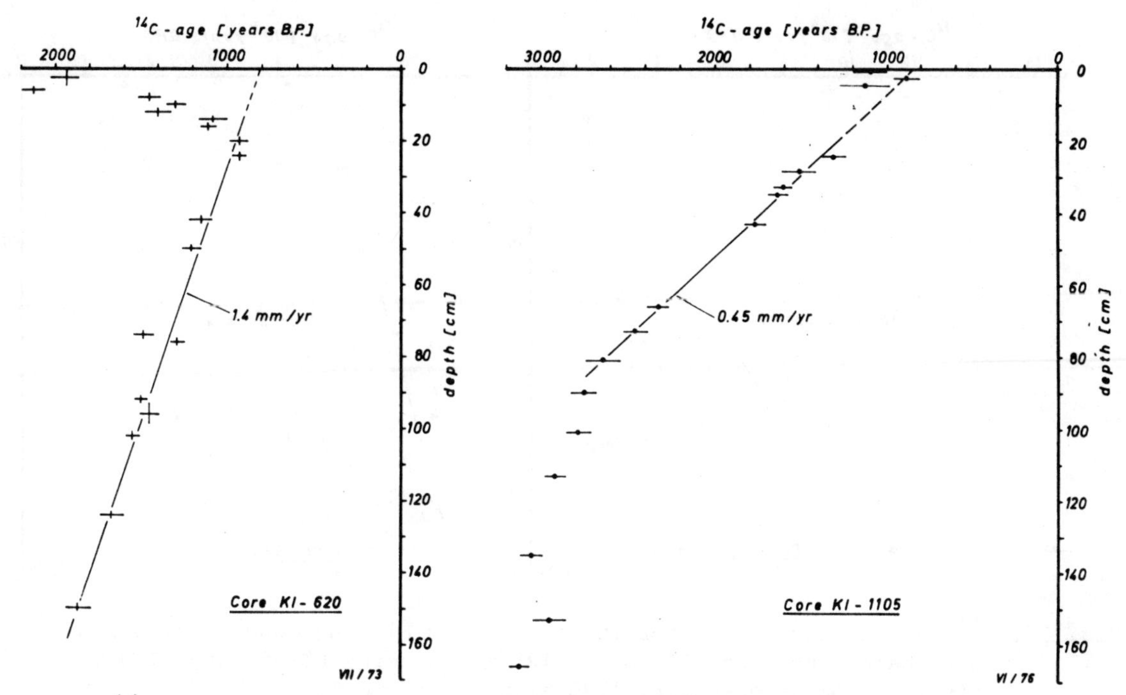


Fig. 1 : $^{14}C$ age depth distributions for two sediment cores from Kiel Bight : (a) core KI-620, outer Eckernförde Fjord, 28 m water depth, (b) core KI-1105 from Vejsnäs submarine channel, medium Kiel Bight, 27 m water depth (from Erlenkeuser, 1976 a).

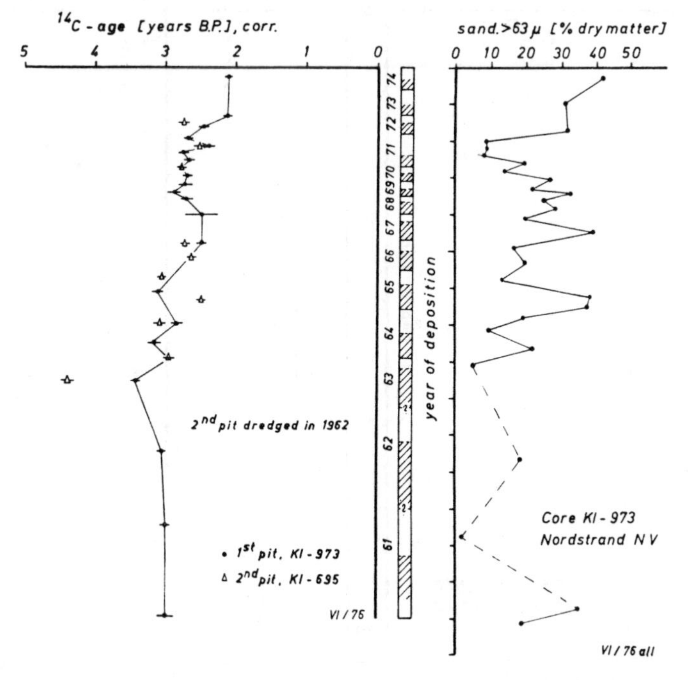


Fig. 2 : $^{14}C$ age and sand fraction > 63μ with depositional age for two cores from the tidal flats south of Nordstrand Isle, western coast of the Federal Republic of Germany. Shaded depth intervals indicate winter layers, light intervals indicate summer deposits. A grain size analysis by G. Unsöld, Kiel (from Erlenkeuser, 1976 a).

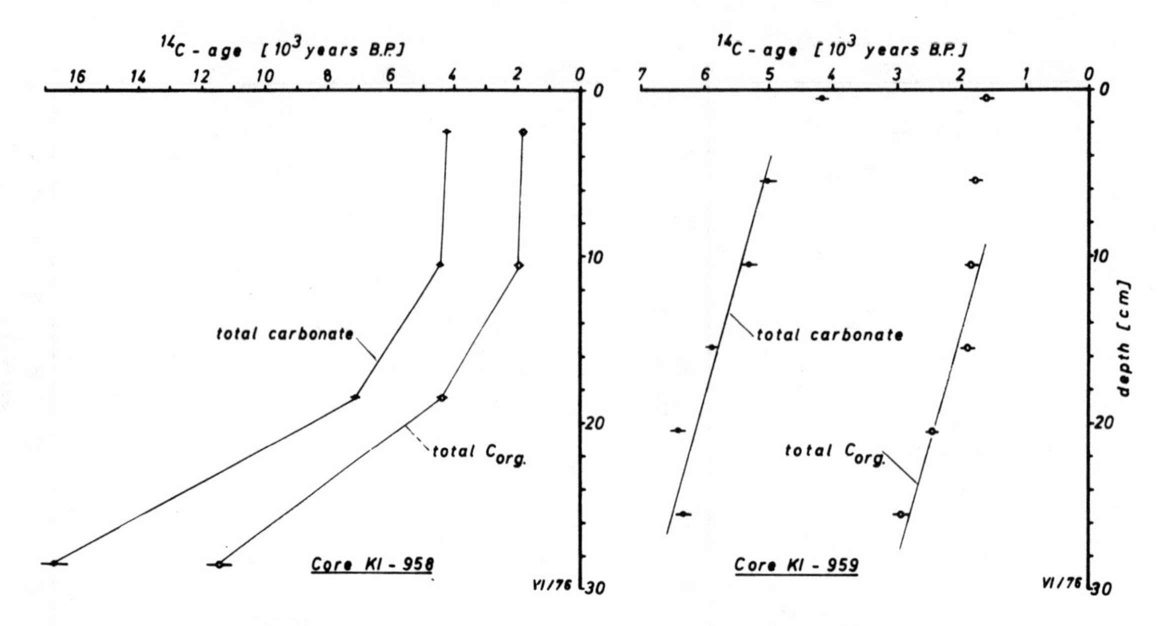


Fig. 3 : $^{14}$C age against depth for two cores from Fladenground area, northern North Sea. Water depth was 146 m and 125 m for cores KI-958 and KI-959, respectively (from Erlenkeuser, 1976 a).

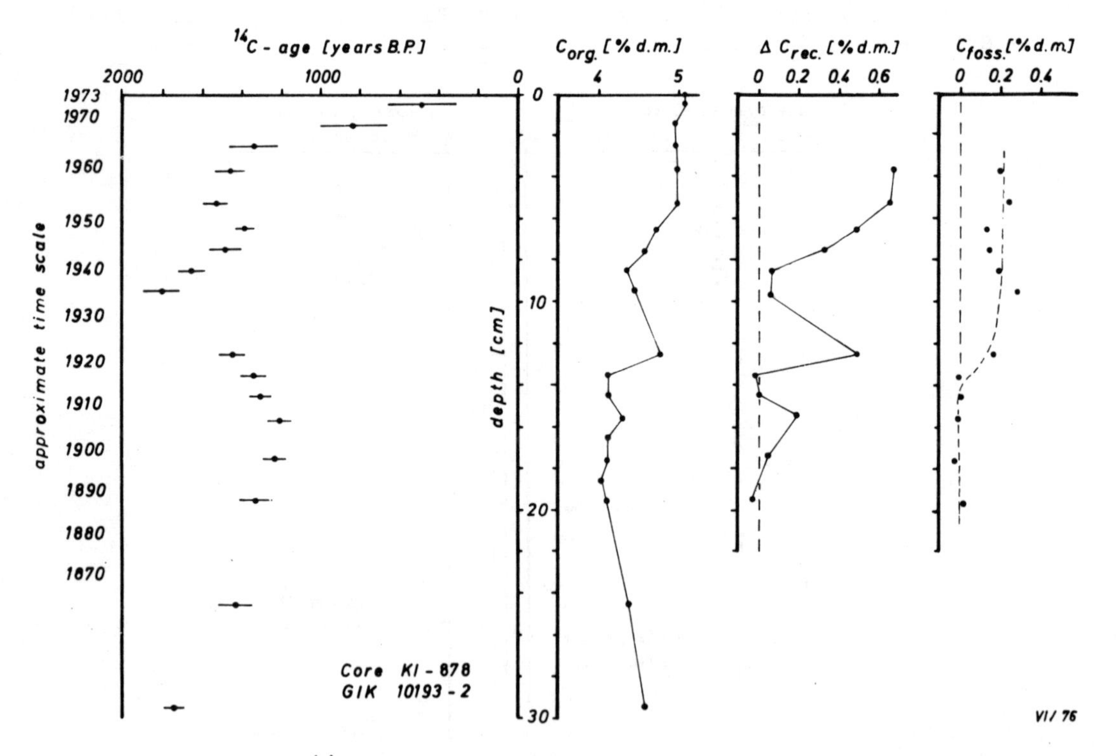


Fig. 4 : Variation of $^{14}$C age and organic carbon constituents with depth in core KI-878 from the Bornholm Basin at 71 m water depth. The variation of the total organic carbon content over that at a reference depth of 17 cm is split into a fossil fraction, $c_{foss}$, and a recent carbon fraction, $\Delta c_{rec}$ (from Erlenkeuser, 1976 a).

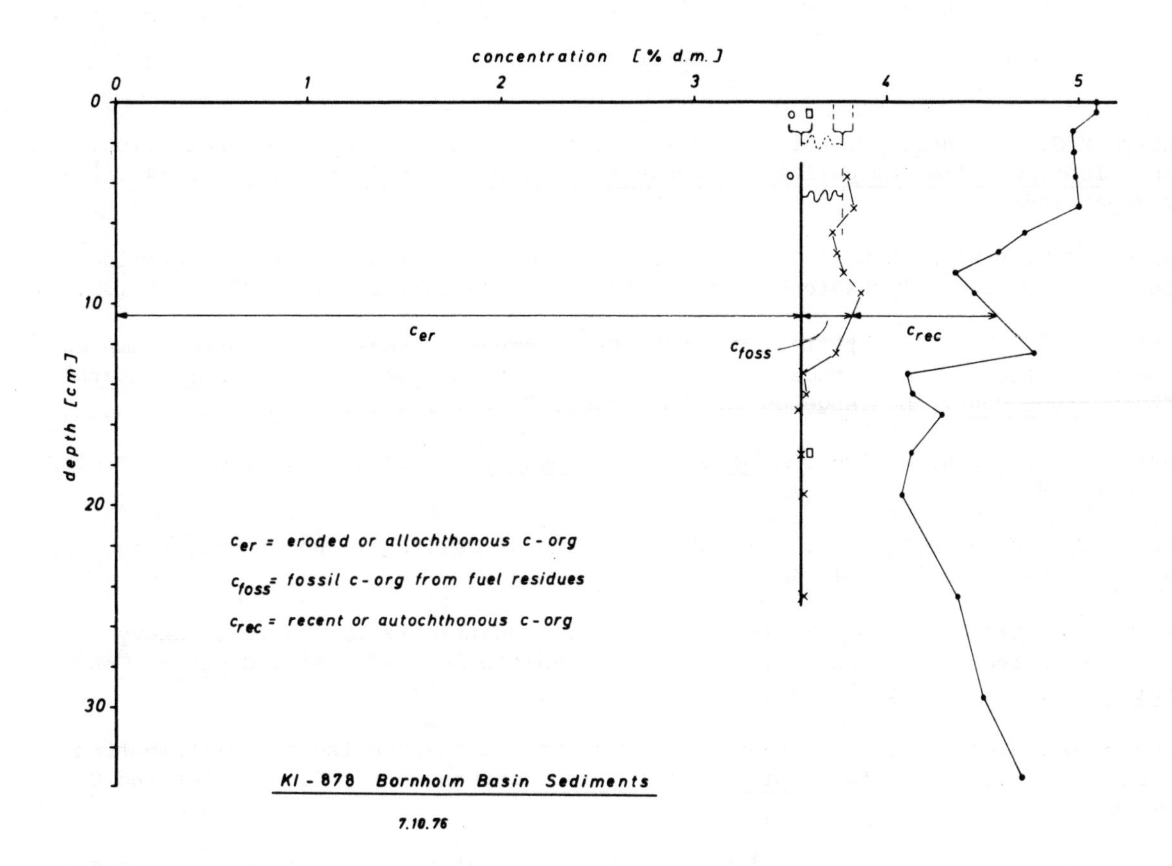


Fig. 5 : Variation of the composition of sedimentary organic matter in core KI-878 from Bornholm Basin, as deduced from $^{14}C$ activity distribution with depth (• = total organic carbon; x = $c_{foss}$ + $c_{er}$ ; o , ◘ = $c_{er}$ as calculated from $^{14}C$ activity difference between surface layer and two deeper strata; heavy line = average of o and ◘ ; total organic C = $c_{rec}$ + $c_{foss}$ + $c_{er}$; $c_{foss}$ of surface layer was extrapolated from below as indicated.)

References

Baxter, M.S.; Harkness, D.D. 1975: $^{14}C/^{12}C$ ratios as tracers of urban pollution. In : Isotope ratios as pollutant source and behaviour indicators, p. 135-41, 1975, Vienna, IAEA.

Beasley, T.M. Fowler, S.W. 1976. Plutonium and americium : uptake from contaminated sediments by the polychaete Nereis diversicolor. Mar. Biol., vol. 38, p.95-100.

Currie, L.; Noakes, J.; Breiter, D. 1976. Measurement of small radiocarbon samples : power of alternative methods for tracing atmospheric hydrocarbons. Ninth internl. radiocarbon conf., Los Angeles and San Diego, June 1976.

Dietrich, G.; Kalle, K. 1965. Allgemeine Meereskinde. p. 234. Berlin, Gebrüder Bornträger, 492 p.

Emery, K.O.; Bray, E.E. 1962. Radiocarbon dating of California Basin sediments. Bull. Am. Ass. Petr. Geol., vol. 46, p. 1839-56.

Erlenkeuser, H.; Suess, E.; Willkomm, H. 1974. Industrialization affects heavy metal and carbon isotope concentrations in recent Baltic Sea sediments. Geochim. Cosmochim Acta, vol. 38, p. 823-42.

Erlenkeuser, H. 1976 a. Environmental effects on radiocarbon isotope distribution in coastal marine sediments. Ninth interntl. radiocarbon conf., Los Angeles and San Diego, June 1976.

Erlenkeuser, H. 1976 b. $^{14}C$ and $^{13}C$ isotope concentration in modern marine mussels from sedimentary habitats. Naturwissenschaften, vol. 63, p. 338.

George, J.D. 1964. Organic matter available to the polychaete Cirrifornia tentaculata (Montagu) living in an intertidal mud flat. Limnol. Oceanogr., vol. 9, p. 453-5.

Gillespie, R.; Polach, H. 1976. Radiocarbon in marine shells. Ninth interntl. radiocarbon conf., Los Angeles and San Diego, June 1976.

Gulliksen, St.; Nydal, R. 1972 : Further calculations on the C-14 exchange between the ocean and the atmosphere. p. 283-96. Proc. 8th interntl. conf. radiocarbon dating. Wellington, The Royal Society of New Zealand.

IAEA, 1970: Isotope Hydrology 1970. Vienna, IAEA, 918 p.

Kölle, W.; Ruf, H.; Stieglitz, L. 1972. Die Belastung des Rheins mit organischen Schadstoffen. Naturwissenschaften, vol. 59, p. 299-305.

Mangerud, J.; St. Gulliksen. 1975. Apparent radiocarbon ages of recent marine shells from Norway, Spitzbergen and Arctic Canada. Quat. Res., vol. 5, p. 263-73.

Oeschger, H. 1976. Recent progress in low level counting with proportional gas counters. Ninth interntl. radiocarbon conf., Los Angeles and San Diego, June 1976.

Rubin, M.; Spiker, E. 1972. $^{14}C$ activity as an indicator of sources of dissolved organic carbon in rivers and lakes. Proc. 8th interntl. conf. radiocarbon dating, Wellington, The Royal Society of New Zealand, p. 324-9.

Spiker, E.; Rubin, M. 1975. Petroleum pollutants in surface and groundwater as indicated by the carbon-14 activity of dissolved organic carbon. Science, vol. 187, p. 61-4.

Suess, E.; Erlenkeuser, H. 1975. History of metal pollution and carbon input in Baltic Sea sediments. Meyniana, vol. 27, p. 63-75.

Unsöld, G. 1974. Jahreslagen und Aufwachsrten in Schlicksedimenten eines künstlichen, gezeiten-offenen Sedimentationsbeckens (Wattgebiet südl. Nordstrand/Nordfriesland) (Annual layering and rates of accrction of muddy sediment in a man-made sedimentary basin of the tidal flats south of Nordstrand.Schleswig Holstein). Meyniana, vol. 26, p. 103-11.

Walger, E. 1974. Einfaches mathematisches Modell der Schlicksedimentation in einem ufernahen, gezeiten-offenen Sedimentationsbecken. Nachr. Deutsch. Geol. Ges., vol. 11, p. 45.

# Evaluation of the exchange fluxes of materials between sediments and overlying waters from direct measurements of bacterial activity and mathematical analysis of vertical concentration profiles in interstitial waters

G. Billen, J. P. Vanderborght*

## I. INTRODUCTION

The evaluation of the fluxes of dissolved materials across the water-sediment interface is one of the main problems in biogeochemistry of sediments. It is very tempting for this purpose to use the information contained in the vertical distribution of dissolved species in the pore water of sediments. However, two precautionary comments have to be made in this connexion :

(i) Static information (vertical distribution of a chemical species) can only be used to evaluate a dynamic magnitude (flux of materials) if, in addition, some auxiliary dynamic parametre is known. Implicitly, in many studies this auxiliary dynamic parametre has been the molecular diffusion coefficient. However, in the upper part of sediments, many other processes than molecular diffusion are responsible for the dispersion of interstitial waters (see article by Vanderborght, Billen and Wollast, in this volume). In sofar as these processes induce only isotropic movements of the interstitial waters, the formalism of Fick's law can still be applied to these situations, but the dispersion coefficient cannot be considered *a priori* to be known.

(ii) If the dispersion coefficient is known, it is theoretically possible to evaluate the flux of a dissolved species to the overlying water directly from the graphically estimated concentration gradient of this species at the sediment-water interface. However, the experimental determination of the concentration gradient is often rather inaccurate, on the one hand because the upper part of sediment cores are easily disturbed during sampling and on the other, because the gradients are often very steep in this zone; the concentration measurements are thus often, not sufficiently numerous for good estimation. In fact, from all the information contained in the concentration profile, this method uses only a small part, and moreover, the part most subject to error and inaccuracy.

We think that the best way to evaluate the fluxes across the sediment-water interface is to perform a complete mathematical simulation of the whole concentration profile, using direct, *in situ* measurement of bacterial activity or chemical

* Université Libre de Bruxelles, 50 ave. F. Roosevelt, 1050 Brussels, Belgium.

reaction rate as auxiliary dynamic parametres.

As an example, we present below an application of this methodology to the problem of evaluating nitrate transfers across the sediment-water interface and their seasonal variations in a shallow marine lagoon, the Sluice Dock at Ostend (Belgium). The two different sediment types of this lagoon - sandy sediments with organic content lower than 2.5%, and muddy sediments with organic content higher than 3% - offer two very different situations which will be compared.

## II. DIRECT MEASUREMENTS OF NITRIFYING AND DENITRIFYING ACTIVITY IN THE SEDIMENTS

### a. Nitrifying activity

Direct estimations of the activity of nitrifying bacteria have been obtained by measurement of nitrogen inhibition of $^{14}C$-bicarbonate incorporation in sediment samples (Billen, 1976). Some typical profiles of nitrifying activity in the sandy sediments are given in Fig. 1. The general pattern exhibited is a roughly constant activity in the few upper centimetres, then a sudden reduction to zero of this activity with increasing depth. Much higher rates of activity are found in summer than in winter (Fig. 2) but the depth of the layer in which the activity occurs decreases considerably from winter to summer; mostly, this depth is very close to the brownish oxidized layer of the sediment which itself shows important seasonal variations (Fig. 3).

### b. Denitrifying activity

Estimation of potential denitrifying activity in sediments has been obtained by adding nitrate to freshly collected sediment samples and following its disappearance as a function of time. Typical results of such an experiment on muddy sediments of the Sluice Dock are given in Fig. 4. From the data of this figure, it is clear that denitrification reasonably fits a Michaelis-Menton kinetics with a Km near 50 µmoles/l. The maximum rate of denitrification (Vmax) in the muddy sediments of the Sluice Dock increases exponentially with temperature ($Q_{10}$ = 2.3) (Fig. 5).

## III. A MODEL OF VERTICAL NITRATE DISTRIBUTION IN THE PORE WATER OF THE SEDIMENTS

Vertical profiles of nitrate concentration in interstitial waters from the Sluice Dock are of two major types. In the first type, A, the concentration of nitrate is lower in the interstitial water than in the overlying water, and decreases with depth. Because the concentration gradient at the sediment-water interface is negative, nitrates diffuse from the overlying water into the sediment, which acts as a nitrate sink for the Dock. In the second type, nitrate concentrations are higher in the upper few centimetres of the sediment than in the overlying water and decrease below that. In this case the concentration gradient at the interface is positive, implying that nitrates diffuse out of the sediments. (Fig. 6 and 7).

### a. Profiles of type A

Type A profiles of nitrate concentration are found in the muddy sediments of the Sluice Dock during spring (Fig. 6). At this time of the year, these sediments, fairly compact and rich in organic matter, are reducing in nature over the first few millimetres. Denitrification predominates because conditions required for nitrification occur only in a thin, unconsolidated oxidized layer of a few millimetres in depth.

The diagenetic equation corresponding to this situation can be written (neglecting the sedimentation term) :

$$\left(\frac{\delta C}{\delta t}\right)_z = D \left(\frac{\delta^2 C}{\delta z^2}\right) - r_{den.}$$

$$\text{with } r_{den.} = \frac{V_{max}}{2K_m} C \quad \text{for } C \leqslant 2K_m \qquad (1),$$

$$V_{max} \quad \text{for } C > 2K_m$$

(this approximation is preferred to the rigorous Michaelis-Menten expression because the latter does not allow the equation to be solved analytically).

Comparison of the profiles of Fig. 6 with the measured values of denitrifying activity shows that the turnover time of nitrate in these conditions is in the order of a few hours. It can therefore be considered that these profiles are stationary with respect to the instantaneous conditions. The stationary solution of equation (1), with the following limit conditions

$$z = 0 \; ; \; C = C_o < 2K_m$$

$$z = \infty \; ; \; C = 0,$$

is simply

$$C = C_o \exp -\sqrt{\frac{V_{max}}{2DK_m}}\, z \qquad (2).$$

This solution is represented in Fig. 8 for the situation measured in the Sluice Dock on 8.4.75 ($C_o = 0.055$ µmoles/cm$^3$; $V_{max} = 3.6 \times 10^{-5}$ µmoles/cm$^3$sec), and for a set of values of the dispersion coefficient. Comparison with the experimental profile shows that the latter is between $10^{-4}$ and $10^{-3}$cm$^2$ sec$^{-1}$.

From the analytical expression (2) of the profile, it is possible to calculate the nitrate flux across the sediment-water interface :

$$F = D \left(\frac{dC}{dz}\right)_{z=0} = C_o \sqrt{\frac{D.V_{max}}{2K_m}} \,.$$

b. Profiles of type B

Type B profiles are found throughout the year in the sandy zone (Fig. 7) of the Sluice Dock, and during winter in the muddy zone (Fig. 6).

We have shown elsewhere (Vanderborght and Billen, 1975) that the existence of a maximum in nitrate concentrations implies the occurrence of nitrification in the upper centimetres of the sediment. This cannot be interpreted as the result of the persistence in a type A sediment, of a transient state keeping a "memory" of its sudden decrease from a higher value of nitrate concentration in the overlying water[+]. Thus, in the Sluice Dock, a type B profile necessarily corresponds to a sediment with a nitrification zone extending to a depth $z_n$ and followed by a denitrification

[+] Such unstationary situations can however occur in sediments subjected to rapid variations of nitrate concentration in overlying water. This is the case for the sediments of a zone in the Scheldt estuary, where tide effects cause periodic variations in nitrate concentration of the overlying water from 0 to 100 µmoles/l within a few hours.

zone. The diagenetic equation for this situation can be written :

$$\left(\frac{\delta C}{\delta t}\right) = 0 = D \left(\frac{\delta^2 C}{\delta z^2}\right) + k_n \, \varepsilon(z) - k_d \, C \, (1 - \varepsilon(z)) \qquad (3),$$

with $k_d = \dfrac{Vmax}{2Km}$

and $\varepsilon(z) = 1$ for $z < z_n$

$0$ for $z > z_n$ .

With the limit conditions $z = 0$ ; $C = Co$

$z = \infty$ ; $C = 0$,

the solution of this equation is :

$$C = -\frac{k_n}{2D} z^2 + \frac{\frac{k_n}{D}\left(\frac{z_n^2}{2} + \sqrt{\frac{D}{k_d}}\, z_n\right) - Co}{z_n + \sqrt{\frac{D}{k_d}}} z + Co \qquad \text{for } z < z_n,$$

$$\text{and } C = C(z=z_n) \ \exp - \sqrt{k_d/D}\ (z - z_n) \qquad \text{for } z > z_n.$$

This solution is represented on Fig. 9 for the situation measured in the Sluice Dock on 7.1.74 ($k_n = 0.5 \times 10^{-5}$ moles/cm$^3$ sec, $z_n = 4$ cm, $k_d = 0.5 \times 10^{-4}$ sec$^{-1}$) and a set of D values. By adjusting the theoretical profile on the experimental one, the value of D is determined as about $2 \times 10^{-4}$ cm$^2$/sec.

The flux of nitrate across the sediment-water interface is given by :

$$F = -D \frac{\frac{k_n}{D}\left(\frac{z_n^2}{2} + \sqrt{\frac{D}{k_d}}\, z_n\right) - Co}{z_n + \sqrt{\frac{D}{k_d}}} \qquad (4).$$

This flux is negative or positive according to the value of Co; when high nitrate concentrations are reached in the overlying water, it can happen that the sediments act as a sink of nitrate, even when nitrification is occurring in the upper layer of these sediments.
Total nitrification rate, integrated on the whole sediment column, is given by :

$$Ni = \int_0^{z_n} k_n \, dz = k_n z_n \qquad (5).$$

Total denitrification rate, integrated on the whole sediment column, is given by :

$$Den = \int_{z_n}^{\infty} k_d \, C \, dz$$

$$\sqrt{k_d D} \left[ \frac{k_n}{2D} z_n^2 + \frac{\frac{k_n}{D}\left(\frac{z_n^2}{2} + \sqrt{\frac{D}{k_d}}\, z_n\right) - Co}{z_n + \sqrt{\frac{D}{k_d}}} z_n + Co \right] \qquad (6).$$

## IV. MODEL OF THE SEASONAL VARIATIONS OF NITRIFICATION, DENITRIFICATION AND NITRATE FLUX ACROSS THE SEDIMENT-WATER INTERFACE.

The observed seasonal variations of nitrate concentration in the interstitial waters of the Sluice Dock's sediments can be entirely explained by the variations of the parametres involved in equation (3), namely Co, D, $k_n$, $z_n$ and $k_d$. The variations of $k_n$, $z_n$ and $k_d$ have been empirically related to the variation of temperature (see Fig. 2, 3 and 5). The seasonal variations of temperature can be approximately simulated by a sinusoïdal function of the time. Nitrate concentration in the waters of the Dock is best simulated by a $\cos^4$ function of the time; the value of D is assumed to remain constant throughout the year. All the empirical relations used are summarized in the following table.

| Commanding parametres | | muddy zone | sandy zone |
|---|---|---|---|
| T | temperature (°C) | $12.5 - 7.5 \cos (2\pi \frac{t - 30}{365})$ | |
| Co | nitrate concentration at the sediment-water interface ( moles cm$^{-3}$) | $0.09 \times \cos^4 (\pi \frac{t - 30}{365})$ | $0.075 \times \cos^4 (\pi \frac{t - 30}{365})$ |
| D | dispersion coefficient (cm$^2$sec$^{-1}$) | $10^{-4}$ | $2 \times 10^{-4}$ |
| **Auxiliary parametres** | | | |
| $k_n$ | nitrification rate (μmoles cm$^3$sec$^{-1}$) | $2.5 \times 10^{-6} \times 10^{T/25}$ | $1.3 \times 10^{-6} \times 10^{T/14}$ |
| $z_n$ | depth of nitrification layer (cm) | $9.5 \times 10^{-T/9}$ | $10 \times 10^{-T/20}$ |
| $k_d$ | denitrification constant (sec$^{-1}$) | $3.2 \times 10^{-4} \times 10^{T/27.6}$ | $0.3 \times 10^{-4} \times 10^{T/27.6}$ |

From these empirical relationships and the solution (3) of the diagenetic equation, the nitrate profiles in the interstitial waters of the sediments can be calculated over a complete annual cycle. This has been done in Figs. 10 and 11 for the two sedimentary zones of the Sluice Dock. Comparison of these calculated profiles with the experimental ones (Figs. 6 and 7) shows that the model indeed accounts for the major trends of the seasonal variations of vertical nitrate distribution in sediments.

By use of the relationships (4), (5) and (6), the total nitrification rate, the total denitrification rate and the nitrate flux across the sediment-water interface can be calculated. The results are shown in Figs. 12 and 13. The different combinations of the parametres $k_n$, $z_n$ and $k_d$, due to the different physical and chemical conditions in the two sedimentary zones, result in two completely different patterns of nitrate circulation, which can be, at least grossly, described by the methodology and the model presented here.

## V. CONCLUSIONS

The analysis of vertical concentration profiles of dissolved substances in the pore water is a powerful method for evaluating the fluxes of matter caused by the various biochemical tranformations in sediments. This analysis can be done with the aid of a mathematical model giving rise to the diagenetic equation, provided that the kinetics of the involved process are sufficiently known.

The example dealt with in this paper shows that it is frequently necessary in such models to distinguish successive layers in the sediment where different microbial metabolisms are active due to different physico-chemical conditions.

Apparent dispersion coefficient, as determined by way of such models and of the direct experimental measurement of biological production or consumption rates, are of the order of $10^{-4} cm^2/sec$ in the shallow sediments studied. This shows that processes other than simple molecular diffusion are responsible for the exchange within the interstitial water and across the water-sediment interface.

## References

Billen, G. 1976. A method for evaluating nitrifying activity in sediments by dark $^{14}C$-bicarbonate incorporation. Water Res., vol. 10, p. 51-7.

Vanderborght, J.P.; Billen, G. 1975. Vertical distribution of nitrate in interstitial water of marine sediments with nitrification and denitrification. Limnol. Oceanogr. vol. 20, p. 953-61.

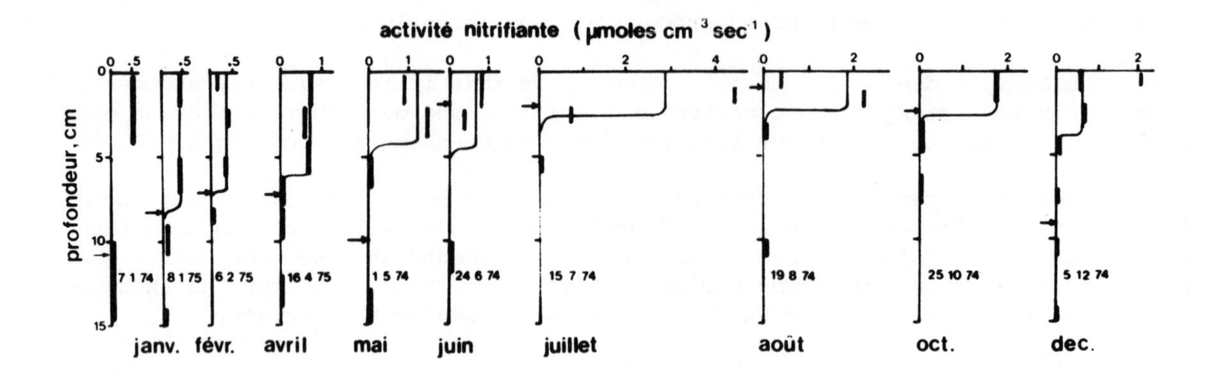


Fig. 1. Vertical distribution of nitrifying activity in the sandy sediments of the Sluice Dock at Ostend over a complete annual cycle.

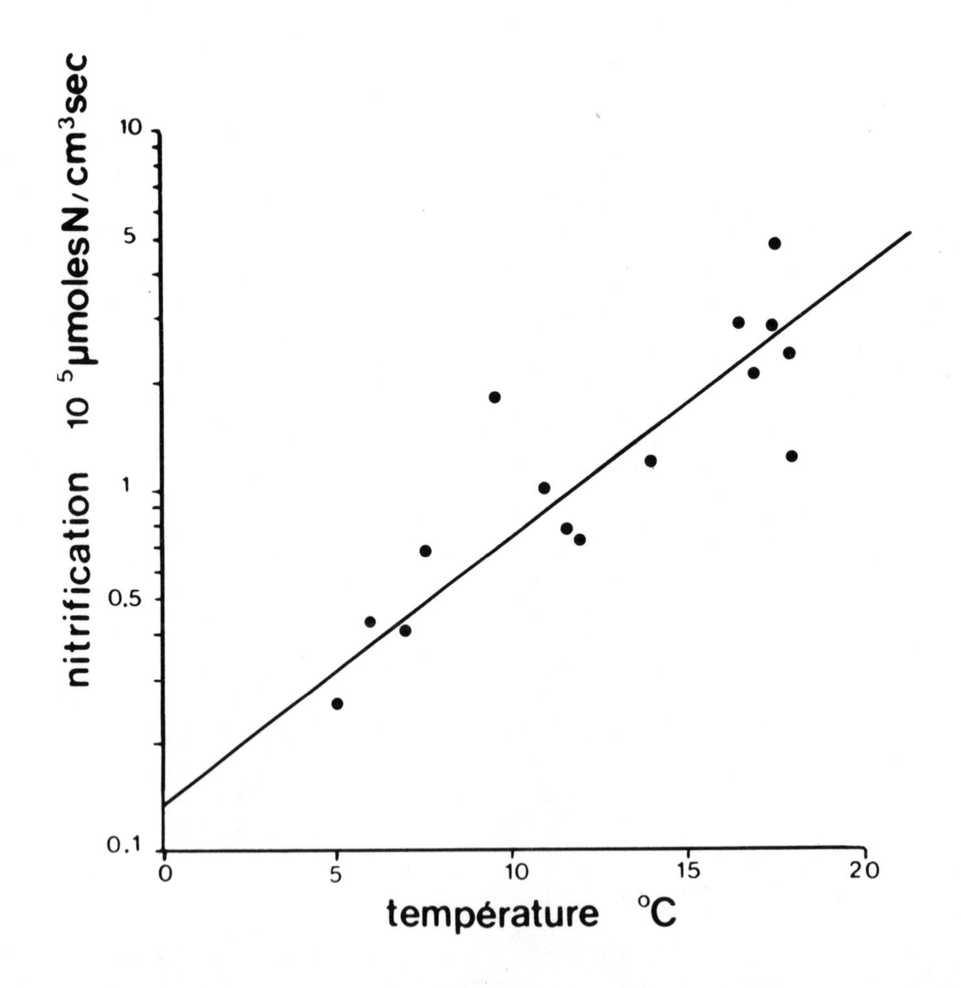


Fig. 2. Relationship between the rate of nitrifying activity in the upper layer of sandy sediments of the Sluice Dock at Ostend and the temperature of the water, over a complete annual cycle.

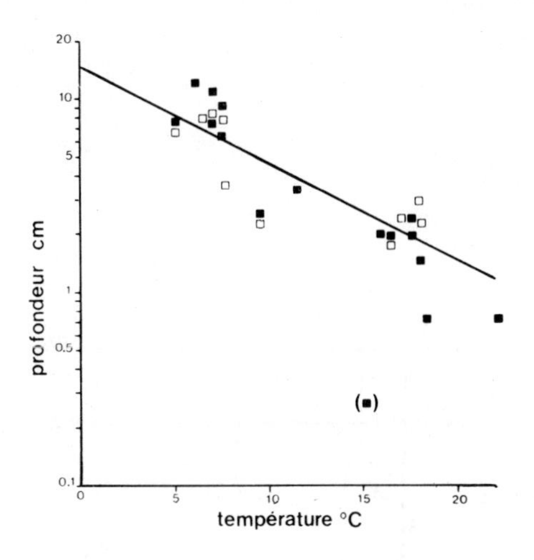


Fig. 3. Dependence of the depth of the nitrification layer (□) and of the brownish oxidized layer (■) on temperature over a complete annual cycle in the sandy sediments of the Sluice Dock at Ostend.

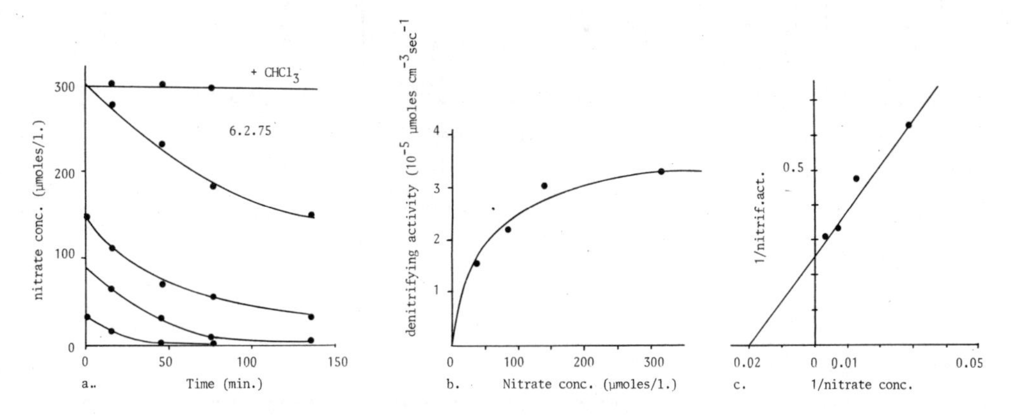


Fig. 4. Potential activity of denitrifying bacteria.

a. Kinetics of nitrate disappearancein sediment samples incubated at in situ temperature (muddy sediments collected in the Sluice Dock at 0-5 cm depth on 6.2.1975) and in a sterile control.

b. plot of initial nitrate consumption rate against concentration.

c. reciprocal plot of initial nitrate consumption rate against concentration.

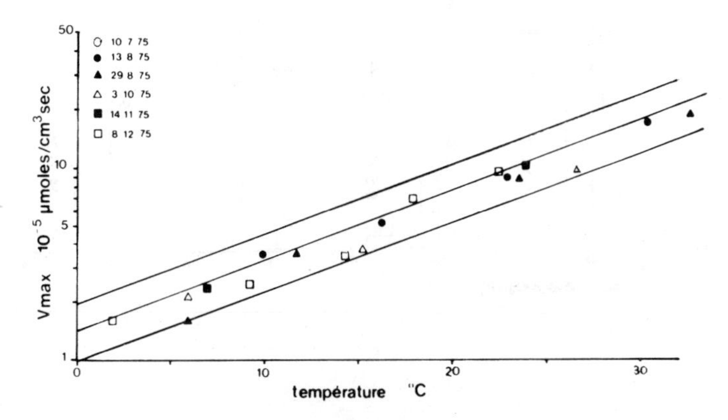


Fig. 5. Effect of temperature on the maximal denitrification rate (Vmax) in samples of muddy sediments from the Sluice Dock at Ostend.

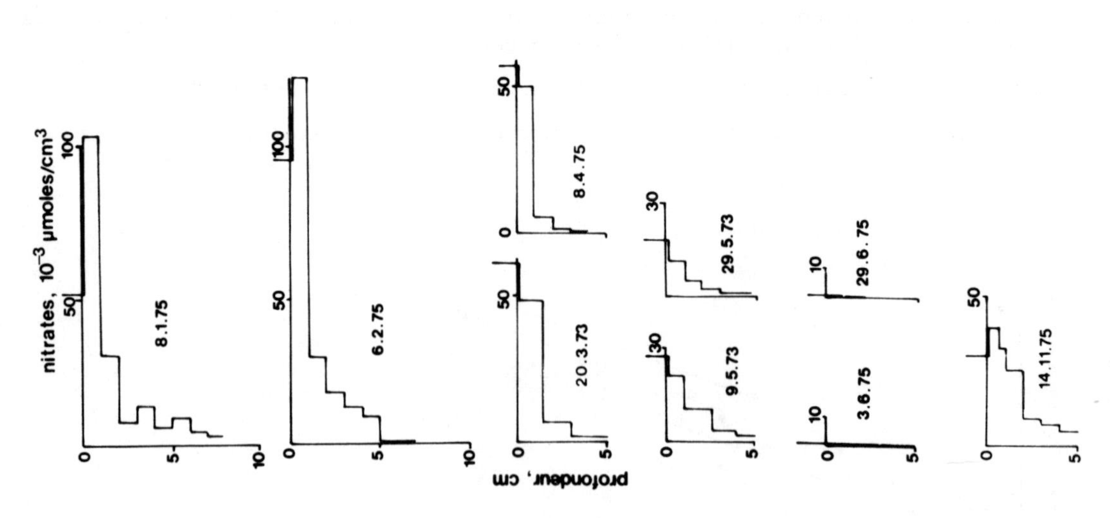


Fig. 6. Seasonal variations of nitrate distribution in the interstitial waters of muddy sediments in the Sluice Dock at Ostend.

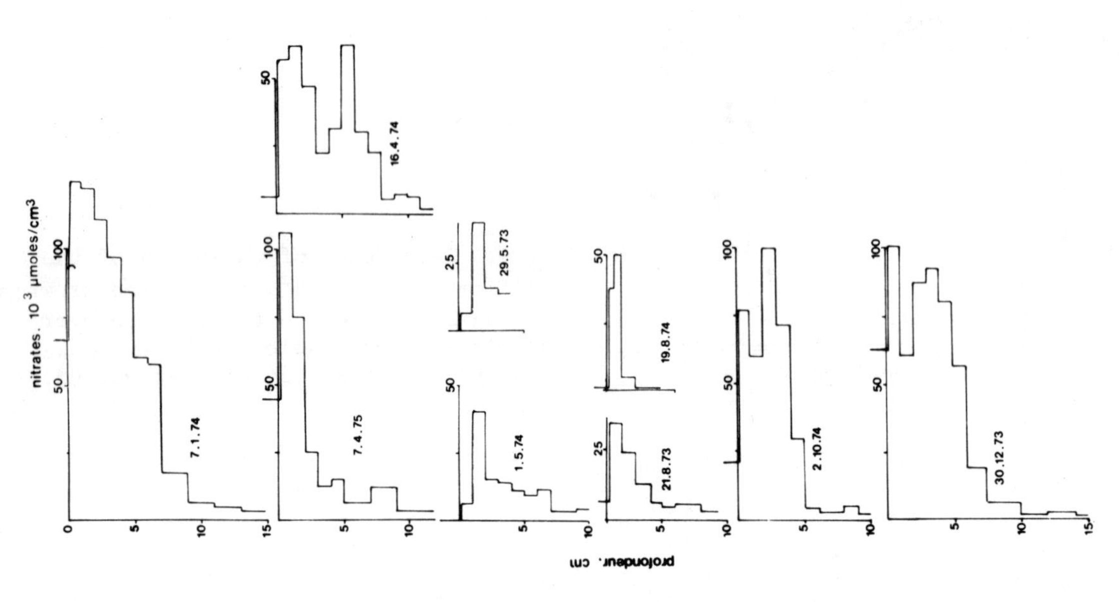


Fig. 7. Seasonal variation of nitrate distribution in the interstitial waters of sandy sediments in the Sluice Dock at Ostend.

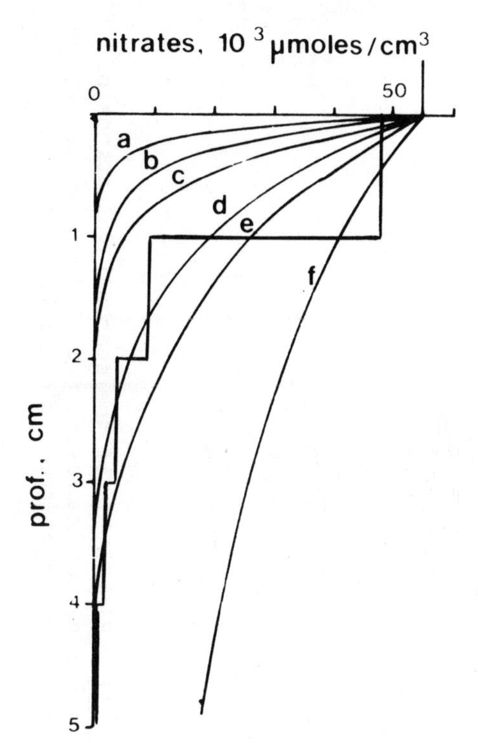


Fig. 8. Experimental and theoretical profiles of nitrate concentration in the interstitial waters of muddy sediments of the Sluice Dock at Ostend (8.4.75). Theoretical curves for

Vmax $= 3.6 \times 10^{-5}$ µmoles $cm^{-3} sec^{-1}$,

Co $= 0.055$ µmoles $cm^{-3}$;

and a set of values of the dispersion coefficient :

(a) $0.1 \times 10^{-4}$ $cm^2 sec^{-1}$
(b) 0.5 "
(c) 1 "
(d) 5 "
(e) 10 "
(f) 100 "

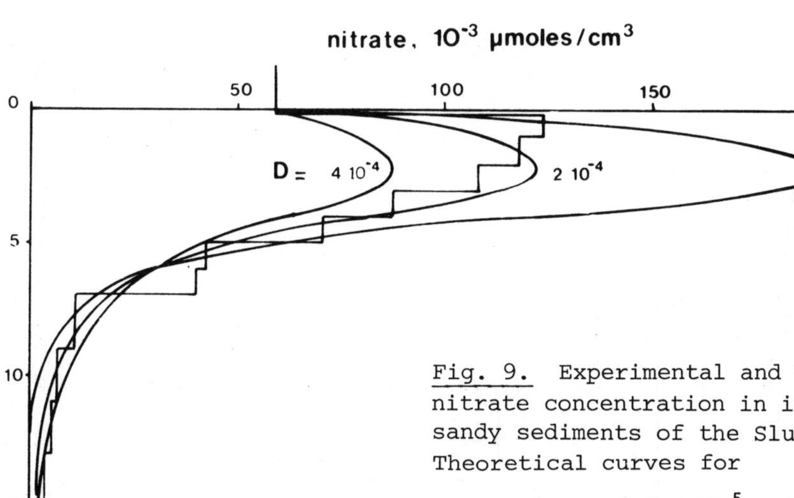


Fig. 9. Experimental and theoretical profiles of nitrate concentration in interstitial waters of sandy sediments of the Sluice Dock at Ostend (7.1.74). Theoretical curves for

$k_n = 0.5 \times 10^{-5}$ µmoles $cm^{-3} sec^{-1}$,
$z_n = 4$ cm,
$k_d = 0.5 \times 10^{-4}$ $sec^{-1}$;

as experimentally determined, and a set of D values :

$D = 4 \times 10^{-4}$ $cm^2$ $sec^{-1}$
$D = 2 \times 10^{-4}$ "
$D = 1 \times 10^{-4}$ "

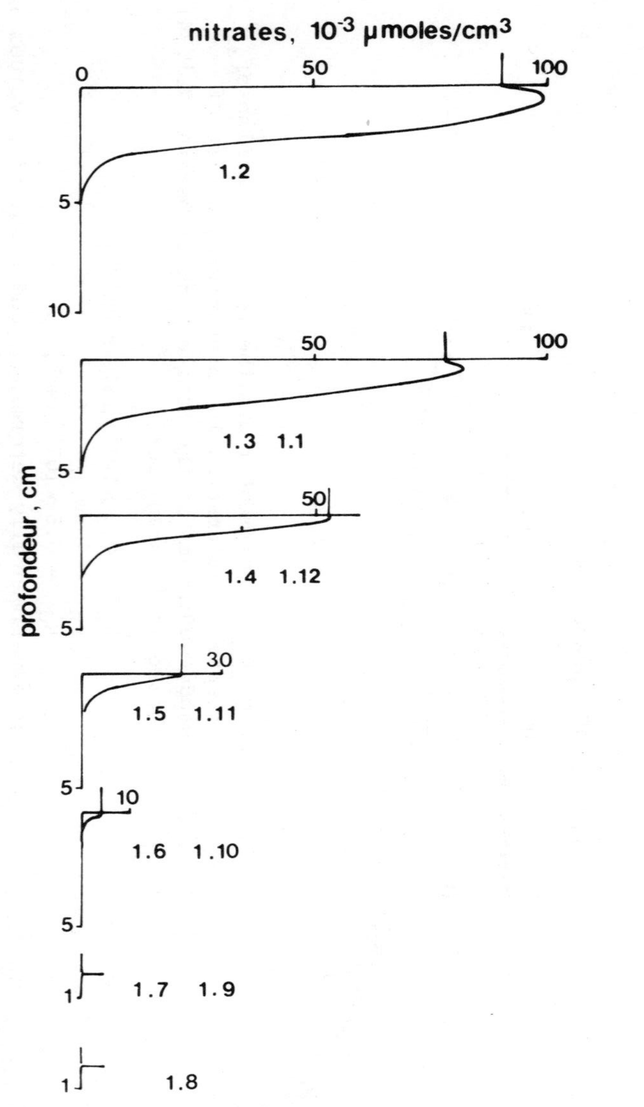


Fig. 10. Simulation of the seasonal variations of the vertical concentration profiles of nitrate in the muddy sediments of the Sluice Dock at Ostend.

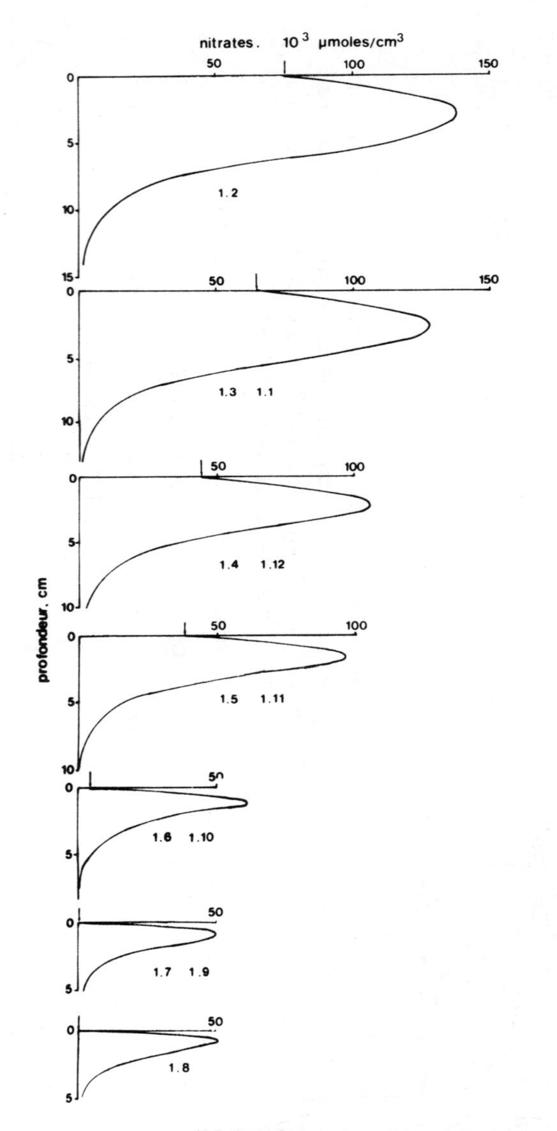


Fig. 11. Simulation of the seasonal variations of the vertical concentration profiles of nitrate in the sandy sediments of the Sluice Dock in Ostend.

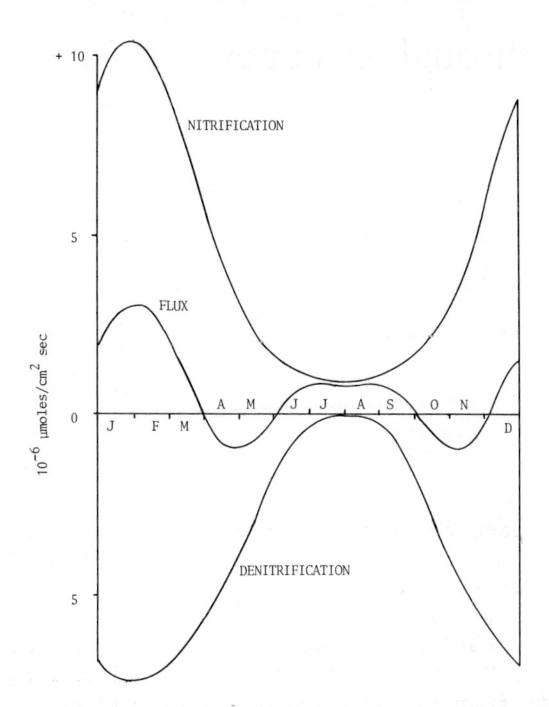


Fig. 12. Calculated evolution of nitrification, denitrification and nitrate flux across the sediment-water interface in the muddy sediments of the Sluice Dock. (For the sake of clarity, denitrification rate has been plotted as negative values, while the flux across the sediment-water interface has been plotted as positive when directed towards the overlying water).

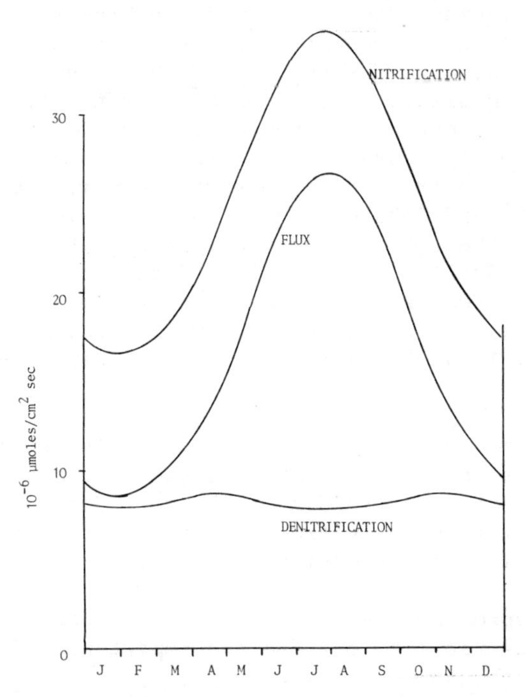


Fig. 13. Calculated evolution of nitrification, denitrification and nitrate flux across the sediment-water interface in the sandy sediments of the Sluice DocK. (Same sign convenction as in Fig. 12 except for denitrification rate, which is plotted as positive values).

# Transfer of chemical species through estuaries to oceans

D. Lal*

## I. Statement of the problem

In natural situations, the interface between the oceans and rivers is usually an extended region, the estuary, where measurable changes occur in the pH, ionic strength and chemical composition of water. A great variety of physical, chemical and biological processes occur in the estuarine environment, e.g. flocculation and deposition of colloidal particles; absorption/scavenging of dissolved components on active surfaces like clay particles, Fe, Mn, hydroxyols, leaching and subsequent release of elements to sea water from fluvial sediments by exchange processes.

These processes and reactions control the net fluvial transport of weathered materials from the continents to the open ocean, as has been firmly established (Bien *et al.*, 1958; Burton *et al.*, 1970; Liss and Spencer, 1970; Wollast and DeBroeu, 1971; Martin *et al.*, 1971, 1973, 1976; Aston and Chester, 1973; Boyle, 1974; Boyle, *et al.*, 1974; Carpenter *et al.*, 1975; Peterson *et al.*, 1975; Windom, 1975; Borole *et al.*, 1976; Graham *et al.*, 1976) by making inter-comparisons of the relative concentrations of major and trace elements in the seawater, river water and in their mixture at different points along the estuary. The results of these investigations have led to discovery that some elements always behave non-conservatively, whilst others may be conservative in one estuary but not in the other ("variable non-conservative"). Elements like nickel, copper and uranium are probably conservative (Borole *et al.*, 1976; Boyle, 1976) whereas iron and rare earth elements behave non-conservatively in most estuaries (Aston and Chester, 1973; Boyle, 1974; Boyle *et al.*, 1974; Martin *et al.*, 1976). Curiously, the total manganese (particulate + soluble) seems to be conservative whereas individually, the soluble and particulate forms are not (Graham *et al.*, 1976). "Variable non-conservative" behaviours are seen in several cases. Silicon and uranium, for example, exhibit both conservative and non-conservative behaviour (Martin *et al.*, 1971, 1973, 1976; Peterson *et al.*, 1975; Borole *et al.*, 1976).

Our understanding of the estuarine reactions is very limited as yet because of paucity of data for a suite of elements in the river-ocean system. It must be borne in mind, however, that our present lack of understanding is largely due to the fact that the chemistry of such systems differs greatly from one estuary to another and that the principal differences have not yet been delineated.

Before going into the various approaches being made towards understanding basic estuarine processes, it seems desirable to outline the importance of such studies.

---

* Physical Research Laboratory Ahmedabad 380 009, India.

Firstly, purely from an academic point of view, the exercise of making precise mass-balance of trace-elements in the hydrosphere is of utmost importance in geochemistry, and determination of input of elements from rivers has always been recognized as of great importance (Goldberg and Arrhenius, 1958). The estimation of residence time of elements in the ocean involves a study of the standing crop of an element, A, and the rate of introduction of the element, dA/dt. In view of what has been said earlier, one should be very cautious in basing estimates of dA/dt on the measured concentrations of trace elements in the river waters. In the case of the normally conservative elements (U, Ni, Cu) there are no obvious problems, but in the case of a host of other elements, e.g. Si, Fe, REE, and others where residence times are $<10^5$ yrs, one faces difficulties in making budgetary balances. Elements which behave non-conservatively may of course be slowly released later on: this process will vary according to their mobilisation from the estuarine sediments and factors of sediment transport between the ocean and the estuary. The process is thus expected to be highly variable from one estuary to another. For several elements, their "hold-up" within the estuaries may be important in regulating trace-element concentrations in the coastal waters. In the case of several heavy toxic metals, the estuarine control, i.e. hold-up and slow release, may even be very necessary from the point of view of harmfulness to marine life.

Secondly, estuarine processes are generally highly variable, with a great deal of reversibility in time scales in the order of or shorter than 1000 years, whereby it becomes difficult to make estiamtes of rates of transfer of chemical species through the estuaries to the oceans in dissolved or particulate phases. As an example of variability, we may cite the case of Storm "Agnes" in Chesapeake Bay; this storm caused more sediment to be discharged during one week in June 1972 than during the previous several decades (Schubel, 1974).

Based on laboratory studies, Sholkovitz (1976) showed that flocculation occurs very rapidly in the case of the elements Fe, Mn, Al, P, organic carbon and humic substances. Does this imply removal of these elements to sediments? It would be necessary to study in detail how elements can be transferred to sediments and whether any of these elements reaching the sediments as particulate phases do get remobilized at a later time. Windom (1975) studied heavy metal fluxes through salt-marsh estuaries. He found (i) removal of Fe and Mn from solution due to precipitation, (ii) recycling of certain elements, otherwise lost to estuarine sediments, due to biological processes, and (iii) near quantitative transfer from the river to the open oceans of certain elements, e.g. Cd and Cu. In view of the few studies carried out so far, our information on such processes is very limited; further work is urgently warranted.

## II. Suggestions for characterizing the nature of reactions in estuarine sediments

In the foregoing sections have been outlined some of the important chemical reactions occurring in the estuarine environment and their importance in geochemical and biological problems. It seems that if a few of the aspects of trace element transfer via estuaries were investigated, the information might provide answers to a host of important questions relating to geochemistry as well as pollution : e.g.

(i) The reasons for the highly variable conservative character of some trace elements, e.g. Si. Uranium has been found to behave as a non-conservative element in the French estuary Gironde but conservatively in the rivers of India (Borole *et al.*, 1976; Martin *et al.*, 1976).

(ii) The concentrations of toxic heavy metals and elements of short residence times ($<10^5$) years in sediments, as a function of depth at a given point in the estuary as well as at the surface, along the river-ocean transect.

These two problems need very careful study in several estuaries, the selection of which has to be made in terms of the prominent chemical differences in the river-ocean systems: type of basement complex of the rivers; chemical composition of rivers, e.g. pH variability in the suspended matter; differences in biological productivity; state of pollution, etc. For example, it is possible that the non-conservative behaviour of uranium in the Gironde is due to heavy pollution.

Ideally, one should obtain data from which a set of simultaneous equations from different estuaries can be derived so that a matrix of the parametres of interest can be constructed.

Of particular interest are the large estuarine systems of India. The uranium concentration in these rivers are highly variable (0.1 - 5 ppb) since these rivers run through different types of basement complexes. The total discharge from the six major rivers flowing into the Bay of Bengal amounts to 2.5 x $10^{15}$g/yr, i.e. about one sixth of the global discharge of river-borne solids into the world's oceans (Holeman, 1968). Further, because of the large variabilities in the levels of industrialization within India, an intercomparison of different Indian estuaries would be of great value. Although the drainage areas are similar for the Ganges, Brahmaputra, Narbada, Tapti and Sabarmati, the Ganges is heavily polluted whereas the Narbada and Tapti are not. The amounts of particulate material in the Indian estuaries also show wide variations.

It becomes quite clear that while basic information on chemical and biological processes should be entirely retrievable from a study of the sediments in the estuarine region, it will be necessary to study the waters as well to define precisely the quantitative delineation of the processes. The reason for this is that the flux of particulate material generally greatly exceeds the efflux of trace elements - due to adsorption. A host of radionuclide tracing and dating methods are available to aid in such studies.

It seems plausible that studies as outlined above will ultimately lead to a useful "chemical" definition of the estuary, even though this naturally implies the making of chemical cand biological generalizations. Such are the uses of estuaries (as source of water and mineral deposits, as channels for dispersal of waste products of industrialization, etc.) the proposed studies quite apart from their academic interest, are of ever-increasing importance.

## References

Aston, S.R.; Chester, R. 1973. The influences of suspended particles on the precipitation of iron in natural waters, Estuar. Coast. Mar. Sci., vol. 1, p. 225.

Bien, G.S.; Contois, D.E.; Thomas, W.H. 1958. The removal of soluble silica from fresh water entering the sea. Geochim. Cosmochim. Acta, vol. 14, p. 35;

Borole, D.V.; Krishnaswami, S.; Somayajulu, B.K.L. Investigations on dissolved uranium, silicon and on particulate trace elements in estuaries. Estuar. Coast. Mar. Sci. 1976.

Boyle, E. 1974. Chemical behaviour of dissolved iron in the mixing zone between rivers and the oceans, Ph. D. thesis M.I.T., Cambridge, Mass.

Boyle, E.; Collier, R.; Dengler, A.T.; Edmond, J.M.; Ng, A.C. and Stalland, R.F. 1974. On the chemical mass balance in estuaries, Geochim. Cosmochim. Acta, vol. 38, p. 1719.

Burton, J.D.; Liss, P.S.; Venugopalan, V.K. 1970. The behaviour of dissolved silicon during estuarine mixing I. Investigation in Southampton water, J. Cons. Explor. Mer, vol. 33, p. 134.

Carpenter, J.H.; Bradford, W.L.; Grant, V. 1975. Processes affecting the composition of estuarine waters, in : Estuarine Research, vol. 1, p. 188, L.E. Cronin (ed.), New York, Academic Press.

Goldberg, E.D.; Arrhenius, G.O.S. 1958. Chemistry of the Pacific pelagic sediments, Geochim. Cosmochim. Acta, vol. 13, p. 153.

Graham, W.F.; Bender, M.L.; Klinkhammer, G.P. 1976. Manganese in Narragansett Bay.

Holeman, J.N. 1968. The sediment yield of the major rivers of the world, Water Resources Res., vol. 4, P. 737.

Krishnaswami, S. 1976. Anthigenic transition elements in Pacific pelagic clays. Geochim. Cosmochim. Acta, vol. 40, p. 425.

Liss, P.S.; Spencer, C.P. 1970. A biological process in the removal of silica from seawater, Geochim. et Cosmochim. Acta, vol. 34, p. 1073.

Martin, J.M.; Jednacak, J.; Pravdic, V. 1971. The physiochemical aspects of trace element behaviour in estuarine environments. Thalassia Jugosl., vol. 7, N° 2,p.619.

Martin, J.M.; Kulbicki, G.; De Groot, A.J. 1973. Terrigenous supply of radioactive and stable elements to the oceans, Proceedings of symposium of hydrogeochemistry and biogeochemistry, vol. 1, p. 463, Washington, D.C., The Clark and Company.

Martin, J.M.; Nijampurkar, V.N.; Salvadori, F. 1976. U and Th isotope studies in estuaries of France.

Peterson, D.H.; Conomos, T.J.; Broenkow, W.W.; Scrivani, E.P. 1975. Processes controlling the dissolved silica distribution in San Francisco Bay, in : L.E. Cronin (ed.), Estuarine Research, vol. 1, p. 153, New York, Academic Press.

Schubel, J.R. 1974. Effects of tropical storm Agnes on the suspended solids of the northern Chesapeake Bay, in : Suspended Solids in Waters, R.J. Gibbs, (ed.), p. 113-32, New York, Plenum Press.

Sholkovitz, E.R. 1976. Flocculation of dissolved organic and inorganic matter during the mixing of river water and sea water, Geochim. Cosmochim. Acta, vol. 40, p. 831.

Windom, H.L. 1975. Heavy metal fluxes through salt-marsh estuaries, in : L.E. Cronin, (ed.), Estuarine Research, vol. 1, p. 137, New York, Academic Press.

Wollast, R.; DeBroeu, F. 1971. Study of behaviour of dissolved silica in the estuary of the Scheldt, Geochim. Cosmochim. Acta, vol. 35, p. 613.

# Chemical variability in estuaries

H. Elderfield*

INTRODUCTION

The unique geochemical interest of the estuary is that it is the site of the mixing of riverwater and seawater. Hence, the objective of the estuarine geochemist is an understanding of the chemical reactions in estuarine waters and sediments that reflect this unique environment. The estuary is also a site for geochemical processes that are common to other environments. For example, estuarine sediments commonly are anoxic and many of the diagenetic reactions which occur in estuarine sediments occur also in anoxic coastal marine sediments. Since the study of estuaries is beset by complex problems caused by temporal and spatial variability, it is clear that processes which are represented elsewhere are best studied elsewhere or in estuaries where variability is minimal. In addition, studies with specific objectives should be matched with estuaries for which the operational variables are small and which optimise the possibility of the sought-after reaction taking place. For example, studies of particle-water interaction will be more profitable in well-mixed estuaries (where the residence times of reactants is long) than in highly stratified estuaries where the majority of particles are transported rapidly out to sea.

Similarly, studies of non-conservative chemical properties using property against salinity tie lines will be more profitable in estuaries where there is a single riverine end member. The unsuitability of certain estuaries for certain studies may account for the apparently variable behaviour of elements in some estuarine processes.

Thus, there is a need to study estuaries where processes might be easily evaluated; logically, estuaries with poorly known hydrodynamics, complex drainage patterns, large temporal variations in discharge, etc., should best be rejected as such sites of study. Unfortunately, it is often estuaries of this latter type of which study is essential. This is because estuarine research is often instigated on "environmental grounds" rather than from a desire to investigate one specific process. For this reason, it is important to be aware of the chemical variability in estuaries. Clearly, this variability must be assessed for budget studies and is critical to evaluations of the importance of anthropogenic material added to estuaries.

VARIABILITY OF INPUT

The variability in the supply of material to estuaries is large and is often underestimated (Gibbs, 1976). This variability is reflected in the flushing time (or water replacement time) which is the average time required for an entering volume of water to pass through the estuary, so :

* Graduate School of Oceanography, University of Rhode Island, Narragawsett Bay Campus, Kingston, RI 02881, U.S.A.

$$\tau_f = \frac{V}{dV/dt} ,$$

Where V = volume of estuary, and dV/dt = discharge rate; or, according to Peterson et al., (1975),

$$\tau_F = \frac{A}{\int_z^0 w/r/dz} ,$$

where A = cross channel area, w = depth variable width of channel, /r/ = absolute current speed, and z = water depth. The value $\tau_f$ is influenced by discharge rate, tidal mixing, density and wind-driven currents (Bowden, 1967) as schematically shown in Fig. 1.

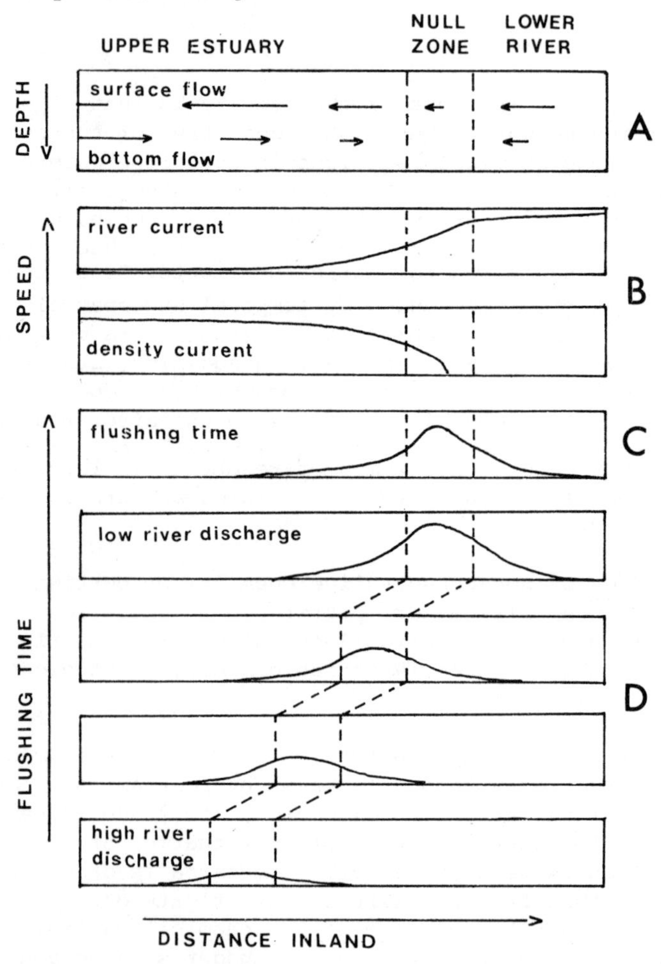


Fig. 1 : Schematic representation of (A) net drift in vertical section through a river-estuarine system, in which the length of the arrows indicates relative current strength; (B) longitudinal variation in average river and density currents within the river, null zone, and estuary; (C) generalised longitudinal variation in average advective water-column replacement time; and (D) seaward movement of null zone and diminishing water column replacement time with increasing river discharge (from Peterson et al., 1975, Fig. 8).

Changes in flushing time directly affect the residence time of chemical species in estuaries, which is the ratio of the standing crop of an element x in an estuary to its rate of introduction; i.e.,

$$\tau_R = \frac{M}{dM/dt} = \tau_F \frac{[x]\ \text{estuary}}{[x]\ \text{river}}$$

where M = mass, and [x] = concentration normalised to zero salinity. In stratified estuaries it may be desirable to specify residence times for surface and bottom waters. It is clear that the identification of non-conservative behaviour (i.e. $\tau_R \neq \tau_F$ ). will be difficult when temporal variations in input produce changes in $\tau_F$ or where complex circulation means that [x] of the estuary cannot be uniquely defined. Similarly, in metal pollution studies, temporal variations in stream supply (see, for example, Fig. 2) produce a shifting baseline of fluxes of metals against which anthropogenic input must be compared.

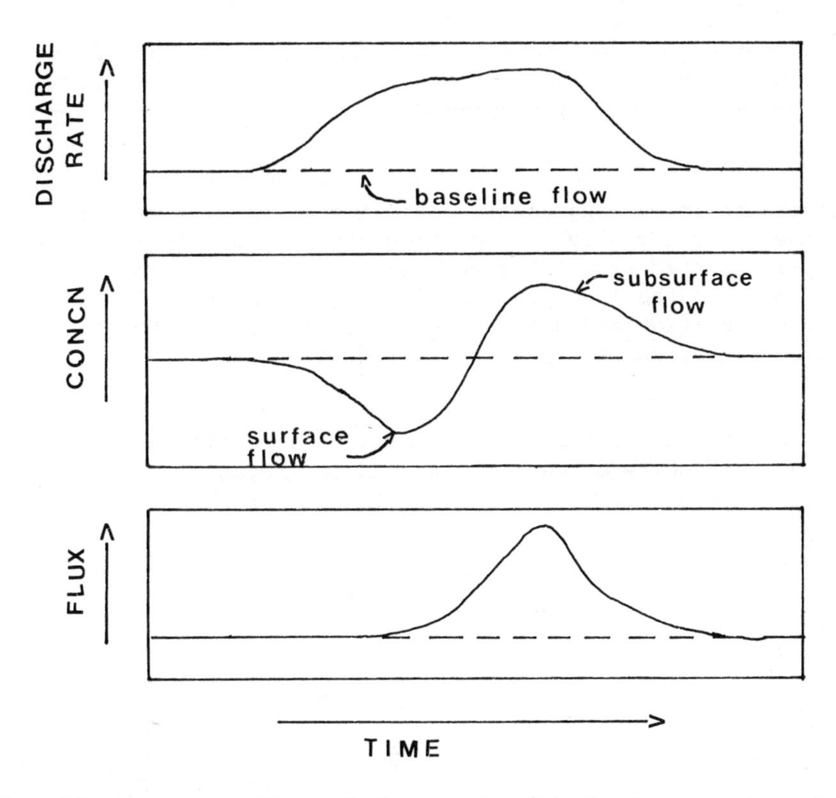


Fig. 2 : Schematic representation of changes in (A) discharge rate and (B) concentration, and (C) flux of species supplied by river to estuary in response to storm discharge. Concentration initially decreases as a consequence of increased surface flow but flux remains constant due to dilution. Sub-surface or groundwater flow into the river then causes both concentration and flux to increase due to flow through mineralised catchment or because of the transfer of "soil water" to the river.

## MULTIPLICITY OF INPUTS

One of the fundamentals in estuarine research is the identification of end members which mix in the estuarine system. As far as dissolved species are concerned, the seawater end member is easily defined, but a characteristic of many estuaries is that inorganic detritus is supplied from marine as well as riverine sources. It is often difficult to distinguish material supplied from these two sources, yet this is an important requirement for budget studies. Similarly, in studies of the halmyrolysis of aluminosilicate debris, it is important to compare clay minerals in an estuary with their appropriate source material (Table 1).

Table 1 Exchangeable cations on river and marine clays from Morphou Bay, Cyprus

| | Na/K | Mg/Ca |
|---|---|---|
| Cyprus rivers | 2.21 | 0.28 |
| Morphou Bay | | |
| inner bay | 1.03 | 0.84 |
| outer bay | 0.95 | 1.78 |

The difference in exchange characteristics of samples from inner and outer bays is that the inner bay clays are derived from Cyprus rivers whereas the outer bay clays are transported from the river Nile by a counter-clockwise gyre in the E. Mediterranean. Whereas the inner bay samples fit an ideal solution model (based on the river clay composition and differences in cation ratios between riverwater and seawater) the outer bay samples do not (from Neal and Elderfield, in preparation).

The multiplicity of riverine end members is a particular problem in the assessment of chemical continuity between rivers and the ocean (Boyle et al., 1974). Natural or artificial tracers are required for each end member where more than one tributary mixes with seawater in an estuary (Fig. 3).

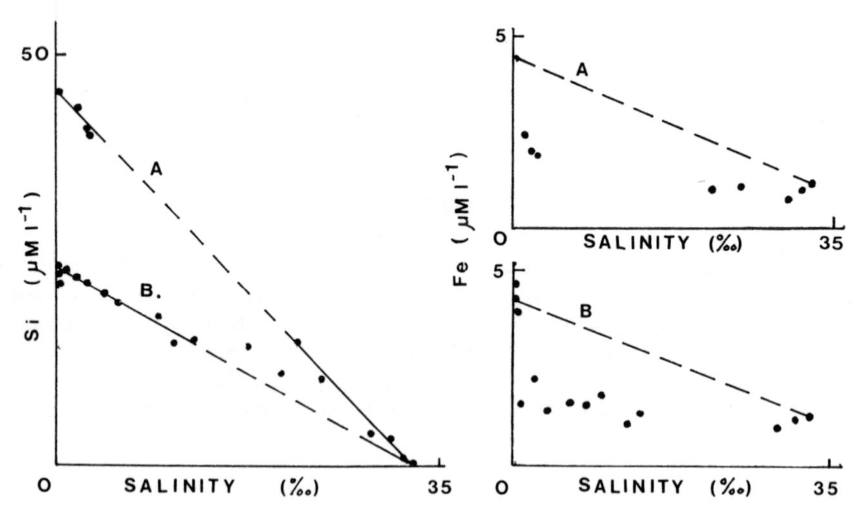


Fig. 3 : A survey of silica varying with salinity for the river Conway reveals two riverine end members and so the data cannot be used to identify non-conservative mixing. However, the different Si concentrations of these end members allows an assessment of the behaviour of iron, which is non-conservative for both mixing regimes (from Elderfield, unpublished data).

## FLUXES AT THE WATER-SEDIMENT INTERFACE

One major problem in the biogeochemistry of estuarine sediments is a determination of the fluxes of dissolved species at the water-sediment interface. One aspect of the problem is that lateral variations in sediment type may result in variable concentration gradients of pore fluids and hence fluxes (Table 2).

Table 2. Concentration gradients ($\mu g cm^{-4}$) of Mn, Fe and Zn in the upper centimetre of seven cores from the Conway estuary

| core | Mn | Fe | Zn |
|---|---|---|---|
| 1 | 11.9 | 0.18 | 0.16 |
| 2 | 5.7 | 0.57 | 0.012 |
| 3 | 0.43 | 0.76 | 0.005 |
| 4 | 2.0 | 1.13 | - |
| 5 | 2.9 | 0.39 | 0.003 |
| 6 | 2.5 | 0.02 | 0.15 |
| 7 | 2.8 | 0.54 | 0.004 |

Results show variations of 1-2 orders of magnitude between individual cores from tidal flat sediments reflecting differences in sediment texture and organic matter contents (from Edwards et al., unpublished data).

Another aspect is that pore fluids can be transferred from the sediment by processes other than molecular diffusion. These include convective replacement (Reeburgh, 1976) and "dispersion" (Billen and Vanderborght, 1976). Storms can flush sediments and can convectively replace relatively metal-rich pore fluids by relatively metal-poor estuarine water at rates about one hundred times that predicted by Fickian diffusion (Fig. 4).

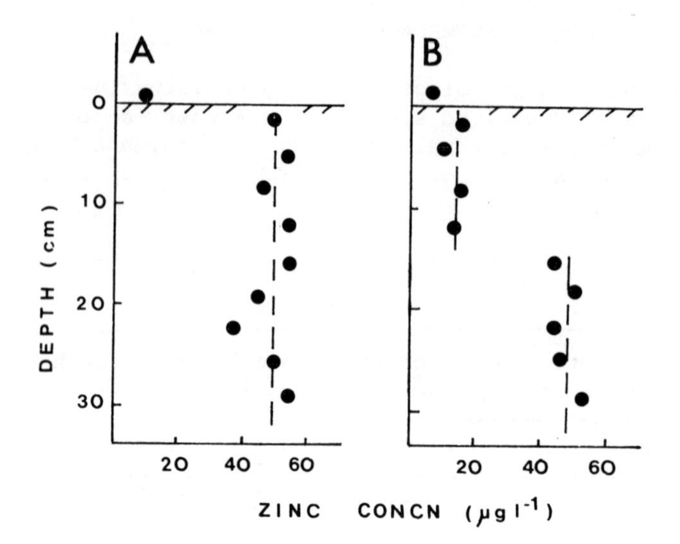


Fig. 4. Concentration profiles of zinc in pore fluids of Conway estuary sediment : (A) normal profile; (B) after storm. In (A) the concentration gradient at the water-sediment interface is 0.035 g $cm^{-4}$ which gives a flux of $\sim 3 \times 10^{-9}$ g $cm^{-2} day^{-1}$, using a molecular diffusion coefficient of $10^{-6}$ $cm^2 sec^{-1}$. In (B) a storm has caused the replacement of the pore fluid with estuarine water down to the base of a sandy horizon at 15 cm giving a net release of 0.035 μg $cm^{-3}$Zn from 15 $cm^3$ of sediment of 70% porosity. The flux via flushing was $3.7 \times 10^{-7}$ g $cm^{-2}$ $day^{-1}$ for one-day storm event (from Elderfield, unpublished data).

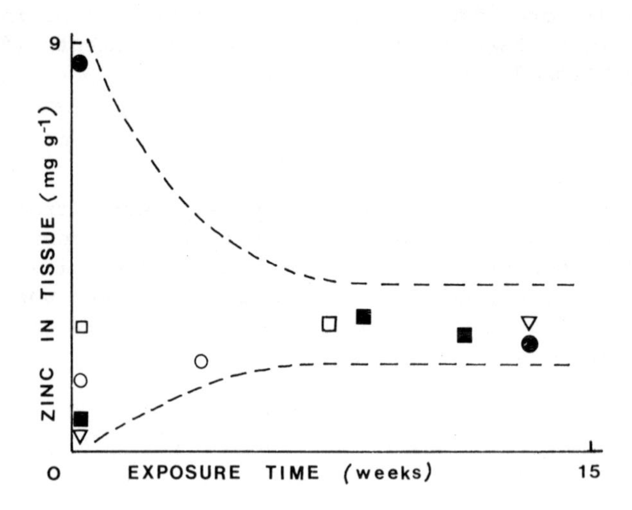


Fig. 5. Temporal changes in zinc content of oyster (Ostrea edulis) brought into Conway estuary from Helford (■ □ ), Colne ( ○ ) and Walton (● ▽ ) estuaries (from (Elderfield, unpublished data).

## METAL ACCUMULATION BY FAUNA

This variability is not unique to inorganic processes. To exemplify this, Table 3 shows how the oyster Crassostrea gigas responds to different metal contents of the estuaries of three English rivers.

Table 3 Changes in metal content of Crassostrea gigas in 3 estuaries

| | | estuary | | Crassostrea gigas | |
|---|---|---|---|---|---|
| | | water (µg/l) | sediment (ppm) | initial (ppm) | eight months later (ppm) |
| Cadmium | 1. | < 1 | 2 | 6 | 2.2 |
| | 2. | < 1 | 1 | 4 | 4.8 |
| | 3. | 1.5 | 12 | 4 | 26.7 |
| Copper | 1. | 11 | 179 | 273 | 643 |
| | 2. | 4 | 35 | 110 | 207 |
| | 3. | 6 | 12 | 155 | 396 |

1. = Helford; 2. = Colne; 3. = Poole (data from Thornton et al., 1975).

There are several examples (e.g. Bryan and Hummerstone, 1971; Pedon et al., 1973) of how metal accumulation by benthic fauna reflect the geochemical status of estuaries (see, for example, Fig. 5).

However, since the factors affecting metal availability are incompletely known, it may be premature to link directly concentrations of all metals in fauna to metal contents of estuarine water and/or sediments. For example, the possibility of competition between Zn and Ca uptake by Ostrea edulis (Coombs, 1972) may produce zinc levels in tissue that reflect salinity ranges to which the fauna are exposed (Thornton et al., 1975). Hence identical species from a single estuary may accumulate some metals in varying degrees simply as a consequence of the different degrees to which they are exposed to freshwater drainage.

## References

Bowden, K.R. 1967. In : G. Lauff, (ed.) Estuaries, Am. Assoc. Adv. Sci. Publ., N°83, p. 15.

Boyle E.; Collier, R.; Dengler, A.T.; Edmond, J.M.; Ng, A.C. and Stallard, R.F. 1974. Geochim. Cosmochim. Acta, vol. 38, p. 1719.

Bryan, G.W.; Hummerstone, L.G. 1971. J. Mar. Biol. Assoc. U.K., vol. 51, p. 845.

Coombs, T.L. 1972. Mar. Biol., vol. 12, p. 170.

Neal, C.; Elderfield, H. (In preparation) Cation exchange reactions during sediment halmyrolysis and effects on seawater composition.

Pedon, J.D.; Crothers, J.H.; Waterfall, C.E.; Beasley, J. 1973, Mar. Pollut. Bull., vol. 4, p. 7.

Peterson, D.H.; Conomos, T.J.; Broenkow, W.W.; Scrivani, E.P. 1975. In : L.E. Cronin, (ed.), Estuarine Research, vol. 1, New York Academic Press, p. 153.

Thornton, I., Watling H. and Darracott, A. 1975. Sci. Tot. Environ., vol. 4, p. 325.

# Migration in the seabed: some concepts

E. K. Duursma*

## INTRODUCTION

Concentration profiles of materials in the seabed are not always easy to explain on the basis of simple migration processes. Migration may be composed of several processes, such as physico-chemical diffusion of dissolved materials in the interstitial waters, turnover of the seabed caused by bioturbation or current and wave actions, interstitial water movements and perhaps redistribution of particles in the sediments due to differences in density.

The separate impact of each of these processes is not easy to predict for most of the seabeds of estuaries, the continental shelf and deep seas, unless detailed studies have been made. This makes it difficult to predict the fate of foreign contaminants in the sea which build up in the seabed. One of the major questions to be answered is whether on short-or long-term bases, steady state concentrations can or will be reached in the seabed, and whether these concentrations will eventually become hazardous to man or the aquatic environment.

For some pollutants, steady states may be reached within a period of a few decades, which simplifies the description of the processes involved. Much more difficult to describe and to predict are those processes which take very long periods to attain steady states of concentration. It is mainly for long-lived, transuranic isotopes and other persistant pollutants that more information is required particularly on the processes that determine the migration and residence of such substances in confined basins. The migration processes of substances into, within or out of the seabed can be divided into two major groups which are connected to the two phases present in the seabed, liquid and solid. For the liquid phase, migration takes place in or with the interstitial waters and the supernatant water of the seabed. The extent of migration will depend on the solubility of the substances concerned either as truly dissolved and or as colloidal material. The migration related to the solid phase depends on the binding and trapping of substances by the solid particles in the seabed, and their migration will be controlled by the sediment mixing and sediment transport.

It is important to know which of these migration processes related to solid-liquid processes regulate the integral displacement of the substances in the seabed. Using recently published information, some facets of this problem will be discussed. Additional material is presented in other papers of this volume.

---

* Delta Institute for Hydrobiological Research, Yerseke, Netherlands.

SOLID-LIQUID PROCESSES

Dissolved and colloidal substances, both organic and inorganic, will be distributed between liquid and solid phases when both phases are present. For the seabed such a distribution is caused by sorption and desorption processes. Sorption causes an immobilisation on solid particles, where desorption favours a mobilisation into the water phase. The rates of these processes influence the migration processes, making a difference whether these are of a short time scale (minutes to a few weeks) or of a long time scale (month or longer).

Typical sorption curves for a number of radionuclides are shown in Fig. 1, and for some it is apparent that within a period of 200 days no real equilibrium has been attained. Uptake by the sediments continues gradually. This uptake may concern a strongly bound form as is shown in Fig. 2 for $^{65}Zn$. The so-called exchangeable fractions are constant from the first days after the start of the sorption. Only the more solidly bound fraction increases over longer periods. Identical results have been found for other elements. The percentage of leachability with either $NH_4$ acetate/acidic acid (pH 5.4) indicates the amount of weakly bound fraction (Fig. 3).

The implications of these figures are, however, not obvious, since the facility of leaching may not always coincide with the exchange processes that occur in the seabed. In soil science, such exchangeable fractions of elements are considered more or less proportional to the availability of the elements to crops.

For short-time processes in a seabed, migration of the non-leachable fraction of sorbed substances will be possible only along with solid particle mixing and transport. The fraction thus transported will increase when solidification is increasing with time. For the soluble and leachable fractions, the migration is less dependent on time and can take place both as diffusion through or with the interstitial waters and as transport or mixing of the sediment particles. In the seabed, particularly of the deep sea, where the mixing and transport of sedimentary particles is absent, diffusion in and with the interstitial waters is the sole migration process possible. In this case, when in addition there exists no interstitial movements, the physico-chemical diffusion is responsible for the total migration process.

Over very long periods, the processes of desorption and sorption have to be regarded also in the context of these long time scales. The solidly bound fractions might then be equally considered as fractions which exchange with the pore waters, and migration along the dissolved phase remains a possibility.

MIGRATION AS A DIFFUSION PROCESS

Fick's diffusion laws have been used by a great number of investigators for the description of migration processes in the seabed. The question posed here is how these diffusion laws can help us to distinguish which migration processes are dominant in the seabeds of confined ocean or sea areas. A mode of approach, illustrated by an example, is presented for a profile of $^{239}Pu/^{240}Pu$ concentration in a seabed as analyzed by Livingston and Bowen (1976), and plotted in Fig. 4, according to the diffusion models of Duursma and Hoede (1967).

Only a straight line could be produced for the plot represented in Fig. 4A, which is connected to the constant source model. For the plot of the instantaneous source model, Fig. 4B, this was not the case. This means two things: apparently that there is a constant source of plutonium at the top of the sediment and that migration into the seabed can be described by diffusion whatever process is causing the transport. Assuming that the constant source has been available as settled material on the seabed in uniform concentrations and this started in the early sixties, for example 1962, then the diffusion in this case occured over 13 years (core was taken in 1975). Hence from Fig. 4A, a diffusion coefficient of $6 \times 10^{-8}$ $cm^2$/sec can be determined.

This is higher than would be expected on the basis of physico-chemical diffusion alone, since for plutonium a diffusion coefficient of $10^{-10}$ - $10^{-9}$ $cm^2$/sec would be expected on the basis of sorption concentration factors of the order of magnitude of $10^4$ - $5 \times 10^4$ (Hetherington, Jefferies and Lovett 1975; Duursma and Parsi, 1976; Hetherington et al., 1975). The more rapid diffusion measured suggests that one of the processes mentioned earlier causes perturbation of the seabed. A striking feature is the regularity, according to the constant source diffusion model. The vertical transport is probably due to regularly occurring perturbation processes of weak character, also noted by Bowen, et al. (1976).

In the case of plutonium the possibility has been mentioned that particles of high density may or may not exist. It was noted by Patterson et al., (1976) that extremely fine particles as small as 0.05 µm were produced from rapidly cooled, hot chunks of $^{238}PuO_2$, which were transported into soils. Little is known about the migration of particles in a seabed as a function of differences in density. For the non-consolidated top surface layers, however, such processes might not a priori be excluded.

In this context it would be interesting to study the distribution in heavy minerals or heavy fall-out particles of natural and artificial radionuclides (or non-nuclear materials), which can be expected to be brought into the sea attached to or incorporated in them. The bromophorm (density 2.82g/$cm^3$) technique used in soil science to separate the heavy minerals can be applied for this purpose. More research in this direction is required to determine whether certain profiles, as observed for various substances in ocean cores, might be partly caused by differences in densities of sedimentary particles and "settling" through unconsolidated layers.

Another point to consider is whether migration in the seabed might be affected by the hydraulic pressures which occur at great depths. Pressure is known to affect equilibrium constants of carbonic acid and water (Hills, 1972) to cause calcium carbonate dissolution at depths greater than 5000 m. Hills also cites a change in the self-diffusion coefficient in water from $2.64 \times 10^{-5}$ $cm^2$/sec at 1 atm, to $3.09 \times 10^{-5}$ and $3.24 \times 10^{-5}$ $cm^2$/sec at 500 and 1000 atm, respectively. Few measurements have been made on the effects of pressure on the diffusion and sorption processes occurring in the seabed. For chloride it was determined that no effect could be observed until a depth of 1000 m was reached (Duursma, in press).

## ESTIMATION OF MAXIMAL ACCUMULATION IN A SEABED

One of the urgent questions to be answered for waste disposal areas concerns the prediction of long-term accumulations of persistent contaminants. For many pollutants the seabed can act as a sink. In such cases, it is of major interest to know how the concentrations in the seabed increase with time, how long this accumulation will proceed, and what finally will be the concentrations when equilibrium is attained. Two methods will be discussed here, the first applicable for radionuclides having stable natural counterparts in the sea, the second for other radionuclides and stable contaminants.

The first method is based on the specific activity evaluation of the radionuclide in the seabed, supposing that this specific activity does not exceed that existing in the seabed into which the radionuclides are released. Specific activity means the radioactivity per unit of stable counterpart. The second supposition for this method is, for liquid radionuclide wastes, that the specific activity in the water of the disposal area is known from the predictable composition of the effluent. The supposition that the specific activity in the sediment will not exceed the specific activity in the seawater is based on the fact that, for a great number of elements of the sedimentary particles, only a small fraction can be exchanged with the element in solution. An attempt to determine this fraction was made by Duursma (1973) based on studies of radionuclide sorption to a number of major ocean sediments, and on the analyses of various elements in these sediments by Rancitelli and coworkers in Richland, Wash., USA.

The fraction exchangeable was determined by tracer experiments for 35 major ocean sediments and 8 radioisotopes. The average values are presented in Fig. 5. The maximum radioactivity possible in a given seabed can be found by multiplying the factor for the fractional exchange by the trace metal content of the sediment (TMC) times the given specific activity in the seawater (SASW). The method presents the extremes which can be expected in certain "hot spots" although, in general, the concentrations would be lower because decay and migration both counter the input.

The second method is a simple box-model approach which might be of help in roughly determining the build-up of contaminants until equilibrium is reached in certain seabeds. For a confined area with an influx, F, to the seabed in gram material/time, the outflow is taken as proportional to the amount present in the seabed of the confined area. This is expressed as $\lambda VC$, V being the volume containing the contaminants, C being the average concentration, where $\lambda$ is the factor of proportionality with dimension $time^{-1}$.

As is illustrated in Fig. 6, the build-up of material according to this model will be given by :

$$\lambda VC = F(1 - e^{-\lambda t}),$$

t being the time at which the influx started. For this model the average residence time of the material in the seabed will be $\bar{T} = 1/\lambda$, and the steady state concentration equal to $F/\lambda V$.

It is of major interest to know the average residence time, since from this value the maximum expected concentration can be evaluated and as well how long it would take to arrive at 90, 95 or 99 per cent of this concentration (Fig. 6).

This box-model approach might be improved for many areas where the conditions at the seabed are known in detail but the objective of such an approach should focus on the factors which lead to the evaluation of maximum concentrations and the time required for a maximum build-up. Such an approach has the advantage that existing knowledge from natural materials present in the seabed, relative to their migration and transport, can be used for predicting the fate of persistant contaminants.

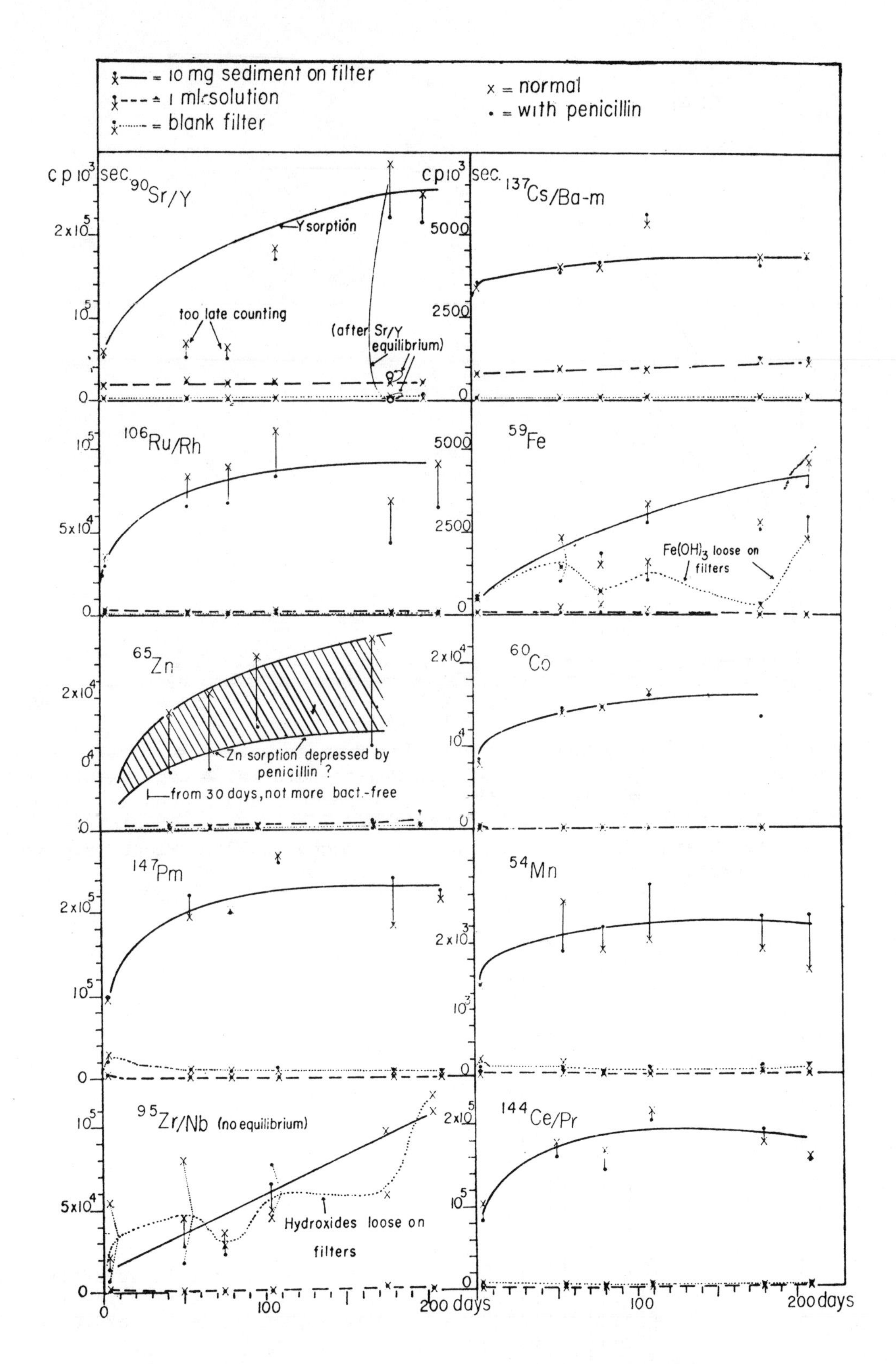


Fig. 1 Sorption results of ten radioisotopes and one Mediterranean sediment, as determined by the thin-layer technique of Duursma and Bosch (1970). (Reproduced from the Netherlands Journal of Sea Research).

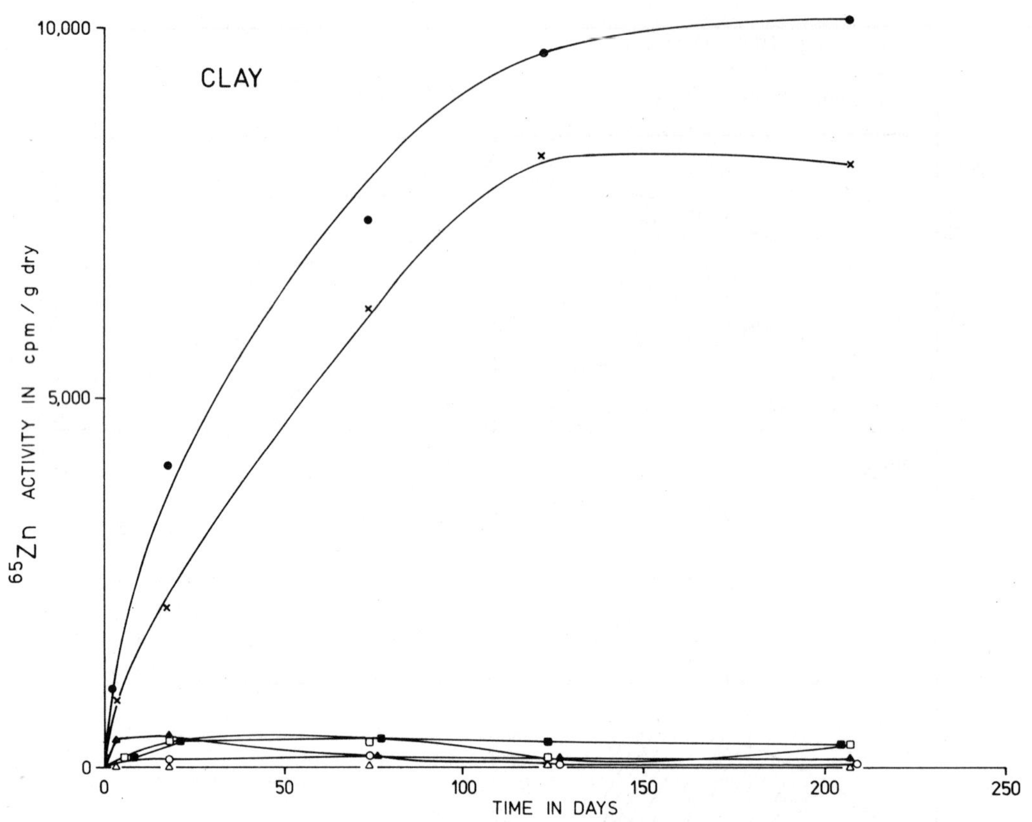


Fig. 2 Sorption of $^{65}Zn$ by Mediterranean clay from seawater; ● total sorbed, △ soluble in $H_2O$, O exchangeable with $NH_4$-acetate 1N, ▲ complexable with Cu-acetate 0.05 N, ■ released with $H_2O_2$ 30 percent and leachable with Cu-acetate 0.05 N, □ leachable with EDTA 1% and x the non-leachable fraction; Results obtained by Ros Vicent and Costa Yagüe (1972). (Figure reproduced from Duursma et al., 1975).

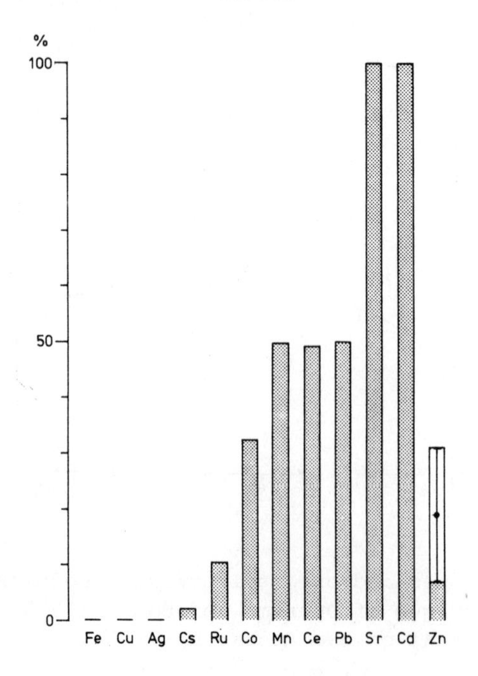


Fig. 3 Each of release after 7 months of sorption of some elements from Mediterranean sediment, excluding the original element present, as determined by leaching with 1N $NH_4$-acetate/acetic acid (pH = 5.4) according to Duursma et al., 1975).

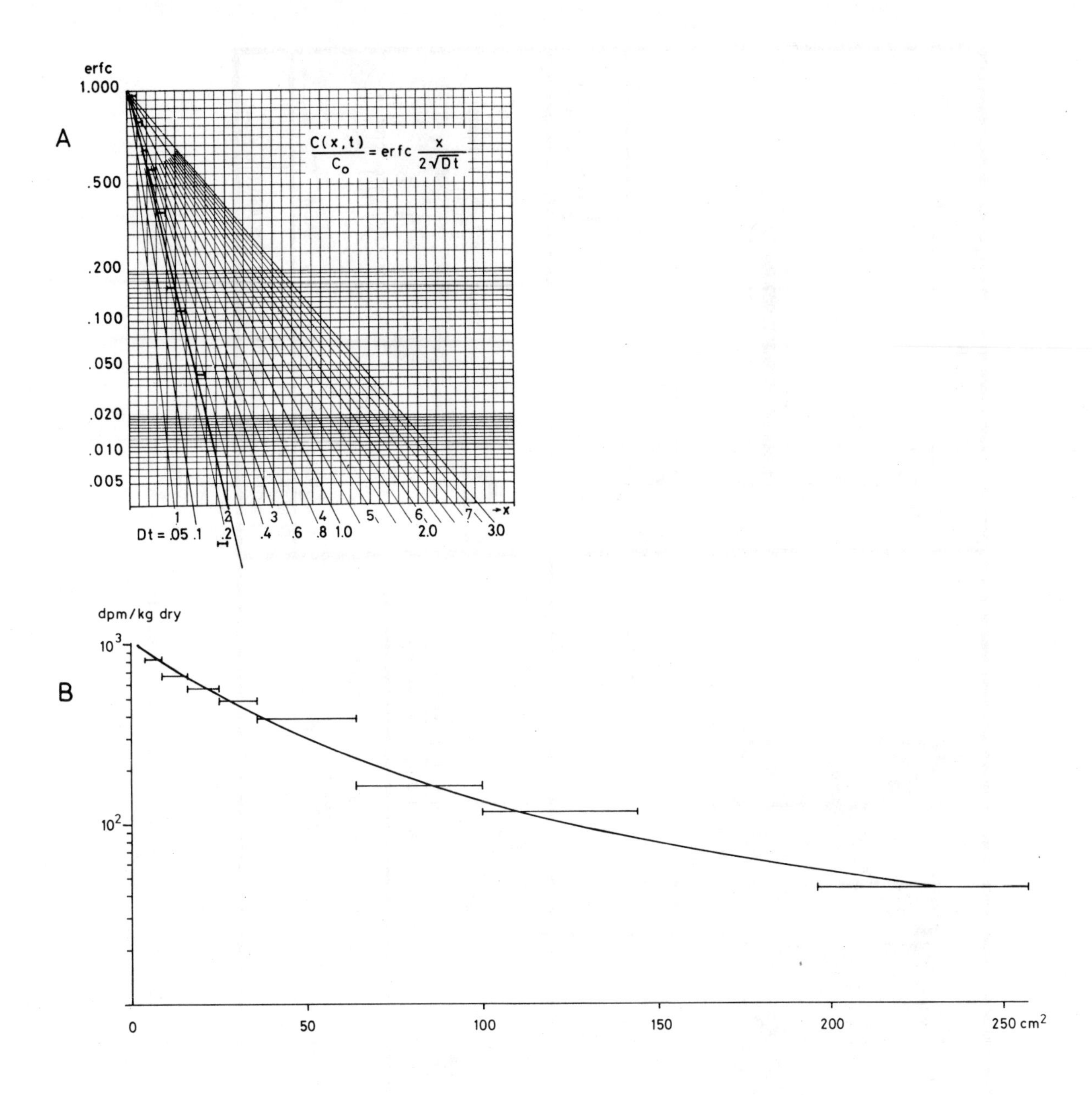


Fig. 4 Plots of $^{239}Pu/^{240}Pu$ concentrations of a North Atlantic core A II 86, No. 6, 42° 25.0'N and 68° 56.76'W from livingston and Bowen (1976) according to two diffusion models of Duursma and Hoede (1967): A, according to a constant source model, B, according to an instantaneous source model.

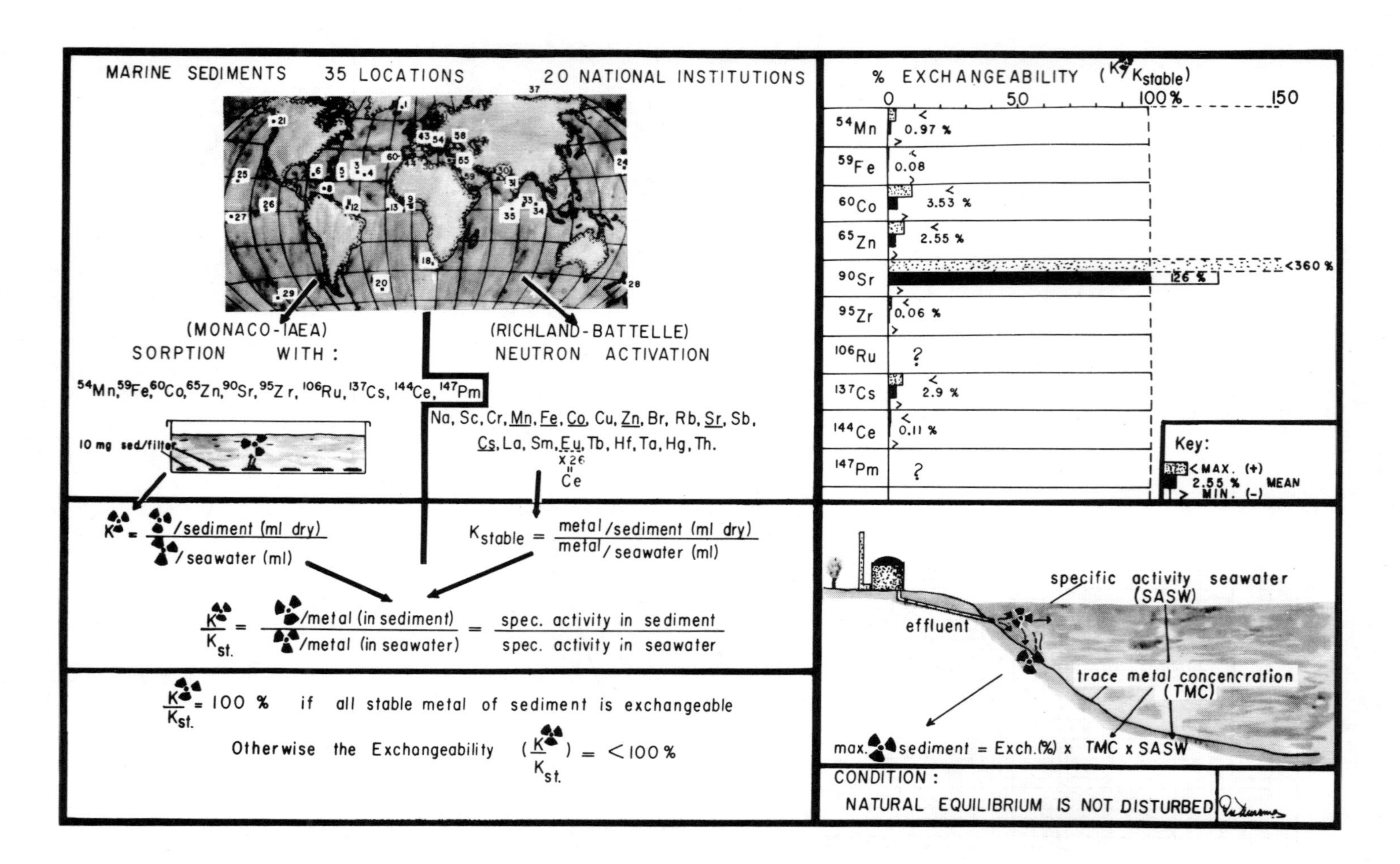


Fig. 5 Schematic representation of methods, final average results and a method of calculation for a maximum radioactivity in a seabed. (Figure reproduced from Duursma, 1973).

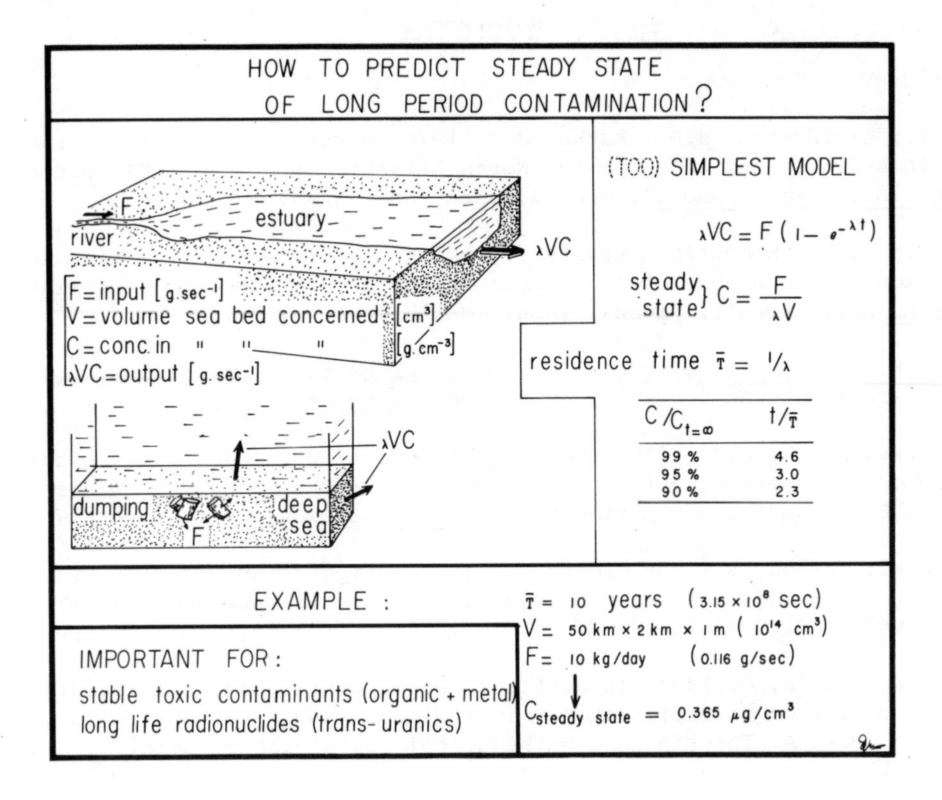


Fig. 6 Box-model example for an evaluation of steady state concentrations of foreign substances in a seabed, e.g. 0-10 cm layer; simplest model.

## References

Bowen, V.T.; Livingston, H.D.; Burke, J.C. 1976. Distribution of transuranium nuclides in sediment and biota of the North Atlantic Ocean. In : "Transuranium Nuclides in the Environment", Vienna, IAEA, SM-199/96, p. 107-20.

Duursma, E.K. 1973. Specific activity of radionuclides sorbed by marine sediments in relation to the stable element composition. In "Radioactive Contamination of the Marine Environment", Vienna, IAEA, SM-158/4, p. 57-71.

Duursma, E.K. ( 1978). Effect of hydrostatic pressure on the diffusion of chloride in marine sediment. An "in situ" experiment. Deep Sea Res.

Duursma, E.K.; Bosch, C.J. 1970. Theoretical, experimental and field studies of radioisotopes concerning diffusion in sediments and suspended particles of the sea. Part B. Methods and Experiments. Neth. J. Sea Res., vol. 4, n° 395-469.

Duursma, E.K.; Dawson, R.; Ros Vicent, J. 1975. Competition and time of sorption for various radionuclides and trace metals by marine sediments and diatoms. Thalassia Jugosl., vol. 11, p. 47-51.

Duursma, E.K.; Hoede, C. 1967. Theoretical, experimental and field studies concerning molecular diffusion of radioisotopes in sediments and suspended solid particles of the sea. Part A. Theories and Mathematical Calculations. Neth. J. Sea Res., vol. 3, p. 423-57.

Duursma, E.K.; Parsi, P. 1976. Distribution of plutonium -237 between sediment and seawater. Monaco, CIESM, XXIVe Congrès Assemblée plénère, dec. 1974.

Hetherington, J.A., Jefferies, D.F.; Lovett, M.B. 1975. Some investigations into the behaviour of plutonium in the marine environment. In : "Impacts of Nuclear Releases into the Aquatic Environment", Vienna, IAEA, SM-198/29, p. 193-212.

Hetherington, J.A.; Jefferies, D.F.; Mitchell, N.T.; Pentreath, R.J. Woodhead, D.S. 1976. Environmental and public health consequences of the controlled disposal of transuranic elements to the marine environment. In "Transuranium Nuclides in the Environment", Vienna, IAEA, SM-199/11, p. 139-154.

Hills, G. 1972. The physics and chemistry of high pressures. In "The Effect, of Pressure on Organisms." Symposia of the Society for Experimental Biology, XXXI, Cambridge, 1-26.

Livingston, H.D.; Bowen, V.T. 1976. Plutonium and cesium -137 distribution patterns in coastal sediments of the Northwest Atlantic Ocean. In "Environmental Chemistry and Cycling Processes", Symposium Atlanta, Georgia.

Patterson, J.H.; Nelson, G.B.; Matlack, G.M.; Waterbury, G.R. 1976. Interaction of $^{238}PuO_2$ heat sources with terrestrial and aquatic environments. In "Transuranium Nuclides in the Environment", Vienna, IAEA, SM-100/100, p. 63-78.

Ros Vincent, J.; Costa Yagüe, F. 1972. Behaviour of radioisotopes concentrated in the surface layer of sediments under quasi-natural conditions. IAEA, Vienna, Final Progress Report, Contract N° 7571/RI/RB, 20 p.

# Convective mixing in sediments

William S. Reeburgh*

Quite recently fragmentary evidence has accumulated to suggest that sediments can be "marked" or invaded by denser water. The data suggest that the process is density-driven and that it operates with a time scale that is fast relative to diffusion. To date, we have only "before" and "after" observations; the "during" observations are non-existent.

Information suggesting convective replacement of interstitial waters comes from a variety of sources. One suggestion is from Sanders et al., (1965), who used an array of chloride-sensitive electrodes to show that the time-average salinity of sediments in the Pocasset River is always higher than the time-average salinity of the overlying waters. More recently, Thorstenson and Mackenzie (1974) observed large summer-winter differences in the bicarbonate profiles in fine-grained sediments from Harrington Sound, Bermuda; composition changes extended to depths of 70 cm in these sediments. They suggested that these changes were related to intersection of the bottom by the thermocline during winter cooling, which explained the distributions by diffusion from above and below the zone. Hesslein (1976) reports penetration of Canadian Shield lake sediments to depths of 10 cm following autumn overturn of the water columns of these lakes. The water column of the lake under study was labelled with tritiated water ($^{3}H_2O$), which served as a convenient and distinctive tracer. This lake overturned only partially before freezing, so sediments from water depths unaffected by the overturn showed no penetration by the labelled water.

Laboratory studies on convective mixing in sediments have been reported by Callame (1961). Callam's experimental conditions were extreme, but the time frame for penetration was shown to be a matter of hours. Okubo and Reeburgh are presently engaged in a study using mathematical models for convection in porous media to determine the extent and importance of this process in nature. Preliminary results indicate that critical Reynolds numbers can be exceeded by density changes comparable to those expected for seasonal temperature changes. We hope to use these models to estimate the depth and rate of penetration for a given density change. The biggest uncertainties are in the values for the permeability of surface sediments, which provides a means of replacing interstitial waters without affecting the solid phase.

The importance of this process as a nutrient source or flushing mechanism depends on the depth to which flushing takes place and how frequently it occurs. Before we can realistically determine reaction time frames, we must assure ourselves that the interstitial waters are not being replace or removed by a non-diffusive pro-

* Institute of Marine Science, University of Alaska, Fairbanks, Alaska, 99701 U.S.A.

cess. In freshwater systems, we can expect the process to occur only once a year; Hesslein's observed penetration is equivalent to 1-2 week's diffusion. Where penetration takes place to depths of 1 m or where the process can occur repeatedly, as in an estuary, much greater contributions may be expected. If the convection is salinity-driven, as in shallow areas subject to intense evaporation, loading the sediments with much denser water might be expected to inhibit fluxes from the sediments.

## References

Callame, B. 1961. Etude sur la diffusion des sels entre les eaux surnageantes et les eaux d'imbibition dans les sediments marins littoraux. Bull. Inst. Oceanog. Monaco, N°. 1181, 13 May 1960.

Hesslein, R.H. 1976. The fluxes of methane, total carbon dioxide, and ammonia from sediments and their consequent distribution in a small lake. Ph.D Thesis, Columbia University.

Lapwood, E.R. 1948. Convection of a fluid in a porous medium, Proc. Camb. Phil. Soc., vol. 44, p. 508-21.

Reeburgh, submitted, A major sink and flux control for methane in marine sediments: anaerobic consumption. In : Benthic Processes and the Geochemistry of Marine Interstitial Waters, F.T. Manheim and K.A. Fanning, eds.

Sanders, H.L.; Mangelsdorf, P.C. Jr.; Hampson, G.R. 1965. Salinity and faunal distribution in the Pocasset River, Massachusetts. Limnol. Oceanogr., vol. 10, R216-R229.

Thorstenson, D.C.; Mackenzie, F.T. 1974. Time variability of pore water chemistry in recent carbonate sediments, Devil's Hole, Harrington Sound, Bermuda. Geochim. Cosmochim. Acta, vol. 38, p. 1-19.

# Group 4
# The role of organisms in estuarine sedimentation processes

# Report

INTRODUCTION

Estuaries are areas of high productivity. Both benthic and planktonic plants contribute to the overall organic budget of the estuary. The magnitude and relative importance of the input from these two communities will depend upon a variety of factors, e.g. topography, hydrology, and accordingly can vary markedly from one estuary to another. A substantial proportion of the products of the plant growth will enter the estuarine sediments, which are known to be important sites for many reactions involving the degradation of organic material and the regeneration of inorganic forms. Estuaries are, in consequence, zones of high biological activity and the sites of important fisheries.

The biology of estuaries has been discussed by many authors (see Lauff, 1967; Barnes and Green, 1972; Perkins, 1974) and it is not appropriate to review the subject in the present account.

The organisms within the sediments may have both biochemical and mechanical effects upon the sedimentary environment. The system is an interactive one and, as a consequence, does not fall nautrally into sections for discussion. We have, however, for the sake of convenience, considered the processes under two headings: (i) those associated with micro-organisms; and (ii) those associated with higher organisms. The important microbiological processes are biochemical whereas the higher organisms play an important role in physical and mechanical processes (e.g. such as bioturbation, irrigation).

2. MICROBIOLOGICAL PROCESSES

The details of the biochemical reactions attributed to micro-organisms are discussed in the three contributions to this volume (Stanley, Martens and Reeburgh) and need not be repeated here. The overall decomposition process of organic material involves the initial hydrolysis of the sedimented organic material into diffusible substances, followed by their assimilation by the organism. The latter process may either be fermentative or oxidative. The oxidative process will require suitable hydrogen accepters. Some of the reduced products (for example $CH_4$, $H_2S$, $NO_2^-$) are essentially transients and when they diffuse out of their zone of formation, may be used by autotrophic organisms as a source of energy.

Most aspects of the biochemical processes in sediments are poorly understood, frequently our knowledge being restricted to the steady state concentrations of the intermediates. In the case of the organic fraction, analyses are too frequently limited to a determination of total organic material which provides little information to the microbiologist. Microbiological processes are, in part, governed by the concentration of low molecular weight compounds and there is a need for more detailed studies of the concentration of relevant organic and inorganic compounds in the interstitial waters.

The measurements of the rates of microbiological processes in sediments present considerable experimental problems. In favourable circumstances it should be possible to measure the integrated activity in the sediments by determining the flux of material, for example, oxygen across the sediment-water boundary layer. This however, provides little insight into the details of the events occurring within the sediments. The measurement of the rates of processes within the body of the sediments is a formidable task, for any attempt to remove the sediment disrupts its structure and thus disturbs the system. At present, there appears to be no means of predicting rates of microbiological processes solely from steady state concentrations. However, it is possible to derive kinetics from isolated systems and use them in conjunction with steady state concentrations in models to predict rates. The paper by Billen and Vanderborght (see this volume) illustrates the potential of this approach. At the same time, it raises the problem of the dispersion coefficients that should be used for such calculations. The ultimate test of these models is that they should be able to reproduce not only the vertical profiles within the sediments but also the fluxes across the water-sediment interface. This indicates the need for reliable measurements of the fluxes of key substrates across the water-sediment interface. The difficulties of making such studies centre around reproducing the hydraulic regime above and within the sediments. This type of study is needed to calculate balances.

There is an array of kinetic and molecular problems associated with the understanding of microbial process in sediments. Williams (in this volume) considers some aspects of molecular diffusion in relation to the factors controlling microbiological activity. The mode of action of extracellular enzymes in sediments is poorly understood, although they catalyze the first step in the breakdown of organic material. It is not known, for example, whether extracellular enzymes present in sediments are in a free state or are bound to surfaces. Surface-bound enzymes are more stable than those in solution and this could have an important bearing on our understanding of biochemical processes occurring over long time scales. Some aspects of the interstitial environment may not be favourable to extracellular enzyme action; humic acids may denature enzymes, some metal ions may inhibit enzyme action. This latter consideration of course raises the question of the speciation of metals such as copper, mercury, chromium etc., in the interstitial waters; this has been discussed elsewhere in the workshop report.

A puzzling aspect of biochemical processes in sediments is the apparent difference between time scales of some processes attributed to micro-organisms, and the probable time scale of microbial processes themselves. Micro-organisms, if they are to remain active, need to expend energy continuously in order to maintain the cell. If the input of energy falls below a certain level, the cell will die or form a spore. This sets a general limit on the minimum possible rate of microbial activity, which in turn sets a maximum time scale to microbial processes in closed systems. It is accordingly difficult to attribute to living micro-organisms reactions occurring in parts of cores that have been isolated for $10^2$ to $10^3$ years. If these reactions are in fact underway (and not purely chemical reactions) then there must be a continual replenishment of material from outside this zone, alternatively it is possible that the reactions are catalysed by fragments of cells, which have retained their enzymic capacity since the reducing nature of the interstitial waters could enhance the stability of enzymes and enzyme systems.

Another problem is the low energy yield of some of the reactions occurring in sediments which are supposedly carried out by micro-organisms. As an example, Reeburgh (in this volume) has provided evidence for the oxidation of methane to carbon dioxide, with sulphate as the hydrogen accepter. The redox potential of these two couples is so close (Billen, this volume) that the likely energy yield associated with the transfer of an electron is less than would be required to synthesise ATP. As a result, it seems unlikely that the reaction could serve as a source of energy for the micro-organisms. It may be argued that it is unlikely that the organism will perpetuate itself, and expend energy in producing enzymes for a reaction from which there is no immediate or foreseeable energy yield. Non-living enzyme systems would be free of such restrictions, but whether they exist is a matter of speculation.

## 3. HIGHER ORGANISMS

The roles of higher organisms have been discussed in other sections (see Day, in this volume). In addition to participating in biochemical processes, higher organisms may be responsible for important physical processes in the sediments.

Vascular plants play a major role in the biology of many estuaries. As well as acting as a major supply of humus, they transport material, such as inorganic nutrients, oxygen, metal ions, etc., through their root systems. This can result in a modification of the chemistry of the sedimentary environment. As an example, the roots take up Ca, Mg, K and Na in ratios quite different from those of the interstitial waters. Some of the estuarine plants (e.g. sea thrift, *Limonium*) are perennials living up to 50 years and may provide an historical record of the estuary.

The benthic fauna of estuaries is best considered under two general categories: the smaller forms of the population (the meio- and micro-fauna); and the larger macro-fauna. The former category is mainly restricted to the oxygenated zone. They have little effect upon the movements of sediment particles; they feed on bacteria and thus promote the flux of material within the surface layer.

The macro-fauna has a variety of effects on the chemistry of the sediments. Many filter feeders have mechanisms for sorting the larger inorganic particles from fine silt and organic matter. Only the latter is ingested and the undigested material is formed into faecal pellets which accumulate on the sediment surface. Some pellets disintegrate rapidly but others are enclosed in membranes and hence persist, being found as fossils. Such latter-type pellets increase the general particle size of the sediments and thus its porosity. The chemical composition of these pellets and the dissolved substances in the interstitial waters of this particular environment requires further study.

Burrowing animals cause perturbation of the sediments but the effect varies with different animal groups. Crustacea, including amphipods, burrowing prawns and crabs, bring excavated material to the surface where it can be oxidized leading to the escape of soluble susbstances. Soft-bodied burrowers, including worms and bivalves, do not excavate sediments but force the sediment particles apart as they burrow down. Suspension feeders make simple vertical burrows which may extend down a metre or more. Deposit feeders and carnivores often plough their way laterally through the sediments while mud swallowers, such as the worms *Arenicola* and *Scoloplos*, eat their way through the soil assimilating subsurface organic particles and voiding indigestible inorganic materials on the surface. The burrowing prawns *Callianassa* and *Thalassina*, which winnow organic particles from subsurface sediments, make branching tunnels and bring large amounts of material to the surface. *Callianassa* often occur in densities of $100/m^2$ and must have important effects on sediment chemistry. Burrowing animals appear to be far more abundant in intertidal banks than below tidal marks where bioturbation is much less.

Burrowing animals draw oxygenated water and change the $E_H$ and the chemistry of the interstitial waters and sediments. Those animals that make U-shaped burrows, for example *Arenicola*, are most effective in ventilating the sediments. The water in the exhalant current must be mixed with interstitial water from the sides of the burrow and this carries the dissolved chemicals and fine particles to the surface.

Estuarine organisms take up and may concentrate trace metals. This is presently an area of considerable interest. Much less is known about the extent and importance of heavy metal accumulation by micro-organisms present in sediment. It is conceivable that they could act as a vector in the process of uptake of heavy metal ions.

## 4. SUMMARY

In summary, benthic organisms will have an important effect on the chemistry of the sediments both through biochemical and mechanical processes. Micro-organisms cause a variety of biochemical transformations and although the descriptive aspects of many of them are understood, there is very little knowledge of the rates of these prcoesses. There is inadequate information on the detailed organic chemistry of the interstitital waters, especially the low molecular weight intermediates of microbial processes. The role played by higher organisms is complex and will vary considerably from one location to another. As well as biochemical transformation, higher organisms, by virtue of burrowing and pumping actions, can have a considerable effect upon the chemical environment by increasing the rate of transport of material within the sediments and by altering its physical nature.

## 5. RECOMMENDATIONS FOR FURTHER STUDY

1. Detailed measurements on better-constrained systems. Depth distributions of low molecular weight organic compounds which are substrates and products of degradation reactions should be determined. Also needed are measurements of more chemical species in the C, N, S and P systems, which should serve to identify reactions and provide a more complete understanding of fluxes and mass balances. In systems where isotope effects are large, such as the C, N and S systems, attempts at obtaining an isotope balance are needed. Greater time resolution than is currently available and depth resolution of 1 cm or less are essential to understanding the zonation of processes. Mirror-image sampling techniques employing *in situ* equilibration samplers or variations employing gels, specific reagents and indicators should be used.

2. Direct measurements of biological reaction rates. Techniques are currently available for measuring the rates of denitrification and sulphate reduction in sediments. Similar approaches should be employed using labelled carbon and nitrogen compounds in field and laboratory experiments. Directly measured rates are needed to test and extend the dynamic models developed for sediments.

3. The role and residence time of easily suspended particulate organic matter. This organic carbon reservoir is expected to be important in interactions with heavy metals. Little is known of its residence time, but bomb-produced radiocarbon appears to offer potential in determining the fate and residence time of this material. Techniques are needed for determining the proportions of living and non-living material in this reservoir and in the sediments.

4. Time scales for biochemical reactions. We have only a vague idea of the biodegradability of freshly deposited organic material in sediments, which is partially discussed above. Biochemical processes like sulphate reduction, methane production, and methane consumption take place on a large scale at depths in sediments where the matrix and, presumably, the organic material is old relative to biological time scales. The energy yield of these reactions is low and maintenance energy requirements of organisms over periods of $10^2$ to $10^3$ years presents a puzzling situation. Are these reactions mediated by stable surface-bound extracellular enzymes in such a way that maintenance energy is not a consideration? Does a low supply rate of suitable substrate molecules produce this situation?

5. Biological transport, irrigation and sediment mixing processes. In considering these processes, it is important to distinguish whether liquid phases, solid phases or both are involved. Rooted vascular plants and irrigating organisms are important in transporting oxidants into the sediments and products from the sediments. This is accomplished with little physical reworking, and results in oxidized microzones and displacement of anoxic reactions to greater depths in the sediments. When transport or irrigation cease, the original conditions are established relatively quickly. The activities of burrowing organisms involve solid phases. Results of these processes are evident in distributions of solid phase associated isotopes, such as $^{210}Pb$, $^{137}Cs$, and $^{90}Sr$. Means of measuring and understanding bioturbation are important in establishing recent chronologies. The activities of organisms also may affect the mechanical properties of sediments. Faecal pellets deposited on the sediment surface are reasonably durable and may affect the susceptibility of surface sediments to mixing by physical processes.

6. Several processes not considered in present mathematical models were discussed at this workshop. Evidence that molecular diffusion cannot be considered the only transport process was advanced. When direct reaction rate measurements are available, a dispersion coefficient one hundred-fold faster than the currently used molecular coefficient best describes distributions showing denitrification. Rapid convective replacement of interstitial waters driven by seasonal density changes also requires attention. The notion that faecal pellets persist and form a more open structure in muddy sediments was discussed in light of physical interaction by waves and currents. Evidence of zonation of reactions in sediments will require a better understanding of product-substrate inter-relationships and will probably require use of more complicated reaction kinetics. If fluxes from sediments are to be understood, models incorporating these processes will be needed.

# References

Barnes, R.S.K.; Green, J. (eds.) 1972. The Estuarine Environment, London, Applied Science Publishers, p. 133.

Lauff, G.H. (editor) 1967. Estuaries, Washington D.C., American Association for the Advancement of Science, Publication N° 83, 757 p.

Perkins, E.J. 1974. The Biology of Estuarine Waters, London, Academic Press, 678 p.

# Stable carbon isotope characteristics of organic sedimentary source materials entering the estuarine zone

Helmut Erlenkeuser*

## 1. Introduction

An estuary may be considered as a temporary reservoir where river-borne material passes through and is subjected to decomposition, diagenesis, and alteration before being "re-shipped to the ocean. During this process, pollutants may be deposited and although accumulation rates are negligible on the geological time scale, certain substances may reside long enough for high concentration levels to build up. This aspect seems especially important with respect to compounds toxic to the biota of the estuary, which is often intensively used as a source of human food. The adjacent marine environment may be likewise affected when heavily-polluted estuarine sediments are swept out to sea.

In order to understand the changes taking place in the estuary, investigations into the present and former distribution patterns of pollutants are called for as they are produced by the variety of interactions between the water and the sediments. The second investigation should be directed at the variability of sediment accumulation and to the time scales of transport processes, which determine the residence time of the particulate matter in the estuary. In view of the large number of physical, chemical, and biological processes involved in the transfer of the riverborne material along the estuary to the sea, the transfer properties of the estuary can be defined only from a general point of view, indicating behaviour and including some probability figures to characterize the variability of the parametres entering the model of the transfer function. From this aspect, the study of the sediments at locations with fluctuating accumulation seems as important as the study of well-developed records in particle-by-particle sediments.

A great number of the pollutants which enter the estuarine zone are associated with fine-grained particulate and organic matter (de Groot et al., 1971). Adsorption to grain surfaces (Gibbs, 1973), enrichment by planktonic organisms (e.g. Groth, 1971), and complexed with humic acids (e.g. Förstner and Müller, 1974) are some of the mechanisms leading to increased heavy metal content for example, in the sediments. The distribution of river-supplied organic matter in the estuary may therefore provide a general basis upon which to discuss the possible fate of certain pollutants and their influence on the estuarine environment.

In this paper the differentiation between organic matter originating from freshwater land or the marine environment is discussed using stable carbon isotope ratios.

---

* Institut für Reine und Angewandte Kernphysik, University of Kiel,- 23, Kiel, Federal Republic of Germany.

## 2. Methods

The stable carbon isotope ratio, i.e. the $^{13}C:^{12}C$ ratio, is easily measured mass-spectrometrically on the carbon dioxide gas prepared from a sample by dry combustion (for techniques, see Craig, 1953 or Degens, 1969). Stoffers (1976) used wet combustion with potassium dichromate and concentrated sulphuric acid in a closed reaction system. Impurities of the oxides of nitrogen, which may interfere with the carbon dioxide mass signals (Craig and Keeling, 1963) must be carefully avoided.

The carbon isotope ratio is measured as the relative difference of the carbon-13 to carbon - 12 ratio from a reference substance (for reference standards, see Craig, 1957 and Wedepohl, 1969) and was formally re-normalized to the PDB standard, a standard of carbonate of Belemnitella americana from the Cretaceous Pee Dee formation in South Carolina U.S.A. This standard source, however, is now exhaustive.

The stable carbon isotope ratio is expressed by :

$$\delta^{13}C = \frac{(^{13}C/^{12}C)_{sample} - (^{13}C/^{12}C)_{standard}}{(^{13}C/^{12}C)_{standard}}$$

The reproducibility of $\delta^{13}C$ analyses is about ±0.1 or ±0.2‰, depending on the sample and the chemical preparation, while the internal accuracy of the mass-spectrometre is generally better than ±0.05‰. Samples containing even a few micrograms of carbon can be analyzed.

## 3. Carbon isotopes in nature

Due to their different atomic masses, the carbon isotopes may become fractionated in transport processes and chemical reactions as a result of kinetic and quantum-mechanical effects of the atomic mass on the transport behaviour and the interaction poetential of the molecules. The natural range of $\delta^{13}C$ variation is about 100‰. Carbon isotope studies can hardly be applied to estuarine systems without understanding the different fractionation mechanisms that affect the isotopic composition of the sedimentary source materials.

### 3.1 Terrestrial biosphere

The $\delta^{13}C$ values of atmospheric $CO_2$, are closely grouped around -7‰ (Craig and Keeling, 1963; Wedepohl, 1969). Considerable fractionation occurs during the different steps of the assimilation process (Degens, 1969). The carbon isotope ratio in higher land plants such as trees, which obey the Calvin or C3 assimilation cycle, is about -25‰ to -26‰ and peat has about the same average (Stahl, 1968). Smaller fractionation effects occur in the Hatch-Slack or C4 cycle ($\delta^{13}C \approx -13$‰; Bender, 1971; Smith and Epstein, 1971; Troughton, 1972) and these plants may dominate in semi-arid climates. A third group of plants follow the Crassulacean Acid Metabolism. These plants can use both C3 and C4 pathways, and their $\delta^{13}C$ covers the range between -13‰ and -26‰ (Lerman, 1972). The bulk of common land plants, however, is centred around an average of about -26‰.

The carbon isotope composition appears to be essentially preserved during decomposition of plant organic matter in the soil (e.g. Stout and Brien, 1972; Tate, 1972). Terrigenic contributions to the particulate load of a river, such as plant litter, peat, or soil humus, have $\delta^{13}C$ values very close to the average isotopic composition of the vegetation; the effect of C4 plants, however, may not always be negligible.

When the ground water comes into contact with the atmosphere, the bicarbonate gradually becomes enriched in $^{13}C$ through carbon dioxide exchange, finally changing to slightly positive $^{13}C$ values - as found in the sea. Thus the actual isotopic composition depends on the fate of the river water. For the Rhine river mouth, for example, Mook (1970) reported $^{13}C$ values of about -11‰, varying seasonally by about 2‰.

It appears from the above discussion that, in general, the freshwater system is isotopically lighter than the marine carbonate system. This relation should be reflected in the respective organic fractions as well. Degens (1969), in summarizing literature data, found $\delta^{13}C$ of freshwater plankton centred at about -30‰, and low $^{13}C/^{12}C$ ratios in the range between -25‰ to -30‰ have frequently been found in lake sediments (Stuiver, 1975). In Great Plön Lake, northern Federal Republic of Germany, where the overwhelming part of the organic fraction is planktonic matter (Ohle, 1973) and the contribution of terrigeneous plant litter is small, we found $\delta^{13}C$ about -28‰ to -30‰ in the sedimentary organic matter ($\delta^{13}C$ of the water : -7‰ to -10‰, Erlenkeuser and Willkomm, 1971, 1973). However, as stable carbon isotope ratios suggest higher subaquatic plants are generally more affected by low $CO_2$ availability than are plankonic organisms. The $^{13}C$ isotope in these plants is generally more concentrated than in the plankton, and $\delta^{13}C$ may exceed that of the average marine sedimentary organic carbon. These effects need attention when fresh or brackish water plants provide an important source of sedimentary organic matter.

## 4. Identification of sedimentary source materials

The above discussion illustrates that the isotopic composition of riverborne organic matter may be expected to range between -25‰ and -30‰, and hence to be lighter than in the marine plankton, which yields $\delta^{13}C$ values of about -20‰ to -23‰ for tropical and temperate zones. Hunt (1970) reports $\delta^{13}C$ of about -25‰ for the deposits of rivers on the eastern coast of the U.S.A., while abyssal sediments off shore possessed a carbon isotope ratio of about -21‰; estuarine sediments fell between these two values. It should be noted, however, that in larger estuaries and when primary production is high, the isotopic composition of organic matter produced in the local brackish water zone varies with salinity and can interfere with the isotopic ratio of sediments mixed with marine and river material.

Although considerable differences in carbon isotope composition occur between different specimens of a plant community (Parker, 1964) and even between the different chemical constituents of organic matter (Degens, 1969), the isotope ratio of organic matter as a whole remains remarkably constant during the early steps of diagenesis and is found to be essentially the same in sedimentary organic matter as in the organic source material (Hedges and Parker, 1976). The variations in isotopic composition as found between individual marine or terrestrial plants or plant species are not observed in the sediments and appear to be smoothed down by the numerous processes of sediment mixing and redistribition. This agrees with results from $^{14}C$ measurements on coastal marine sediments, which show highly reproducible $^{14}C$ activities although they are mixed from differently-aged fractions and are deposited under widely differing conditions (Erlenkeuser, 1973).

Stable carbon isotopes have been successfully used to study the distribution of Mississippi sediments in the coastal zones of the Gulf of Mexico (Sackett and Thompson, 1963). Newman *et al.* (1976) extended this work and studied the $\delta^{13}C$ distribution with depth in sediment cores from all over the Gulf. They found a varying supply of terrigeneous organic matter in relation to the major climate cycle during the late Quaternary era. Recently, Hedges and Parker (1976) have demonstrated the relevance of stable carbon isotope ratios with respect to the fraction of terrigeneous organic carbon in marine sediments by analyzing lignin oxidation products as inde-

Similar isotopic ratios are also found in the effluents of petrochemical plants and in the organic fraction of municipal wastes and sewage sludges which mainly consist of decomposed terrigeneous matter. This isotopic fingerprint may be used for tracing these pollutants if they are discharged directly to the marine environment (Tan and Pearson, 1975).

### 3.2 The ocean

The isotopic composition of the dissolved molecular $CO_2$ in seawater is nearly identical to that in the atmosphere (Deuser and Degens, 1967; Wendt, 1968; Vogel *et al.*, 1970) and the fractionation effect during the passage of carbon dioxide from the atmosphere into the sea appears to be small. The carbon - 13 isotope, however, becomes enriched during the hydration stage, and $\delta^{13}C$ of the bicarbonate is about + 1‰. The latter fractionation is temperature dependent.

Due to isotopic fractionation during assimilation, marine plankton has $\delta^{13}C$ ratios of about -22‰ (Degens, 1969). Newman *et al.*, (1973) assume a $\delta^{13}C$ of -19‰ (on the PDB scale) for marine plankton of the Gulf of Mexico, and Hunt (1970) reports carbon isotope ratios in the deeper sediments off the eastern coast of the U.S.A. to be about -21‰. A greater fractionation effect was observed in cooler waters: $\delta^{13}C$ values of about -27‰ were found in plankton from arctic bottom waters (+1°C; Sackett *et al.*, 1965). These isotopic variations with temperature are probably not an effect of the assimilation process itself: Degens *et al.* (1968) demonstrate that the availability of the dissolved molecular $CO_2$ in the immediate environment of the assimilating plant may effectively influence the isotopic composition. The drainage of the local molecular $CO_2$ pooled by the assimilating organism disturbs the chemical equilibrium of the carbonate system. Molecular $CO_2$, then, is derived from the bicarbonate - with the isotopic composition of this species - and is rapidly consumed by the organism before the equilibrium fractionation between the molecular $CO_2$ and the ionized carbonic acid species is re-established.

Environmental conditions, such as temperature and pH, which affect the quantity ratios of the different constituents of the carbonate system in the water, the turbulence structure, which affects the rate of water renewal at the surface of the organism, and other environmental and biological factors controlling the rate of assimilation, may thus influence the final carbon isotope composition of the plant. For example, in the Kiel Bight (western Baltic Sea), where water transparency is low and pH is often below 8, marine macrophytes from the water at 10 m depth showed large fractionation effects yielding isotope ratios as low as -34‰ (Stoffers, 1976; $\delta^{13}C$ of the water : ∿0‰), whereas the bulk of higher marine plants, as reported in literature, have $\delta^{13}C$ ranging between -10‰ and -20‰.

### 3.3 Freshwater

The carbon isotope ratio in freshwater is affected by the dissolution of fossil carbonates during groundwater formation (cf. Munnich, 1968; Vogel, 1970; Pearson and Hanshaw, 1970). Although the geochemistry of ground water carbonates may be very complex in any particular case, the general scheme, which is valid at least for temperate, non-arid climate zones with well developed soil profiles, is as follows. The average $\delta^{13}C$ of common land plants is about -26‰ and is nearly the same for soil-$CO_2$ (Galimov, 1966). During infiltration, one mole of carbonate - in general fossil marine carbonate with $\delta^{13}C$ close to 0‰ - is dissolved by one mole of soil-$CO_2$ resulting in two moles of bicarbonate with $\delta^{13}C$ about -12‰. Subsequent exchange with excess soil-$CO_2$ causes the isotopic ratio to decrease towards -18‰, the final equilibrium composition of bicarbonate in contact with soil-$CO_2$.

pendent indicators of land-derived organic matter.

Sackett and Thompson (1963) and Newman *et al.* (1973) obtained lines of constant $\delta^{13}C$ ("isodels") which show the distribution of terrigenic organic matter both at present and in the past. Such a distribution could perhaps be taken to represent the steady state distribution pattern of the river-supplied sedimentary organic carbon, as it results from the action of redistribution and transport processes that cause the spreading of particulate matter from the coastal depositional zones out to greater depths. Since the time scales of these processes greatly exceed the 50 - 100 years period when large-scale pollution occurs, the steady state distribution of land-derived organic matter might provide relevant information on what the long-term fate of certain pollutants might be particularly those which at present are found at high concentration levels in the estuarine or coastal marine environment.

The discharge of anthropogenic waste materials such as municipal wastes, sewage sludges, and petrochemical wastes to natural water systems may also be recognized from the isotopic composition of the bicarbonate, which becomes affected by the isotopically light $CO_2$ released through decomposition of the organic fraction of these materials (Tan and Walton, 1975). Because of the high mobility of the water, the effects are small and restricted mainly to the locality of the waste outfall. Over greater distances its seems difficult to discriminate against natural biogenic $CO_2$ contributions. However, the stable carbon isotope ratios offer the possibility of *in situ* decomposition rate studies. Finally, the power of stable carbon isotope studies is greatly enhanced if combined with measurements of the radiocarbon isotope.

## References

Bender, M.M. 1971. Variations in the $^{13}C/^{12}C$ ratios of plants in relation to the pathway of photosynthetic carbon dioxide fixation. Phytochemistry, vol. 10, p. 1239-44.

Craig, H. 1953. The geochemistry of the stable carbon isotopes. Geochim. Cosmochim. Acta, vol. 3, p. 53-92.

Craig, H. 1957. Isotopic standards for carbon and oxygen and correction factors for mass spectrometric analysis of carbon dioxide. Geochim. Cosmochim. Acta, vol. 12, p. 133-49.

Craig, H.; Keeling, C.D. 1963. The effects of atmospheric $NO_2$ on the measured isotopic composition of atmospheric $CO_2$. Geochim. Cosmochim. Acta, vol. 27, p. 549-51.

Degens, E.T.; Guillard, R.R.L.; Sackett, W.M.; Hellebust, J.A. 1968. Metabolic fractionation of carbon isotopes in marine plankton - I. Temperature and respiration experiments. Deep Sea Res., vol. 15, p. 1-9.

Degens, E.T. 1969. Biogeochemistry of stable carbon isotopes. In: G. Eglington and and M.T.J. Murphy (eds.), Organic Geochemistry. Berlin, Springer.

Deuser, W.G.; Degens, E.T. 1967. Carbon isotope fractionation in the system $CO_2$ (gas) - $CO_2$ (aqueous) - $HCO_3^-$(aqueous). Nature, vol. 215, p. 1033-5.

Erlenkeuser, H.; Willkomm, H. 1971. University of Kiel radiocarbon measurements VI. Radiocarbon, vol. 15, p. 113-39.

Erlenkeuser, H.; Willkomm, H. 1973. University of Kiel radiocarbon measurements VII. Radiocarbon, vol. 15, p. 113-26.

Erlenkeuser, H. 1976. Environmental effects on radiocarbon isotope distribution in coastal marine sediments. Ninth Interntl. radiocarbon conf., Los Angeles and San Diego, June 1976.

Förstner, U.; Müller, G. 1974. Schwermetalle in Flüssen und Seen. Berlin, Springer, 225 p.

Galimov, E.M. 1966. Carbon isotopes of soil $CO_2$. Geochemistry intern., vol. 3, p.889-97.

Gibbs, R.J. 1973. Mechanisms of trace metal transport in rivers. Science, vol. 180, p. 71-3.

Groot de, A.J.; Goeij de, J.J.M.; Zegers, C. 1971. Contents and behaviour of mercury as compared with other heavy metals in sediments from the rivers Rhine and Ems. Geol. Mijnbouw, vol. 50, p. 393-8.

Groth, P. 1971. Unter suchungen über einige Spurenelemente in Seen. Arch. Hydrobiol. vol. 68, p. 305-75.

Hedges, J.I.; Parker, P.L. 1976. Land-derived organic matter in surfaces sediment from Gulf of Mexico. Geochim. Cosmochim. Acta., vol. 40, p. 1013-29.

Hunt, J.M. 1970. The significance of carbon isotope variaticns in marine sediments. In : G.D. Hobson and G.C. Speers (eds.) Advances in Organic Geochemistry, Proc. 3rd Congr. Oxford Pergamon, p. 27-35.

Lerman, J.C. 1972. Carbon 14 dating : origin and correction of isotope fractionation errors in terrestrial living matter. Proc. 8th interntl. conf. radiocarbon dating. Wellington, The Royal Society of New Zealand, p. 613-24.

Mook, W.C. 1970. Stable carbon and oxygen isotopes of natural waters in the Netherlands, Isotope Hydrology, 1970, Vienna IAEA, p. 163-88.

Münnich, K.O. 1968. Isotopen-Datierung von Grundwasser. Naturwissensch, vol. 55, p. 158-63.

Newman, J.W. Parker, P.L.; Behrens, E.W. 1973. Organic carbon isotope ratios in Quaternary cores from the Gulf of Mexico. Geochim. Cosmochim. Acta, vol. 37, p.225-38.

Ohle, W. 1973. Die Rasante Eutrophierung des Grossen Plöner Sees in frühgeschichtlicher Zeit. Naturwissensch., vol. 60, p. 47.

Parker, P.L. 1964. The biogeochemistry of the stable isotopes of carbon in a marine bay. Geochim. Cosmochim. Acta., vol. 28, p. 1155-64.

Pearson, F.J.; Hanshaw, B.B. 1970. Sources of dissolved carbonate species in ground water and their effects on carbon-14 dating. Isotope Hydrology 1970, Vienna, IAEA, p. 271-86.

Sackett, W.M.; Eckelmann, W.R.; Bender, M.L.; Bé, A.W. 1965. Temperature dependence of carbon isotope composition in marine plankton and sediments. Science, vol. 148, p. 235-7.

Sackett, W.M.; Thompson, R.R. 1963. Isotopic organic carbon composition of recent continental derived clastic sediments of Eastern Gulf Coast, Gulf of Mexico. Bull. Am. Ass. Petrol. Geol., vol. 47, p. 525-8.

Smith, B.N.; Epstein, S. 1971. Two categories of $^{13}C/^{12}C$ ratios for higher plants. Plant. Physiol., vol. 47, p. 380-4.

Stahl, W. 1968. Die Verteilung der $C^{13}/C^{12}$ - Isotopenverhältnisse von Torf, Holz und Holzkohle. Brennstoff-Chemie, vol. 49, p. 69-72.

Stoffers, H. 1976. Untersuchungen zum Isotopenverhältnis des Kohlenstoffs bei Rotalgen aus der Kieler Bucht. Staatsexamensarbeit, Univ. Kiel.

Stout, J.D.; Brien, B.J. 1972. Factors affecting radiocarbon enrichment in soil and the turnover of soil organic matter. Proc. 8th interntl. conf. radiocarbon dating. p. 394-407. Wellington, The Royal Society of New Zealand.

Stuiver, M. 1975. Climate versus changes in $^{13}C$ content of the organic component of lake sediments during the late Quaternary. Quaternary Res., vol. 5, p. 251-62.

Tan, F.C.; Pearson, G.J. 1975. Stable carbon isotope ratios as water quality indicators. In : Water Quality Paramenters, ASTM STP 573, Am. Soc. for Testing and Materials, p. 543-549.

Tan, F.C.; Walton, A. 1975. The application of stable carbon isotope ratios as water quality indicators in coastal areas of Canada. In : Isotope ratios as pollutant source and behaviour indicators. p. 35-47, Vienna, IAEA.

Tate, K.R. 1972. Radiocarbon dating in studies of soil organic matter - vegetation relationships. Proc. 8th interntl. conf. radiocarbon dating., p. 408-19, Wellington, The Royal Society of New Zealand.

Troughton, J.H. 1972. Carbon isotope fractionation by plants. Proc. 8th interntl. radiocarbon conf., p. 420-38, Wellington, The Royal Society of New Zealand.

Vogel, J.C.; Grootes, P.M.; Mook, W.G. 1970. Isotopic fractionation between gaseous and dissolved carbon dioxide. Z. Physik, vol. 230, p. 225-38.

Vogel, J.C. 1970. Carbon-14 dating of ground water. Isotope Hydrology 1970, Vienna, IAEA, p. 225-239.

Wedepohl, K.H. (ed.), 1969. handbook of geochemistry. Section 6-B : Isotopes in Nature, Berlin, Springer.

Wendt, I. 1968. Fractionation of carbon isotopes and its temperature dependence in the system $CO_2$ - gas - $CO_2$ in solution and $HCO_3$ - $CO_2$ in solution. Earth and Planetary Science Letters, vol. 4, p. 64-8.

# Environmental impact of lead and zinc in recent sediments

N. B. Price*, S. J. Malcolm* and J. Hamilton-Taylor[x]

INTRODUCTION

In recent years there has been a growing awareness that certain anthropogenic constituents released by industry and domestic effluents are incorporated into accumulating sediments, both marine and non-marine. In many instances these constituents, especially artificially produced radioisotopes and halogenated hydrocarbons, clearly delineate the pathways of substances from their release to their incorporation into a sediment. The rates of input of these types of constituents into the environment can in some instances be satisfactorily assessed by measuring their concentration in a sediment whose accumulation rate is known.

It is well known too that man's discharge of heavy metals can be incorporated into sediments, the amounts of which have been historically assessed by measuring their concentrations above a natural background (Erlenkeuser *et al.*, 1974; Bruland, *et al.*, 1974; Förstner and Reineck 1974; Goldberg *et al.*, 1976). The premise of these and allied investigations relies on the approach that any concentration of a metal above certain levels, normally a value found at depth within a sediment core, is designated anthropogenic. However, assessments like these and especially in more dubious situations - need more careful consideration before their application to baseline studies of pollution. In particular, it is essential to understand the various systems involved with the transport and modes of incorporation of the metals and their diagenetic readjustment in sediments.

BASELINE STUDIES

The distinction between heavy metals from natural and anthropogenic sources has been widely inferred from the departure of metal concentrations, usually within the upper most 20-30 cms of sediment, from those at depth. Any increase is usually assumed to be anthropogenic (e.g. Chow *et al.*, 1973) being confined to a limited number of metals, i.e. Pb, Zn, Cd, Cu and Hg. Other elements, for example Ni and Co, show little evidence of surface enrichments. Criteria used to support the contention that these increases are anthropogenic relate to the history of the sediments and to their contents in industrial wastes.

In a limited number of studies where metal enrichments occur, the measurement of the $^{210}$Pb and, more indirectly, the $^{14}$C in the sediments show that the onset of

* Grant Institute of Geology, University of Edinburgh, West Mains Road, Edinburgh, Scotland. United Kingdom.

[x] Department of Environmental Studies, University of Lancaster, Bailrigg, Lancaster, England, United Kingdom.

metal increase usually coincides with the beginning of industrialization in the mid-ninteenth century (Erlenkeuser et al., 1974; Bruland et al., 1974). The increasing metal concentrations above this stratum level in a constantly accumulating sediment has been coupled to the increase of metal outputs from industry. For instance Erlenkeuser et al. (1974) have related the increasing trend of Cd, Pb, Zn and, to a lesser degree, Cu in Baltic Sea sediments to the rise in production of coal in Europe over the last 170 years. They claim that the introduction of metals to the sediments was from a direct fallout of fly-ash and other coal products; from 'fossil' carbon determinations they calculate that the content of coal residues in the sediments required to provide the necessary increase in metals is approximately 7 per cent by weight.

Further evidence of anthropogenic inputs of metals into coastal sediments is provided by the patterns and values of certain Pb isotopes in surface sediments (Bruland et al., 1974; Bertine, 1977). For instance Bruland et al. (1974) have assessed from the distribution and change in lead isotopes that basin sediments of the California Borderlands are probably contaminated with petroleum leads. Bertine (1977) has observed Pb isotopic ratios from Lake Whitney, Connecticut. Here, as the contents of Pb increase towards the sediment-water surface, so also do Pb isotopic ratios, and these tend towards the ratios measured in gasoline. These and numerous other examples (e.g. Goldberg, 1976) show plentiful evidence that many coastal sediments are subject to anthropogenic influences. Examination of sediments for the same metals in regions further removed from industry and human population have attracted much less attention. For this reason it is appropriate to examine certain features of the metal distributions in sediments situated in areas which have no apparent metal contamination, at least with respect to an effluent discharge in water. Such sediments may illustrate some of the problems that arise in the attempts to understand the processes of transport and incorporation of metals in sediments.

## HEAVY METAL DISTRIBUTIONS IN MARINE AND NON-MARINE SEDIMENTS

Surface sediments in fjord-like estuaries in western Norway and north-western Scotland, as well as in some Scottish lakes and areas that are very sparsely populated and that lack industry, also show metal enrichment in the uppermost few decimetres. These are illustrated in Figs. 1-3 for Pb and Zn. Copper also shows some enrichment but this is not so obvious as for Pb and Zn. These sediments show, in general, a very constant lithology, as denoted by the constancy of their aluminium contents (Fig. 2), and represent redox environments ranging from oxic to anoxic (Fig. 3). Within the oxic sediments, the manganese enrichment is usually small (0.1 - 0.5 per cent) and confined to the uppermost few centimetres only. In these circumstances, any heavy metal associations with manganese oxides/hyroxides will be small and are unlikely to obscure or influence the general pattern of increased metal contents towards the sediment-seawater surface.

The contents of both Pb and Zn in the upper sediment horizons in Figs. 1-3 are of the same order of magnitude, or even higher, than most of the published data for environments believed to be subject to anthropogenic metal inputs. While variations in the overall metal contents of sediments from different areas can reflect the differences in their supply from natural and anthropogenic sources, they can also be explained by differences in the total sediment accumulation rate of each area. Nevertheless, these metal profiles do demonstrate certain features that relate to the modes of introduction and incorporation of metals which have been overlooked in many investigations.

POSSIBLE MECHANISMS OF METAL INCORPORATION IN SEDIMENTS

In Loch Etive (Fig. 2) lead enrichment relative to that of zinc is high in cores from the more saline environment of its inner basin (core A). Note also that at depth, core A does not assume constant low natural background levels as seen in core B and other sediments in the estuary. It implies in this instance that surface enrichments extend more than 55 cm below the sediment-seawater surface. Within Loch Morar (Fig. 3), situated in the same area as Loch Etive, the Zn enrichments exceed those of Pb and Zn:Pb ratios for the metal enriched zone are very different from that found in Loch Etive and different again from those found in Bolstadfjord in western Norway (Fig. 1).

Metals transported into an estuary by rivers, from the sea and by the precipitation of wind-blown materials can all influence the heavy metal content of a sediment. Many of the arguments presented for surface metal enrichments over a wide geographical area imply that the major proportion of metals are held in precipitated airborne dusts, and assume that the airbone flux of metals has increased over the last century or so (Erlenkeuser et al., 1974). If the enriched heavy metal contents of Loch Morar sediments are anthropogenic, they must have been caused by precipitation as airborne particles. Some support of this theory, that airborne particles transport anthropogenic metals, is provided by the analyses of atmospheric dusts, which show pronounced enrichments of Zn and Pb such that the Zn:Pb ratio is near unity (Peirson et al., 1973; John et al., 1973).

The geographical situations of the estuaries and lakes described in Figs. 1-3, particularly the examples from northwestern Scotland, are not likely to be unduly influenced by winds carrying metals derived from the south and southeast, the most likely sources of industrial contaminants, as they are greatly influenced by westerly winds from the Atlantic. However, should the metal enrichments be anthropogenic and wind-carried, and not natural constant metal ratios, values approximating to those of atmospheric dusts (Zn:Pb$\sim$1) which Goldberg (1976) believes to be compositionally uniform, should be found. Our results for marine and lacustrine sediments show that this is not the case. Besides showing a considerable variation between basins, the Zn:Pb ratios change within a small estuary such as Loch Etive. Clearly if the metals are anthropogenic, they are not incorporated into the sediments as unaltered particulate matter in the manner suggested by Erlenkeuser et al. (1974), but are more likely to interact within the water column before accumulating in the sediments.

It is known that plankton incorporate Zn and Pb into their systems. For instance, both elements are enriched in particulate matter in Norwegian fjords only in the euphotic zone, and especially where biological productivity is high, such that they correlate with the concentrations of particulate phosphorus (Price and Skei, 1975). Cadmium shows a similar effect; Boyle et al. (1976) have shown that dissolved Cd in ocean seawater is strongly affected by biogenic substances and is closely related to the distribution of phosphorus. While it is known that variations are found both in the quantity and diversity of organisms in estuaries, there is a singular lack of information on the controls that different organisms exert on the fate of introduced metals. Further, the relative importance of either the uptake of metals by organisms or uptake on seston and their fallout to and incorporation in an accumulating sediment, has not been established. Some elements not associated with pollution, such as the minor halides, seem to be mostly scavenged onto seston at the sediment-water interface (Price et al., 1970).

A similar phenomenon could also occur with the heavy metals. Such fundamental reactions must be critically examined before we can understand the fate of metals, anthropogenic or natural, in both marine and lacustrine environments. Only after these have been studied can some understanding of the changes in metal ratios in surface sediments be achieved and baseline levels of pollution assessed.

Sediment cores, especially from lakes, often show increased organic matter content in their surface horizons which may have a direct association with the enrichment of heavy metals. This is overlooked in much of the literature on metal pollution in sediments. The cause of this increase can be related to a change in biological production caused by either an increase in natural or anthropogenic nutrients inducing eutrophism, and/or an increase in the load of terrigenous organic matter. In many estuarine sediments, the organic matter is often dominated by a terrestrial supply and any increase in the fallout of organic matter from biological activity is likely to be masked.

## DIAGENESIS OF METALS IN SEDIMENTS

While the surface enrichment of metals can be explained by recent changes in the environment affecting inputs of metals, it is also possible that metals held in association with organic matter can be subject to diagenesis during burial, so causing some change in their profiles. The redistribution of manganese in the upper layers of sediments is well known, being caused by certain redox reactions mediated by microorganisms. These also induce reactions within sediments which are likely to affect the more labile constituents of organic matter. For instance, the patterns of released nutrients in sediment pore waters are well known. Degens (1965) has illustrated the patterns of loss of certain organic constituents, notably carbohydrates and amino-acids, in certain sediment profiles. Should much of the enriched metals in surface sediments be bound to organic substances, they may also suffer some degree of loss or, at least, redistribution during burial.

That metals are released from sediments during burial is shown by the presence of both Zn and Pb in sediment pore waters in concentrations much greater than is found in normal seawater (Duchart et al., 1973). High concentrations occur in manganese-enriched and manganese-poor sediments, indicating that their release is not solely governed by redox reactions redistributing the Mn. Further, the investigations of Krom (1976) have shown that, unlike Mn, most of the Zn concentrations in sediment pore waters are likely to be chelated by unspecified organic substances.

The amounts of metals mobilised relative to that retained by the sediment at depth is as yet unknown. However, the distributions of element such as iodine (which incidentally show a similar concentration profile as Pb and Zn) show that, with burial, most of the element is lost from the sediment. Should much remobilisation of heavy metals also occur, their upward migration to the sediment surface could result either in a loss from the sediment to the overlying waters and/or its recycling in the surface layer (in a way that is analogous to Mn), thus reinforcing the metal contents there.

## CONCLUSIONS

While the heavy metal enrichment in the surface layers of many sediments can be brought about by anthropogenic influences, the origin of the same metals in sediments further removed from obvious sources of industrial and domestic effluents is more open to question, even though they often display similar metal contents. Regardless of the source of introduced metals (that is natural or anthropogenic), there is evidence to suggest that their accumulation in sediments is not solely due to the unaltered fallout of metal-containing particles.

The variability of Zn and Pb in a number of small estuaries and lakes in northwestern Scotland and western Norway implies that there is some interaction of heavy metals with organic matter either as plankton or seston or both. This association of heavy metals with organic matter is sometimes found in sediment profiles, suggesting that they are incorporated into a sediment in this state and perhaps increase towards the sediment-seawater surface as a response to an increase in biological productivity or terrestrial inputs with time. Such changes can be induced by natural or anthropogenic changes within the environment. It is also suggested that the diagenesis of organic substances during burial can provide a means of redistributing heavy metals in sediments.

Only after these suggested reactions are thoroughly investigated can any baseline estimates of pollution be reliably assessed.

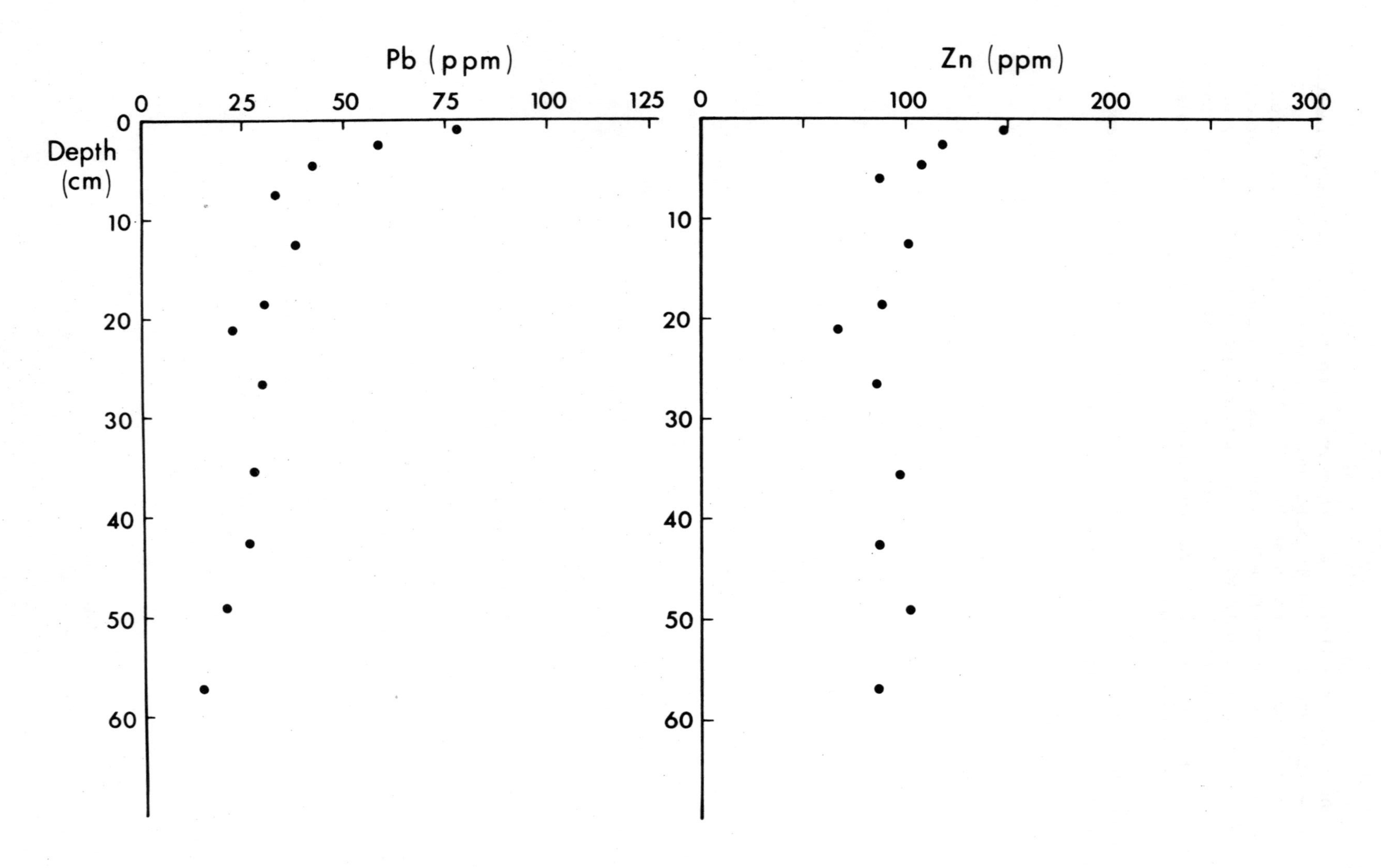


Fig. 1 Distributions of lead and zinc with depth in a core taken from the inner parts of Bolstadfjord, situated 50 km to the northeast of Bergen, W. Norway. The fjord is anoxic and the sediments contain 7% (by weight) of organic carbon at their surface, falling to 4% at depth. Non-silicate Mn is absent.

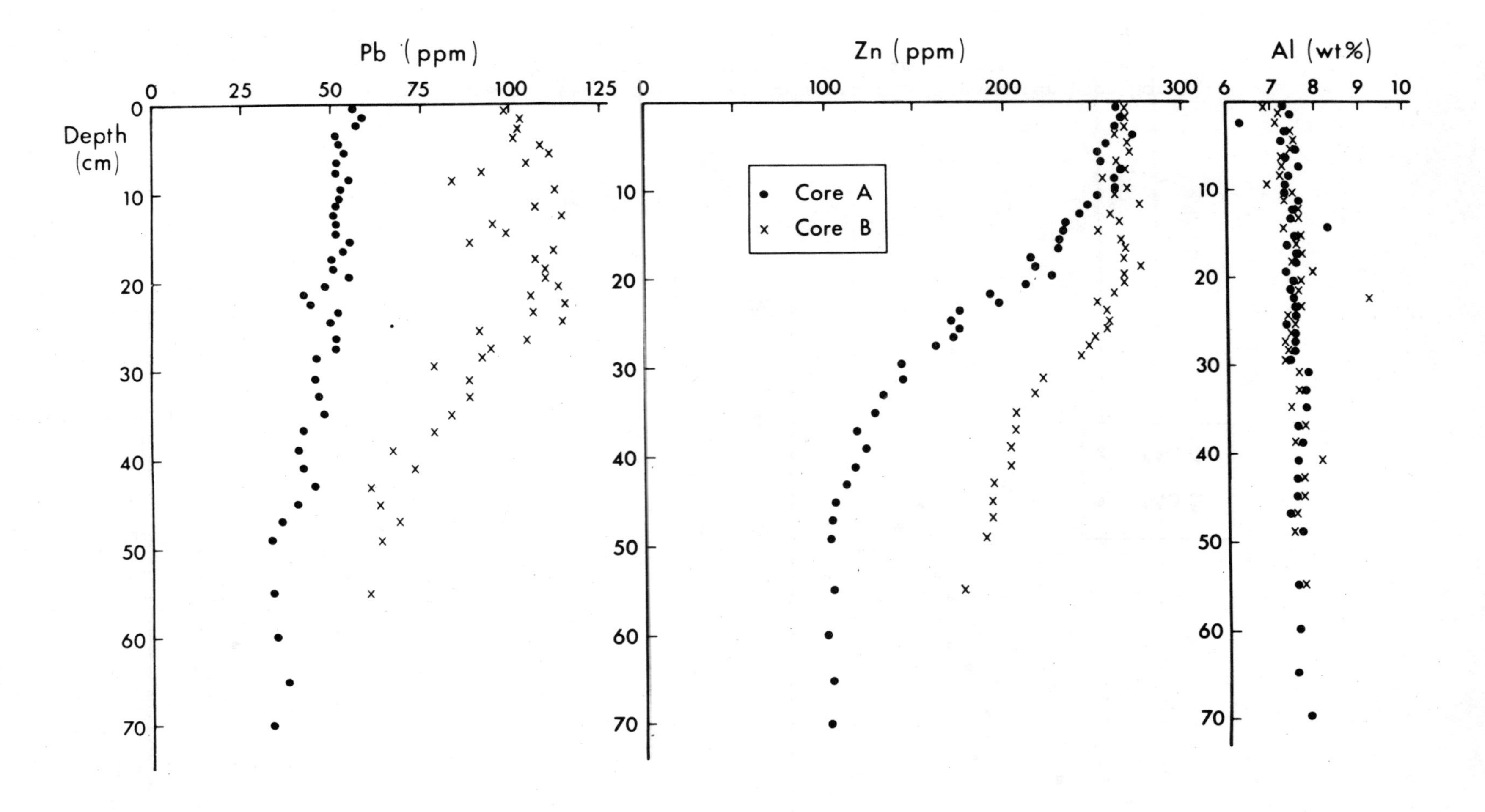


Fig. 2 Distribution of lead and zinc with depth in sediments from the inner basin of Loch Etive, situated some 30 km north of Oba, N.W. Scotland. Core A represents sediments from the innermost part of the estuary while core B is situated in a region of higher salinity and represents sediments of a more oxic nature.

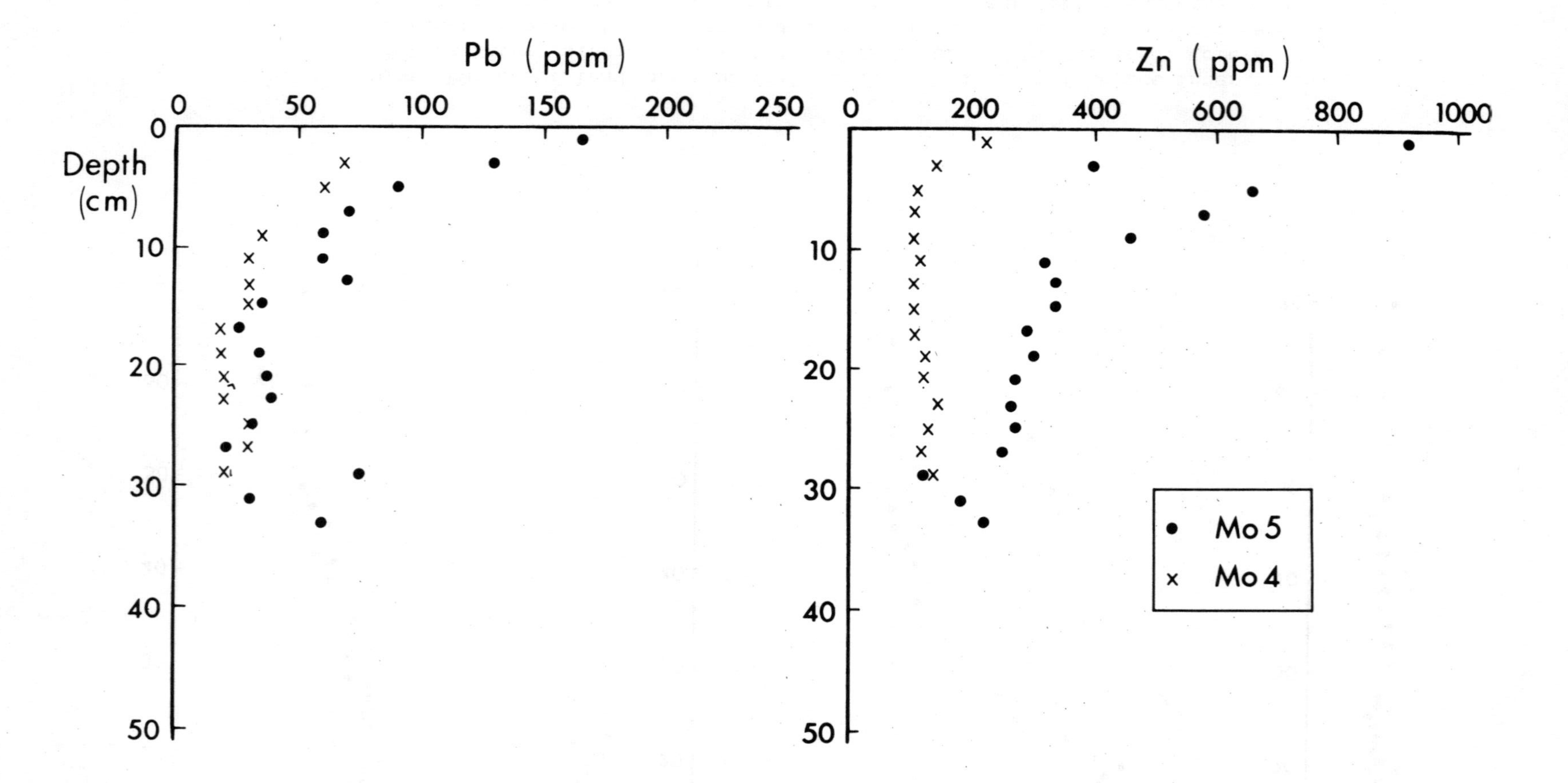


Fig. 3 Distributions of lead and zinc in sediments from freshwater Loch Morar, N.W. Scotland. Mo5 was collected from the central parts of the lake (320m). Mo4 is of shallower depth (170 m).

## References

Boyle, E.A. Sclater, F.; Edmond, J.M. 1976. On the marine geochemistry of Cadmium. Nature, vol. 263, p. 42-4.

Bruland, K.W.; Bertine, K.K.; Koide, M.; Goldberg, E.D. 1974. Hisotry of metal pollution in southern California coastal zone. Environ. Sci. Technol., vol. 8, p. 425-32.

Chow, T.J.; Bruland, K.W.; Bertine, K.K.; Goldberg, E.D. 1973. Lead pollution : records in southern California coastal sediments. Earth Planet Sci. Letters, vol. 1, p. 397-400.

Degens, E.T. 1965. Geochemistry of sediments : A brief survey. Englewood Cliffs, N.J. Prentice Hall.

Duchart, P.; Calvert, S.E.; Price, N.B. 1973. Distributions of trace elements in the pore waters of shallow water marine sediments. Limnol. Oceanogr., vol. 18, p. 605-10.

Erlenkeuser, H.; Suess, E.; Willkomm, H. 1974. Industrialization affects heavy metal and carbon isotope concentrations in recent Baltic Sea sediments. Geochim. Cosmochim. Acta, vol. 38, p. 823-42.

Förstner, U.; Reineck, E.-E. 1974. Die Anreicherung von spurenelementen in den rezenten sedimenten eines profilkerns aus der Deutschen Bucht. Senckenbergiana marit., vol. 6, p. 175-84.

Goldberg, E.D. 1976. The health of the oceans. Paris, Unesco Press, 172 p.

Goldberg, E.D.; Hodge, V.; Koide, M.; Griffin, J.J. 1976. Metal pollution in Tokyo as recorded in sediments of the Palace Moat. Geochim. J., vol. 10, p. 165-74.

Hamilton-Taylor, J. 1974. The geochemistry of fjords in south-west Norway. Unpub. Ph.D. Thesis, University of Edinburgh.

John, W.; Kaifer, R.; Rahn, R.; Wesolonski, 1973. Trace element concentration in aerosols from San Francisco Bay area. Atmos. Envir., vol. 7, p. 107-18.

Krom, M.D. 1976. The geochemistry of Loch Duich, Scotland. Unpub. Ph.D. Thesis, University of Edinburgh.

Peirson, D.H.; Cawse, P.A. Salmon, L.; Cambray, R.S. 1973. Trace elements in the atmospheric environment., Nature, vol. 241, p. 252-6.

Price, N.B.; Skei, J.M. 1975. Areal and seasonal variations in the chemistry of suspended particulate matter in a deep water fjord., Estuar. Coast. Mar. Sci., vol. 3, p. 349-69.

Price, N.B. Calvert, S.E.; Jones, P.G.W. 1970. The distribution of iodine and bromine in the sediments of the South Western Barents Sea., J. Mar. Res., vol. 28, p. 22-34.

# The effect of plants and animals on the chemistry of estuarine sediments

J. H. Day*

Several other contributors to this volume have discussed the important effects that bacteria have on the biogeochemistry of estuarine sediments and the resulting flux of materials to the water column. Therefore, in this paper, attention is focussed on the effects of higher plants and animals. These are, of course, a source of organic detritus, including the substrates upon which the bacteria depend, and in the sediment the rate of biodegradation is more rapid than in the water column since the extracellular enzymes are more concentrated. Complex organic molecules are oxidized or reduced and simple substances are released including plant nutrients - which are taken up by the roots of vascular plants or diffuse into the water column where they can be absorbed by algae. There are many other changes as well. Both plant roots and burrowing animals carry down oxygen into the sediments and simple organic compounds diffuse into the interstitial waters. Trace metals are absorbed and usually concentrated by living organisms, including migratory fish and birds which transport them out of the estuary. Possibly the most important changes are due to bioturbation by burrowing animals which bring up material from the reduced subsurface layers to be oxidised on the surface and thus replenish continuously the intersitial waters.

## The effects of plants

Organic detritus is derived partly from the river, partly from the estuary itself, plus the addition of small quantities from the sea. River water contains about ten times more soluble organic matter than seawater; however, the contribution to the estuary depends not only on the volume of river flow but also on seasonal changes. In high latitudes where the spring thaw results in heavy flooding, and in arid regions where the rains are restricted to a few months of the year, the river spreads over the flood plain and the run-off carries great quantities of plant fragments into the estuary. In areas where the annual rainfall is more uniform, the water seeps through the soil of the drainage basin and leaches out soluble and colloidal materials. The organic fraction includes simples compounds, such as organic acids, manganese acids, monosaccharides, amino-acids and vitamins, such as biotin, thiamine, and vitamin B12, as well as humic material. "Humus" includes several compex organic compounds of large molecular size which are very resistent to bacterial decomposition and which absorb trace metals. According to Phillips (1971) they are "composed of large, aromatic nuclei with phenobic and carboxylic functional groups, some of which contain nitrogen.

* Department of Zoology, University of Cape Town, South Africa.

When river water mixes with estuarine water with a salinity of 1-4 ‰ at the head of the estuary, the colloidal humus and silt flocculate and this material adsorbs much of the soluble organic matter and trace metals. Flocculation is largely complete at a salinity of 5 ‰ and the turbidity reaches a maximum. The floccules gradually sink, carrying down the organic matter and trace metals, and are deposited as mud in the upper reaches of the estuary. Peters and Wollast (1976) determined mass balances for the suspended and dissolved solids carried by the Scheldt. A total of 1.52 x $10^6$ tons/yr of suspended solids including silt, organic matter and metal pollutants was carried by the river and of this, 1.2 x $10^6$ tons/yr was deposited in the upper reaches. More was deposited in the lower reaches of the estuary and only finally 1.2 x $10^5$ tons/y entered the North Sea. Individual mass balances for organic matter, Pb, Cu and Zn show that deposition and biodegradation in the upper reaches is an effective means of removing pollutants from suspension or solution and trapping them in the sediments.

The nature and amount of organic matter contributed by estuarine plants and animals depends on the depth of the estuary and the velocity of the residual current over a complete tidal cycle. In deep, slow-flowing estuaries the long residence time of estuarine water allows several generations of planktonic organisms to develop during the spring and summer, and phytoplankton blooms often occur. In such estuaries as the Scheldt, the Mersey and the Thames, where intertidal flats have been reclaimed and pollution boosts the concentration of plant nutrients, plankton provides more organic matter than benthic plants and animals. Indeed, Peters and Wollast (1976) have shown that in the Scheldt the growth of planktonic diatoms absorbs 95 per cent of the dissolved silica brought down by the river whereas large amounts of nitrates and phosphates are discharged into the sea. This is most unusual since in unpolluted estuaries, dissolved inorganic phosphate is only present at low concentrations and nitrogen compounds fall to such low levels during the spring that they limit algal growth. In shallow, marine-dominated estuaries the residence time is too short for the development of high plankton densities and the benthic plants which grow on the extensive tidal flats are the main source of organic detritus, producing about 1000 $g/m^2/yr$ of organic carbon. Sea grasses, such as _Zostera_, salt marsh plants such as _Spartina_ in temperate estuaries, or mangroves such as _Avicennia_ and _Rhizophora_ in tropical estuaries, produce most of the organic matter. Microscopic algae growing on the surface of the sediments are extremely productive and in Britain, Jefferies (1971) estimated that algae contribute 25 per cent of the organic estuarine matter. The productivity of estuarine vegetation is now being investigated and, from the work of Haines (1976) it appears possible that the C3 and C4 pathways of photosynthesis, as measured by the $^{13}C$ ratio, may provide a means of estimating the contributions of the several components.

Apart from benthic diatoms which migrate in and out of the surface sediments, algae obtain nutrients from the water column and not from the sediments. Benthic algae do, however, produce vitamins necessary for phytoplankton growth and the blue-green algae fix atmospheric nitrogen. It is the rooted, vascular plants that take part in the flux of chemical substances between the sediments and the water column. Sea grasses, such as _Ruppia_ and _Zostera_ in termperate estuaries, and _Halodule_, _Thalassia_ and _Cymodocea_ in tropical estuaries, grow at and below the low-tide mark and obtain part of their nutrients from the sediments. They also transport oxygen and simple organic compounds, such as monosaccharides and amino acids, down to the roots and some of these substances diffuse into the sediments. Stephens (1967) has identified twelve amino acids in interstitial waters and further studies are in progress. Salt-marsh plants, such as _Spartina_, _Salincornia_ and _Limonium_ in temperate estuaries, and mangroves including _Avicennia_ and _Rhizophora_ in tropical estuaries, all grow above the mid-tide mark and are absorbed water and nutrients from the sediments. The major ions $Na^+$, $K^+$, $Ca^{2+}$, $Mg^{2+}$, $Cl^-$ and $SO_4{}^{2-}$ are sbsorbed in concentrations which differ from those in seawater. The exact concentrations vary from species to species but the ratio of ions in the interstitial waters surrounding the roots must be affected. The roots penetrate down into the black reducing layers and absorb the soluble $PO_4^{3-}$ and $NH_4^+$. Nitrogen is usually the limiting factor during the growing season. In the absence of soluble nitrogen compounds in the interstitial waters around mangrove roots, Disulphovibrio which lives symbiotically in the rhizosphere, is capable of fixing nitrogen gas. Trace metals, including Fe, Mn, Zn, Cu and Mo are required for healthy growth in minute concentrations but Cu, Zn and Mo are toxic in high concentrations. In addition to these metals, many others of no known biological significance are absorbed from the sediments and part of the oxygen and organic compounds formed in the leaves during photosynthesis are transported to the soil. For all these reasons the chemical composition of the interstitial waters in areas dense in vascular plants must differ from that of bare mud. It is also of interest that while some salt-marsh plants are annuals, many others - including the sea thrift Limonium and the mangroves - are perennials and live for fifty years or more, so that an analysis of the trace elements they contain provides an historical record of recent changes in the sediments.

## The effects of animals

The most obvious animal effect on the sediment is bioturbation but feeding activities also contribute to the flux of materials between the sediments and the water column. While a number of animals feed on microscopic algae, very few feed on living vascular plants. However, 80 per cent of the animal population are detritus feeders so that it is in this form that most of the plant materials produced in the sea-grass beds, the salt marshes and the mangroves are converted into animal tissue.

Zooplankton and other suspension feeders, such as oysters and mussels, take in detritus drifting in the water along with the phytoplankton. Large amounts of suspended materials are filtered from the water by these animals so that the turbidity is reduced. After digestion, the remains are formed into faecal pellets and fall to the bottom. Some of the pellets disintegrate rapidly but others produced by bivalves and consolidates with mucus will persist and have been recognized in fossil deposits. This aggregation of fine particles into large pellets increases the average size of the sediments in which they accumulate and must affect the permeability.

Most of the pelagic fishes and other nektonic animals, including the penaeid prawns and swimming crabs that are abundant in tropical estuaries, breed in the sea and enter the estuaries to feed and find shelter. When they approach maturity they return to the sea and thus transfer the organic matter and the trace metals they have consumed out of the estuary. Many wading and fish-eating birds also migrate elsewhere to breed. These losses of estuarine materials are substantial, shown by work in the Ythan estuary in Scotland, and so must be included in any estimate of mass balance.

The epibenthic fauna includes many small crustacea, such as amphipods, isopods and shrimps, and enormous numbers of small gastropods, such as Assiminea or Hydrobia. The majority are detritus feeders and accelerate biodegradation partly by the mechanical breakdown of plant litter and partly by the ingestion of bacteria so that the products of bacterial action are removed. Apart from this they have little effect on the sediments.

The infauna includes a variety of burrowing animals, some of which cause a marked disturbance of the sediments while others do not. Enormous numbers of microscopic animals - particularly nematodes and harpacticoid copepods - are present in the oxygenated surface layers and a few penetrate deeper into the reducing layers. These animals, known as the meiofauna, are sufficiently small to move through the pores of the sediment without disturbing the particles . They feed on detritus and the associated micro-organisms and void excretory products into the interstitial waters. The larger burrowers, or macrofauna, have different effects according to their method of burrowing and feeding, but all of them draw down oxygenated surface water - particularly if they live in the black reducing layers - and all of them void faecal material on the surface.

Polychaete worms are the most abundant members of the macrofauna, and in highly polluted sediments Capitella is the dominant inhabitant. During burrowing, all worms thrust their headsinto the sediment and then distend the proboscis or anterior segments to grip the sediment while the rest of the body is drawn downwards. This process is repeated many times as waves of contraction pass along the body. The essential point is that the sediment particles are merely thrust aside as the worm progresses and there is no excavation of material to disturb the original layering. Mucopolysaccarides are secreted to form a lining to the burrow and in many sedentary species indigestible food particles, sand grains and shell fragments are incorporated to form a tough permanent tube. Suspension and deposit feeders make simple tubes, carnivores make irregular burrows while searching for prey, while mud swallowers eat their way through the soil ingesting sediment particles along with bacteria and organic detritus. Arenicola, which is one of the largest, makes U-shaped burrows and, in the case of A. loveni, this extends down almost a metre into the black sulphide layers. It pumps down oxygenated surface water and when the tide is down it draws down air. At intervals, the undigested sediment in the gut is voided on the surface from the tail shaft while surface sediment sinks down the head shaft. In this way Arenicola and other mud swallowers cause a considerable turnover of the sediments and increase the oxygen content in the reducing layers.

The burrowing methods of bivalves were investigated by Trueman et al. (1966), who report that they are essentially similar to those of worms and other soft-skinned animals with the foot merely thrusting aside the sediment particles as the shell is drawn down. However, during the process, oxygenated surface water is drawn down through the inhalent siphon and ejected from the shell to facilitate its passage. Suspension feeders such as Mya, Solen and Cardium make permanent burrows and have little effect on the sediment structure. Deposit feeders such as Macoma, Tivela and Scrobicularia use the longer inhalent siphon as a vacuum cleaner and extend it over the surface to suck up detritus. After filtration, undigested material is ejected at the surface through the exhalent siphon and the bivalve ploughs its way through the sediment in search of fresh detritus. These bivalves thus cause more disturbance of the sediments than suspension feeders.

Crustacea have a completely different way of burrowing. They use their appendages to excavate material and bring it to the surface; here, reduced substances, such as sulphides, are oxidised and soluble substances in the interstitial waters diffuse into the water column. Burrowing amphipods, such as Corophium, are reported to increase the concentration of soluble phosphate in this way. In tropical estuaries where burrowing crabs, including Uca, Sesarma and Macrophthalmus are extremely abundant, large volumes of sediment are turned over and in mangrove swamps their burrows are so numerous that water squirts out of the surface at each step. Moreover, the burrows extend down to the permanent water table and the air water interface rises and falls with the tide so that the interstitial waters are continually refreshed. Most of the crabs and gastropods in the mangroves feed on detritus and microscopic algae on the sediment surface but the large, gregarious burrowing prawn, Thalassina, feeds on organic particles buried in the sediments. It makes extensive burrows and draws down water to winnow the particles from the sediment and makes large mounds of excavated material. Callianassa, which often reaches densities of 100/m$^2$ in African and Australian estuaries, brings large volumes of sediment to the surface and the lower layers are so riddled with burrows that the concentration of oxygen in the interstitial water is high to a depth of 0.5 metres.

There is increasing evidence that detritus feeders obtain their protein requirements not from the detritus but from the bacteria and other micro-organisms that grow on it. Odum and de la Cruz (1967) show that during biodegradation of Spartina into finer and finer particles, there is only a slight reduction of the carbohydrate content. The protein content at first decreases slightly and then rises to higher values. They suggest that these changes are due to nitrogen fixing bacteria using the carbohydrate as an energy source. Newell (1965) collected the faecal pellets of the deposit feeding gastropod Hydrobia ulvae. Analysis of freshly-produced pellets showed that organic carbon formed 10 per cent of the dry weight and organic nitrogen 0.25 per cent. Pellets which had been cultured in seawater for three days contained 8 per cent of organic carbon and 1.75 per cent of organic nitrogen. When this material was fed back to Hydrobia and the resulting faeces were analysed the organic nitrogen content was only 0.075 per cent. It is also known that Arenicola and several other detritus feeders that have been tested experimentally will grow well on a pure diet of bacteria. From all these observations it would appear that organic detritus may be recycled several times through bacteria and animals with the gradual reduction of organic carbon and the formation of bacterial protein which is assimilated by the animals.

## The concentration of trace metals by living organisms

Aquatic organisms are well known to concentrate most elements to higher levels than occur in the surrounding waters. The only exceptions are F, Br, Mg and S, while Na and Cl are absorbed in lower concentrations than are present in seawater. The affinity of organic surfaces for cations increases with their valency but the affinity is not the same for all groups of organisms. According to Perkins (1974) the redox state of an ion is also important and electonegative metals such as Cu, Hg and As, which have an affinity for amino and sulphydryl groups and those which form insoluble sulphides, such as As, Mo and Sb, are absorbed very rapidly. Transportation through biological membranes is an active process demanding an expenditure of energy but, due to their lithophylic properties, the process is facilitated for certain cations. Those metals which are present in the water in insoluble forms require chelators to bring them into solution for their passage through membranes. Thus Fe combines with ATP, Al combines with $PO_4$ and methyl mercury is more readily absorbed than inorganic mercury.

Some of the metals that are absorbed are essential for healthy growth in minute concentrations but become toxic in higher concentrations. Copper is a well-known example. Plants require traces of Fe, Co, Mn, Cu and Zn and some require Mo, Si, and V. All animals require Fe and Cu, molluscs require Zn and some ascidians, but not all, require V. It seems possible that some of the heavy metals with no known biological significance but that are nevertheless absorbed, such as Pb, Hg and Cd, substitute for essential metals with similar properties.

These disconnected fragments of information provide an indication of the complexity of the problems that urgently require solution. Meanwhile toxic metal pollutants are being continuously concentrated in estuarine sediment with unpleasant consequences for the future.

## References

Haines, E.B. 1976. Stable carbon isotope ratios in the biota, soils and tidal waters of a Georgia salt-marsh. Estuar. Coast. Mar. Sci., vol. 4, p. 609-16.

Jefferies, R.L. 1971. Aspects of salt-marsh ecology with particular reference to inorganic plant nutrition. In : R.S.K. Barnes, J. Green (eds.), The estuarine environment. London, Applied Science.

Odum, E.P.; Cruz de la, A. 1967. Particulate organic detritus in a Georgia salt-marsh estuarine ecosystem. In : G. Lauff, (ed.), Estuaries. Washington, Am. Ass. Advmt. Sci.

Perkens, E.J. 1974. The biology of estuaries and coastal waters. London, Academic Press.

Peters, J.J.; Wollast, R. 1976. Role of sedimentation in the self purification of the Scheldt estuary. Proc. 3rd federal inter-agency sedimentation conference.

Phillips, J. 1971. Chemical processes in estuaries. In : R.S.K. Barnes, J. Green, (eds.), The estuarine environment. London. Applied Science.

Stephens, G.C. 1967. Dissolved organic material as a nutritional source for marine and estuarine invertebrates. In : G. Lauff, (ed.), Estuaries. Washington, Am. Ass. Advmt. Sci.

Trueman, E.R.; Brand, A.R.; Davies, P. 1966. The dynamics of burrowing of some common littoral bivalves. J. Exp. Biol., vol. 44, p. 665-77.

# Biodegradation of organic matter in marine sediments

S. O. Stanley*

The objective of this paper is to discuss the parametres that determine the rates of biological degradation at various levels in the sediment.

The critical environmental factors which are likely to control the rate of organic degradation in near-shore marine sediments can be summarized as : (i) the presence of an active microbial flora capable of degrading the individual organic substrates present in the sediment; (ii) the oxygen regime within the sediments and in the immediate overlying waters; (iii) the availability of nutrients other than oxygen; and (iv) bioturbation.

1. The Microbial flora: Under ideal environmental conditions the rate of breakdown of a newly introduced organic substrate depends initially on the build-up of an active microbial flora capable of breaking down that particular substrate. After a lag period, whose length depends on the activity of the micro-organisms and the nature of the substrate, a steady state is reached where the rate of breakdown is related to the activity of the microbial flora and to the input levels of that particular organic substrate. This means that analysis of levels of a particular organic substrate gives no indication of the rate of input into the sediment unless it is also related to the microbial activity associated with that particular substrate. It is important to emphasize that individual organic compounds frequently require specific microbial species for their degradation and that the species involved also depends on the other environmental conditions, particularly the oxygen regime.

2. Oxygen : The level of dissolved oxygen determines the types of microbial populations which exist in the sediments as well as the pathway of organic degradation. As the dissolved oxygen concentration is lowered, there is an increased tendency for incomplete oxidation of organic substrates with the formation of substances such as volatile fatty acids and finally - in the complete absence of oxygen - of methane. The lowering of the redox potential within the sediment with increasing oxygen depletion results in the step-wise, microbially mediated processes of denitrification, sulphate reduction and methane formation. These steps are brought about by different groups of micro-organisms and the extent to which they occur is likely to depend on the biological oxygen demand of the substrate, its rate of deposition and its initial rate of breakdown. The relationship between the biological oxygen demand of the substrate, the level of dissolved oxygen in the water overlying the sediments and the diffusion rate of oxygen within the sediment determines the extent to which the redox levels are altered by microbial processes.

---

* Dunstaffnage Marine Research Laboratory, P.O. Box 3, Oban, Argyll, Scotland. United Kingdom.

3. Nutrient availability: The input of carbon-enriched organic material such as cellulose may lead to nutrient limitation, with a consequent change in the rate of organic breakdown. However, mineralization processes are likely to be quite rapid at the sediment-water interface and the controlling steps in any possible nutrient limitation are the exchange rate between the sediment and the overlying waters and the diffusion of nutrient within the sediment matrix. Most of the mineralization processes occurring in the sediment are microbiological and are carried out by specific groups of organisms whose activity is again dependent on the physico-chemical environment.

4. Bioturbation: The activity of the benthic fauna in turning over the upper sediment layers is an important factor in nutrient exchange reactions and in the oxygenation of the upper sediment. Also important is that during the passage of material, such as leaf fragments, through the gut the material becomes macerated and more susceptible to further degradation. These animals may also be important with respect to the resuspension of sedimented material into the water column.

Acknowledgements

Some of the information contained in thçs paper was obtained as part of the Loch Eil project, a cooperative study involving the Scottish Marine Biological Association, the Department of Biological Sciences, University of Dundee, and the Department of Civil Engineering, University of Strathclyde.

# Distinction between natural and anthropogenic materials in sediments

Erwin Suess*

INTRODUCTION

For the purpose of this study we shall refer only to those materials which have both natural and anthropogenic sources, thereby excluding synthetic organic chemicals such as insecticides, plastics and detergents which are used extensively in tracer studies. To study these problems we strongly rely on, what unfortunately has become a current slogan in environmental research, "baseline studies". This type of approach is based on determining high concentration of certain constituents above a natural background and attributing the excess to anthropogenic sources. Careful consideration should be given, however, before applying the baseline approach to estuarine environments, as will be shown below. For greater refinement, the study of entire assemblages of constituents often helps in distinguishing between natural and anthropogenic materials. A third, still further refined approach can be the study of the physical and chemical states of the various materials in the sediments. This is essential in determining and predicting post-depositional behaviour. Finally, a synthesis of the various approaches is seen in budget or mass-balance studies. This last approach has the benefit of allowing predictions to be made for man's future additions to the environment.

In the following considerations, examples of each of these four approaches are presented and will be biased towards the geochemical behaviour of transition metals and carbon isotopes in near-shore sediments.

Baseline approach

The differentiation of materials from natural and anthropogenic sources has been widely inferred from above-baseline concentrations encountered in innumerable samples of the earth's sediment. They include marine and lacustrine sediments (Chow *et al.*, 1973; Aston *et al.*, 1973; Förstner, 1976), fluvial bed loads (Banat *et al.*, 1972; Förstner and Müller, 1974), terrestrial peat bogs (Tyler, 1972) and even glacial ice (Murozumi *et al.*, 1969).The inferred anthropogenic input has not been wholly accepted by other workers in all instances, particularly when the concentrations above baseline were low or when alternative natural enrichment processes were known. Therefore, these natural processes will be briefly considered here. Some further refinement in distinguishing natural from anthropogenic constituents is added when characteristic assemblages that identify the immediate source are studied; in one case a "coal-residue-assemblage" consisting of fossil carbon and of the trace metals Cd, Pb, Zn, Cu will be discussed below in detail.

---

* School of Oceanography, Oregon State University, Corvallis, Or. 97331 USA.

Table 1. Trace metal contents of manganese oxide layers near the sediment surface and from reducing sediment column immediately below.

| | Depth (cm) | Fe (%) | Mn (ppm) | Cu (ppm) | Pb (ppm) | Zn (ppm) |
|---|---|---|---|---|---|---|
| A. Oslo Fjord, water depth 200 m | | | | | | |
| | 0 - 5 | 5.28 | 11,230 | 35 | 158 | 265 |
| | 15 - 20 | 5.63 | 2,480 | 20 | 64 | 165 |
| B. Deutsche Bucht, water depth 22 m | | | | | | |
| | 0 - 2 | 4.5 | 1,600 | 40 | 200 | 400 |
| | 50 - 100 | 3.0 | 400 | 20 | 20 | 100 |
| C. NW Africa, water depth 1800 m | | | | | | |
| | 0 - 5 | 1.75 | 1,600 | 32 | - | 65 |
| | 15 - 20 | 1.42 | 480 | 37 | - | 61 |

A and B are from coastal environments of the North Sea, C is from the continental slope off NW Africa. It is believed that Pb and Zn in near-surface sediment layers from the North Sea area is from anthropogenic sources, whereas the NW African sediments are not affected by this input. Data from : Price (1967); Förstner and Reineck (1974); Hartmann *et al.* (1976).

According to Table 2, fossil fuel residues as well as products from ore smelting processes utilizing coal as energy source could be recognized by increased enrichments of Cd > Pb > Zn > Cu (Bertine and Goldberg, 1971; Gluskoter and Lindahl, 1973). Typical composition of airborne particulates also appears to be characterized by similar enrichment factors for Pb and Zn (Peirson *et al.*, 1973; Linton *et al.*, 1976). Almost completely unknown, however, are post-depositional fractionation processes which might obscure or accentuate the original assemblage pattern.

Natural processes for trace metal enrichment: manganese migration and others

(a) Manganese migration. The best known diagenetic recycling mechanism involves the reduction of Mn oxides, the migration of soluble $Mn^{2+}$ and the reprecipitation of oxides under suitable oxidation-reduction conditions (Lynn and Bonatti, 1965; Weijden et al., 1970). This process is illustrated by the depth distribution of manganese in pore waters and sediments from the continental rise off N.W. Africa (Hartmann et al., 1976; Fig. 1). The Zn and Cu contents of these sediments appear unaffected by the post-depositional mobilization of Mn. However, in the pelitic fraction of sediments from the North Sea (Förstner and Reineck, 1974) and in sediments from the Oslo Fjord (Price, 1967) Zn and Cu concentrations paralleled those of Mn (Table 1). It seems here that two processes affect the metal distribution in near-surface sediments: diagenetic mobilization of Mn in response to changes in redox potential upon which is superimposed an increased input of transition metals from anthropogenic sources. It is essential that the effect of diagenetic recycling of Mn on the distribution of other metals be known before anthropogenic sources are more strongly implicated.

(b) Breakdown of organo-metal chelates. Price (1973, and this volume) and, more recently, Krom (1976) have suggested a recycling mechanism involving mobilization of trace metals due to diagenetic breakdown of organic chelating agents. They imply that this process takes place both in oxidizing and reducing environments and may result in trace metal and iodine concentration-depth profiles which resemble those from polluted environments. Fig. 2 show the variations in Pb:C-org, Zn:C-org, Cu:C-org and I:C-org ratios with varying depth in sediments from Loch Fyne, Scotland (Price, 1973). According to both Price and Krom, the high relative metal concentrations in the sediment layers between the surface and 20 cm into the sediment are due to organic compound chelation. It appears, however, that for reference purposes, unpolluted environments are difficult to locate; also, changes with depth in the iodine:C-org ratios might well indicate such a breakdown process but other trace metal distributions may not necessarily be related to it. Studies of metal-organic interactions should have high priority in order to evaluate this process.

(c) Heavy mineral enrichment. It is likely that in the coastal and estuarine environment, heavy mineral concentrations develop due to changing transport directions or fluctuating current intensities, as for example, has been reported for the sediments of the western Baltic Sea (Walger, 1966; Cordes, 1966). Natural sorting processes could therefore lead to high metal concentrations in coastal sediments, and without further knowledge of the physical and chemical states in which the trace metals are contained, such placer-type mineral enrichments could well resemble metal polluted environments.

Assemblage studies of anthropogenic trace metals

In this approach the distribution of a number of constituents is determined in order to recognize a characteristic assemblage which might be related to the same anthropogenic source material. Specifically, the trace metal composition of fossil fuel residues and of airborne particles from certain industrial processes appear to serve this purpose. Assemblages where the enrichment factors of the components differ by a wide margin from those in natural materials are well-suited for identifying anthropogenic material.

Table 2. Enrichment of trace metals in coal ash and in airborne particles compared to materials from the natural weathering cycle.

| Elements | Natural Weathering | Coal Ash | | Airborne Particles | |
|---|---|---|---|---|---|
| | | Ratio | Factor | Ratio | Factor |
| Fe/Al | 0.7 | 4.6 | 6 | 5.9 | 8 |
| Mn/Al | $1.4 \times 10^{-2}$ | 0.124 | 8 | $2 \times 10^{-2}$ | 1 |
| Cd/Al | $1.4 \times 10^{-6}$ | $3.1 \times 10^{-4}$ | 217 | - | - |
| Pb/Al | $1.5 \times 10^{-5}$ | 0.019 | 130 | 0.04 | 260 |
| Zn/Al | $5.7 \times 10^{-4}$ | 0.064 | 112 | 0.08 | 140 |
| Cu/Al | $5.7 \times 10^{-4}$ | $1.6 \times 10^{-2}$ | 30 | - | - |

Elements with large enrichment factors may be used to identify anthropogenic materials in sediments; data from: Bertine and Goldberg (1971), Erlenkeuser *et al.* (1974) and Linton *et al.* (1976).

*Carbon isotope distribution*. The introduction of large amounts of fossil carbon into the sedimentary cycle by man's use of fossil fuel has resulted in an "ash-effect", dilution of the natural $^{14}C$-distribution of sediments through dead carbon (Suess and Erlenkeuser, 1975; Erlenkeuser, 1976). Increasing amounts of $^{14}C$ from fusion bomb tests have also entered the environment and have produced high radiocarbon activities known as the "bomb effect" in the marine sedimentary record (Vogel, 1972). Erlenkeuser has shown how these two man-made effects can be recognized in the coastal deposits of the North and Baltic Seas and can be used to estimate the proportions of recent, eroded and fossil carbon input (Fig. 3c). The recent and eroded carbon fractions are natural in origin and consist of carbon from the biomass and redeposited dead carbon. The amounts are highly variable and depend upon the specific conditions of the environment of sedimentation. They produce a $^{14}C$-distribution resulting in a recent age for surface sediments of between 800 and 1200 years (Fig. 3a). The fossil carbon fraction is anthropogenic and responsible for the apparent age reversal in the near-surface sediments. In a typical example from the Bornholm Basin (central Baltic Sea) Erlenkeuser estimated that during the last 70 years 1.3 per cent of recent carbon, 3.5 per cent of eroded carbon, and 0.3 per cent of fossil carbon were deposited. The concentration of 0.3 per cent of fossil carbon corresponds to an annual flux of about 100 $\mu g\ cm^{-2} y^{-1}$ which is 1000 times more than the carbon input estimated from the natural burning of vegetation (Smith *et al.*, 1973). In Figs. 3b, c are illustrated the $^{14}C$-distributions with depth and the calculated amounts of anthropogenically mobilized fossil carbon mixed with the recent and eroded carbon fractions.

*Coal-residue assemblage*. Erlenkeuser *et al.* (1974) and Suess and Erlenkeuser (1975) also reported an enrichment of the coal-residue assemblage (Cd,Pb, Zn, Cu) from the surface sediments of the Western and Central Baltic Sea, which is reproduced in Fig. 4. The interpretation that the enrichment reflects anthropogenic mobilization was largely based on the concomitant fossil carbon content as discussed in the previous section. Sediments from the Bay of Gdansk, another major sedimentary basin in the SE Baltic Sea, apparently contain the same characteristic coal-residue assemblage as seen here from the Zn contents (Fig. 5). The $^{14}C$ age-reversal - indicative of fossil carbon input - is found about 10 cm below the sediment surface. Suess and Djafari (1976) observed the same assemblage concentrated even in the outer layers of ferromanganese concretions (which were formed in the same basin) and concluded that this too was caused by the increased metal input during the past century (Fig. 6). Meanwhile Förstner and Reineck (1974) had shown that the coal-residue assemblage and Hg were also found in near-surface sediments from the Deutsche Bucht of the southern North Sea (Fig. 7, also see Table 1).

The numerous examples suggest, therefore, that the coal-residue assemblage identified here can be used to distinguish between natural and anthropogenic metals. These arguments are largely based on the absence of any diagenetic concentrating mechanism and on the association with fossil carbon. There is little evidence for post-depositional changes in the trace metal assemblage because its characteristic distribution is observed in oxidizing and reducing sediments and even in ferromanganese concretions.

Physical and chemical state

Confidence in identifying anthropogenic materials by assemblages increases as the number of components increases and will supposedly be greatest when the physical and chemical states of anthropogenic materials become fully known. For example, the porous appearance, concave fractures, and rounded outer surfaces of smelted and solidified slag particles are easily identified in the coarse silt and fine sand fractions from sediments of the western Baltic Sea where they may be present in significant quantities. The same characteristics apply to airborne particles (Doyle et al. 1976) and their identification as sediment constituents would be possible if their grain size ranges and transport properties were known. Identification of native carbon particles in sediments resulting from forest fires are another example for the applicability of this approach (Smith et al., 1973). In general, however, identification of physical and chemical states of anthropogenic particles appears to be the least advanced and there is a need for new techniques to succeed in this kind of morphological identification. Adaptations of conventional size and density separation methods used in sedimentology might possibly be considered for preconcentrating anthropogenic particles.

Mass-balance approach

In the absence of any data on a complete mass-balance of metals in an estuarine environment, flux estimates for anthropogenic metals as reported by Chow et al. (1973), Bruland et al. (1974), Suess and Erlenkeuser (1975) and Hansen et al. (1976) might be used to understand man's additions to coastal sediments. Based on the coal-residue assemblage, we have distinguished between anthropogenic and minerogenic (= natural) Zn fluxes and concluded that the present-day anthropogenic Zn input is twice the natural input in the case of the main sedimentary basins of the Baltic Sea (Fig. 8). Chow and Bruland and coworkers have estimated anthropogenic and natural fluxes to the sedimentary basins of the Southern California Borderland; they have shown that there the anthropogenic Pb flux is about 3 times greater than natural flux, and that the anthropogenic Zn-flux is only about one half as large. Table 3 summarizes the respective data and it is apparent that the natural fluxes from these widely separate areas are comparable but that anthropogenic Zn is prevalent in the Baltic Sea deposits and Pb in those of the California Borderland.

As pointed out before, such flux estimates are only one attempt in understanding man's addition to the coastal marine environment. If applied to estuarine systems, it is highly desirable to establish a mass-balance, as recently attempted in great detail for the entire Baltic Sea by Hansen et al. (1976). The flux to the sediments then represents only one segment of the total output which, in relation to the dissolved and particulate input, allows certain predictions to be made.

Finally, for a summary of the methods discussed, Table 4 below gives an overall viewpoint with some anotated remarks on each method.

Table 3. Present-day trace metal fluxes to the sediments of various depositional basins: data from Bruland *et al.* (1974) and Suess and Erlenkeuser (1975).

| Basin | Source | Flux (mg $m^{-2}yr^{-1}$) | | | |
|---|---|---|---|---|---|
| | N = natural<br>A = anthropogenic | Cd | Pb | Zn | Cu |
| San Pedro | N | - | 2.6 | 31 | 12 |
| | A | - | 17 | 19 | 14 |
| Santa Monica | N | - | 2.4 | 28 | 10 |
| | A | - | 9 | 21 | 11 |
| Santa Barbara | N | 1.4 | 10 | 97 | 26 |
| | A | 0.7 | 21 | 22 | 14 |
| Kieler Bucht | N | 0.1 | 5.5 | 70 | 7 |
| | A | 0.6 | 22 | 80 | 12 |
| Bornholm Basin | N | 0.4 | 26 | 78 | 23 |
| | A | 1.0 | 43 | 75 | 11 |
| Gotland Deep | N | - | - | 73 | 31 |
| | A | - | - | 75 | 11 |

Table 4 : Summary

| Approach | Indicators | Remarks | References |
|---|---|---|---|
| Tracer studies | PCB's | not discussed here | |
| | stable Pb isotopes | not discussed here | Chow et al, 1973 |
| | radiocarbon | "bomb effect" | Erlenkeuser, 1976; Vogel, 1972. |
| Baseline approach | "high" concentrations of individual elements | interferences from: | |
| | | diagenetic Mn-mobilization | Hartmann et al, 1976; |
| | | breakdown of organo-metal chelates | Lynn & Bonatti, 1965; Van der Weijden et al., 1970 |
| | | heavy mineral accumulations | Krom, 1976; Smith et al, 1973 |
| Assemblage studies | fossil carbon | dilution of recent $^{14}C$ activity; "ash effect" | Suess and Erlenkeuser 1975; Erlenkeuser, 1976; Erlenkeuser et al., 1974. |
| | fossil fuel residue (Cd, Pb, Zn, Cu) | source identification | |
| | | coal ash | Bertine & Goldberg, 1971; Suess & Erlenkeuser, 1975 Erlenkeuser et al. 1974. |
| Physical and chemical state | slag and coal ash particles | morphological and compositional classification | Erlenkeuser et al., 1974. |
| | airborne particulates | new techniques required, magnetic spherules, size distribution | Linton et al., 1976; Doyle et al., 1976; Johnson, 1976. |
| Mass-balance approach | anthropogenic metal fluxes | sedimentary basins off Southern California and Baltic Sea | Bruland et al., 1974 Suess and Erlenkeuser, 1975; Hansen et al., 1976. |

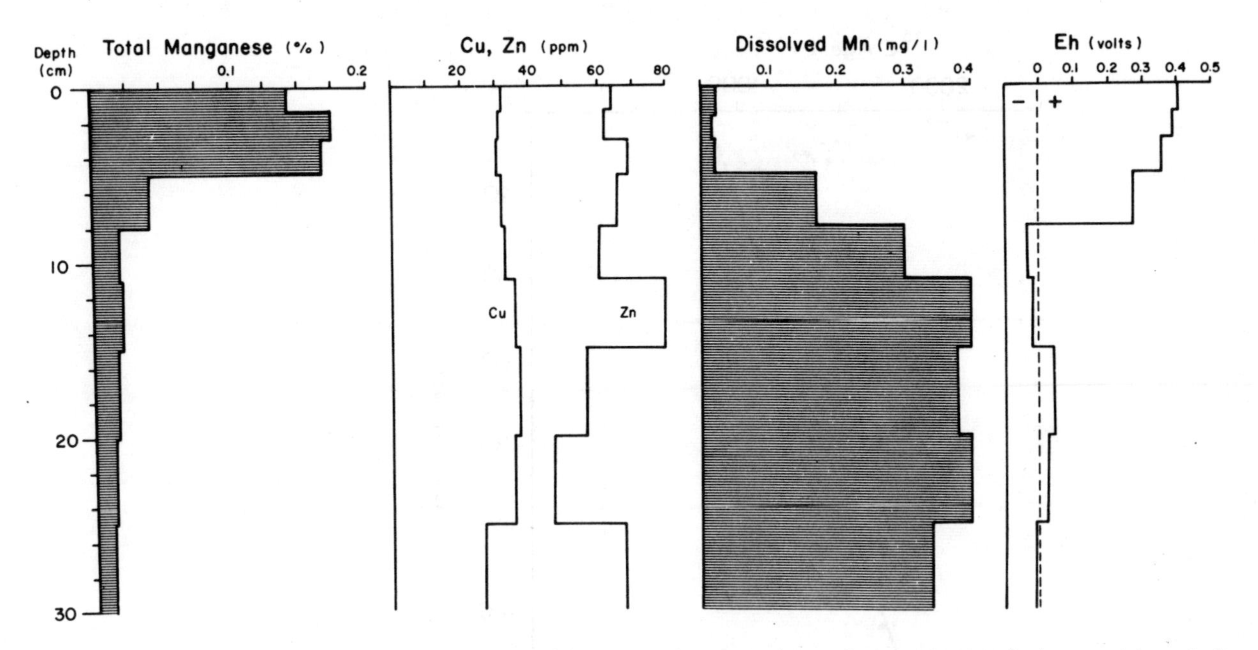


Fig. 1. Diagenetic recycling of Mn due to changes in redox potential in continental rise sediments off NW Africa; Cu and Zn contents appear to be unaffected by manganese mobilization, data from Hartmann et al. (1976).

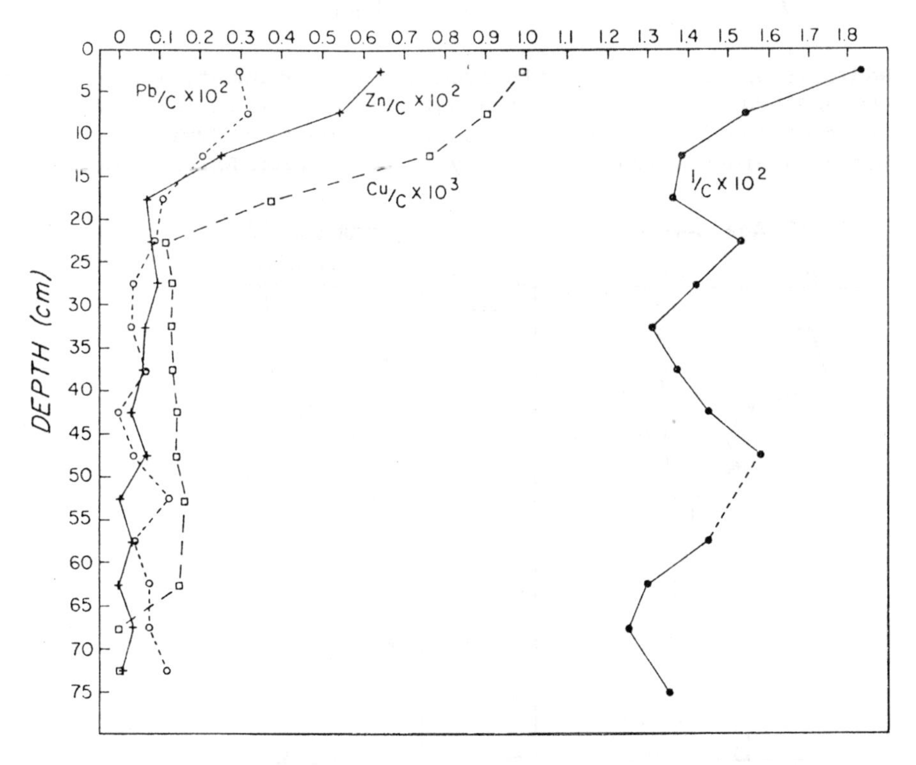


Fig. 2. Variation in Pb/C-org, Zn/C-org, Cu/C-org and I/C-org ratios with depth in oxidizing sediment from Loch Fyne, Scotland; the relative concentrations of metals are thought to reflect diagenetic breakdown of organo-metal chelates as indicated by a change in quality of organic matter, i.e. by a change in I/C-org ratio; data from Price (1973).

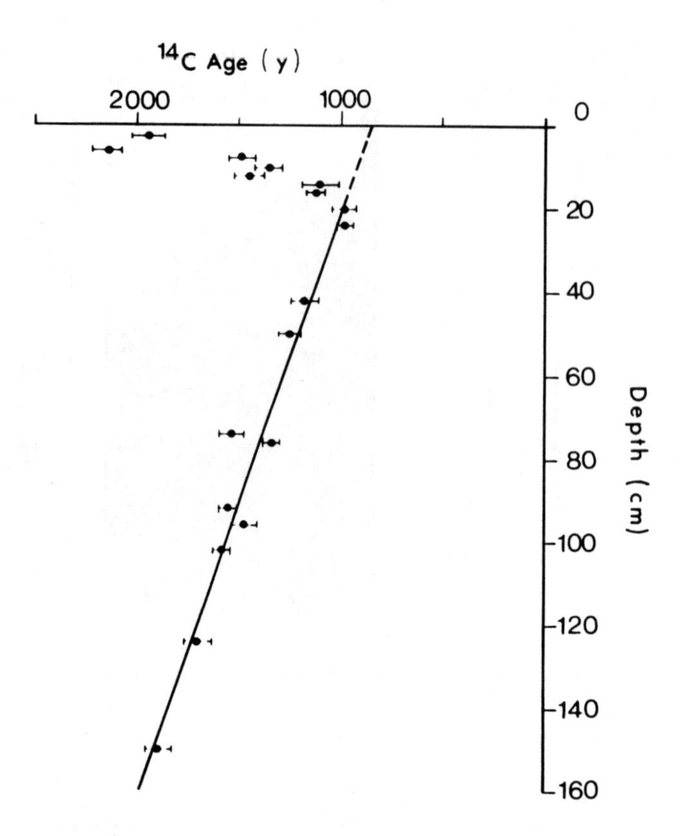


Fig. 3a. Radiocarbon age of sediments from the Kieler Bucht, western Baltic Sea; the apparent age reversal at about 20 cm of depth is thought to reflect input of "dead" carbon from fossil fuel burning during the past century, linear sedimentation is 1.4 mm $y^{-1}$; data from Erlenkeuser <u>et al</u> (1974).

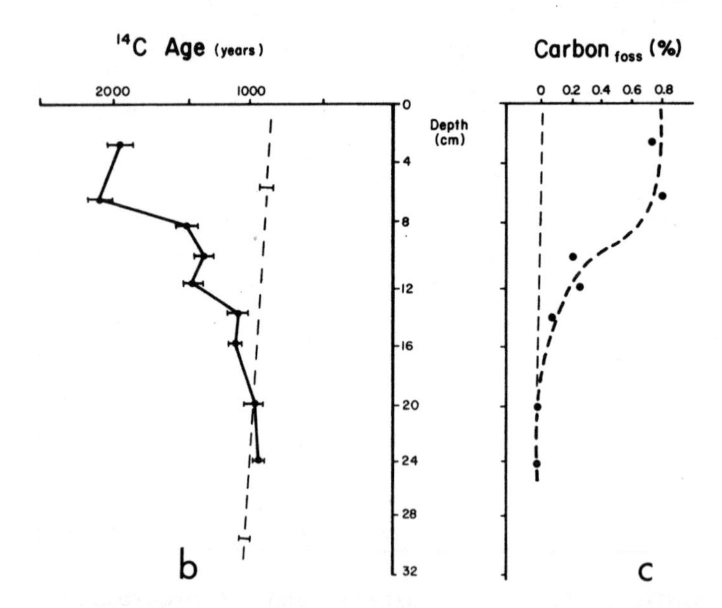


Fig. 3(b,c) Input of fossil carbon into sediments of the Kieler Bucht (c) as calculated from the apparent age reversal (b), where the broken baseline indicates extrapolation of constant sedimentation rate to the surface; data from Suess and Erlenkeuser (1975).

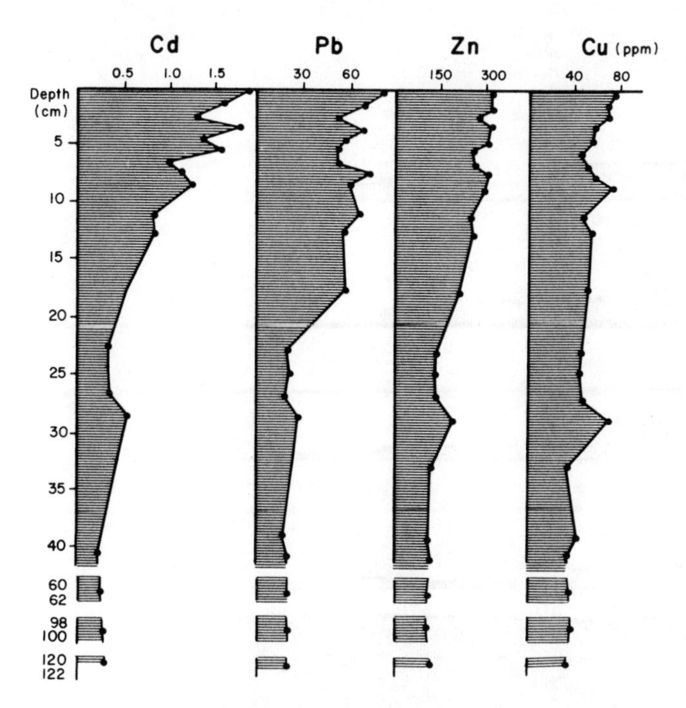


Fig. 4. Increase of coal-residue-assemblage (Cd, Pb, Zn, Cu) in sediments of the Kieler Bucht, the depth of increase as indicated by the dashed line coincides with increased input of fossil carbon (see also Fig. 3a); data from Erlenkeuser et al (1974).

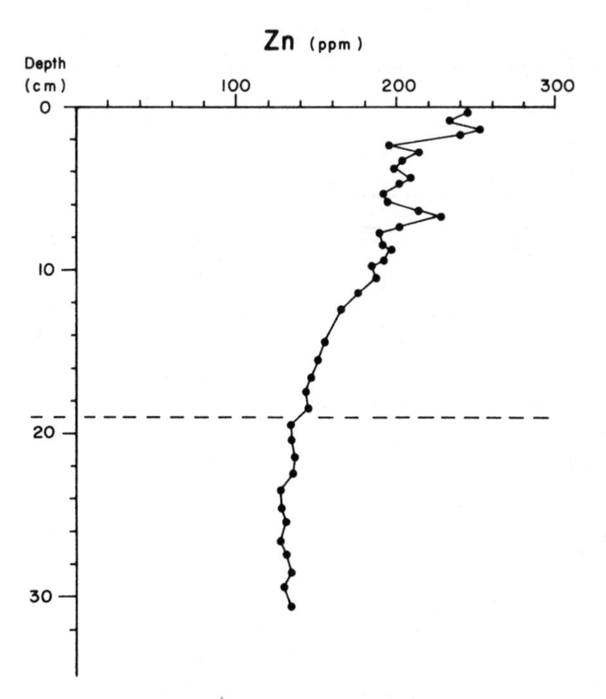


Fig. 5. Increase of Zn, one of the constituents of the coal-residue-assemblage - in the sediments of the Bay of Gdansk, central Baltic Sea; the broken line indicates depth of apparent age reversal from radiocarbon distribution.

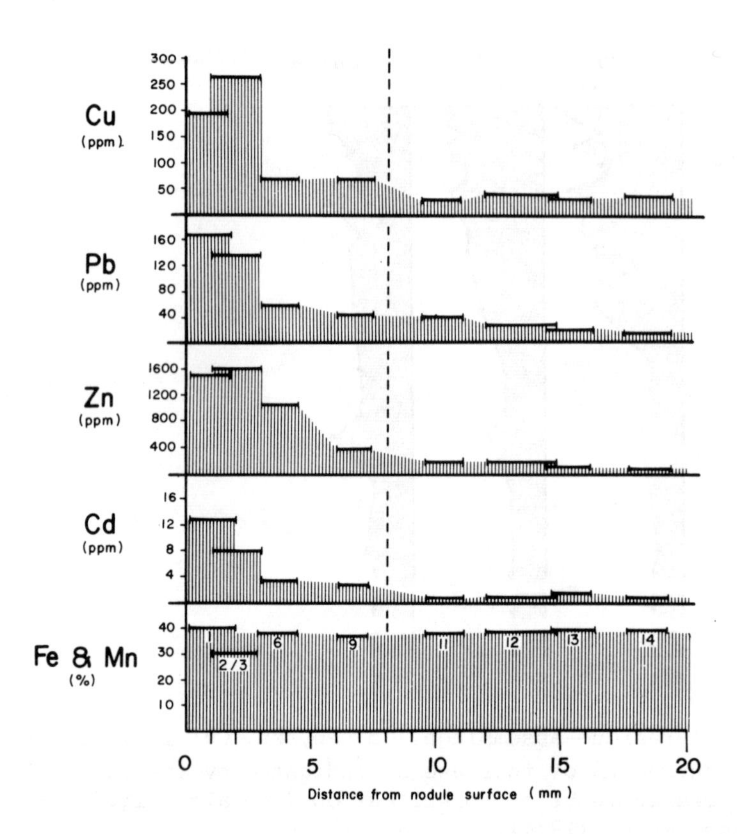


Fig. 6. Increase of trace metals in most recently accreted portion (0-8 mm from nodule surface) of ferromanganese concretion of the western Baltic Sea; it is thought that the increase is due to coal residues and coincides with the onset of trace metal input into surrounding sediments; data from Suess and Djafari (1976).

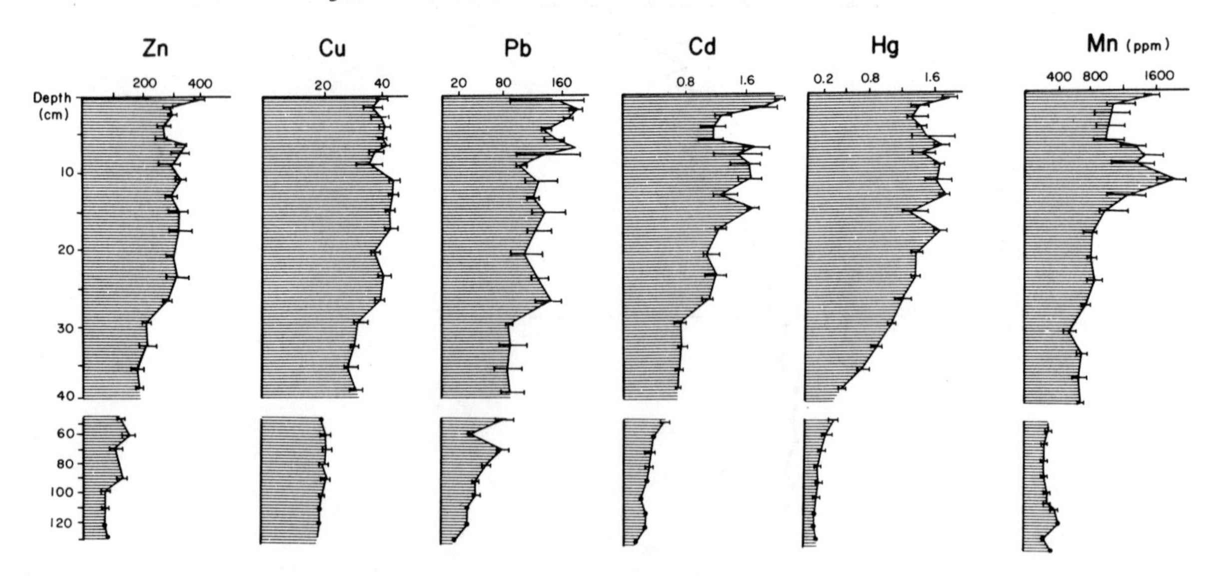


Fig. 7. Increase of metals in pelitic fraction of oxidizing sediments from the Deutsche Bucht, southern North Sea; the increase in Mn is thought to be due to diagenetic recycling whereas the increase in coal-residue metals and Hg is attributed to anthropogenic sources; bars indicate range of tripilicate sampling; data from Förstner and Reineck (1974).

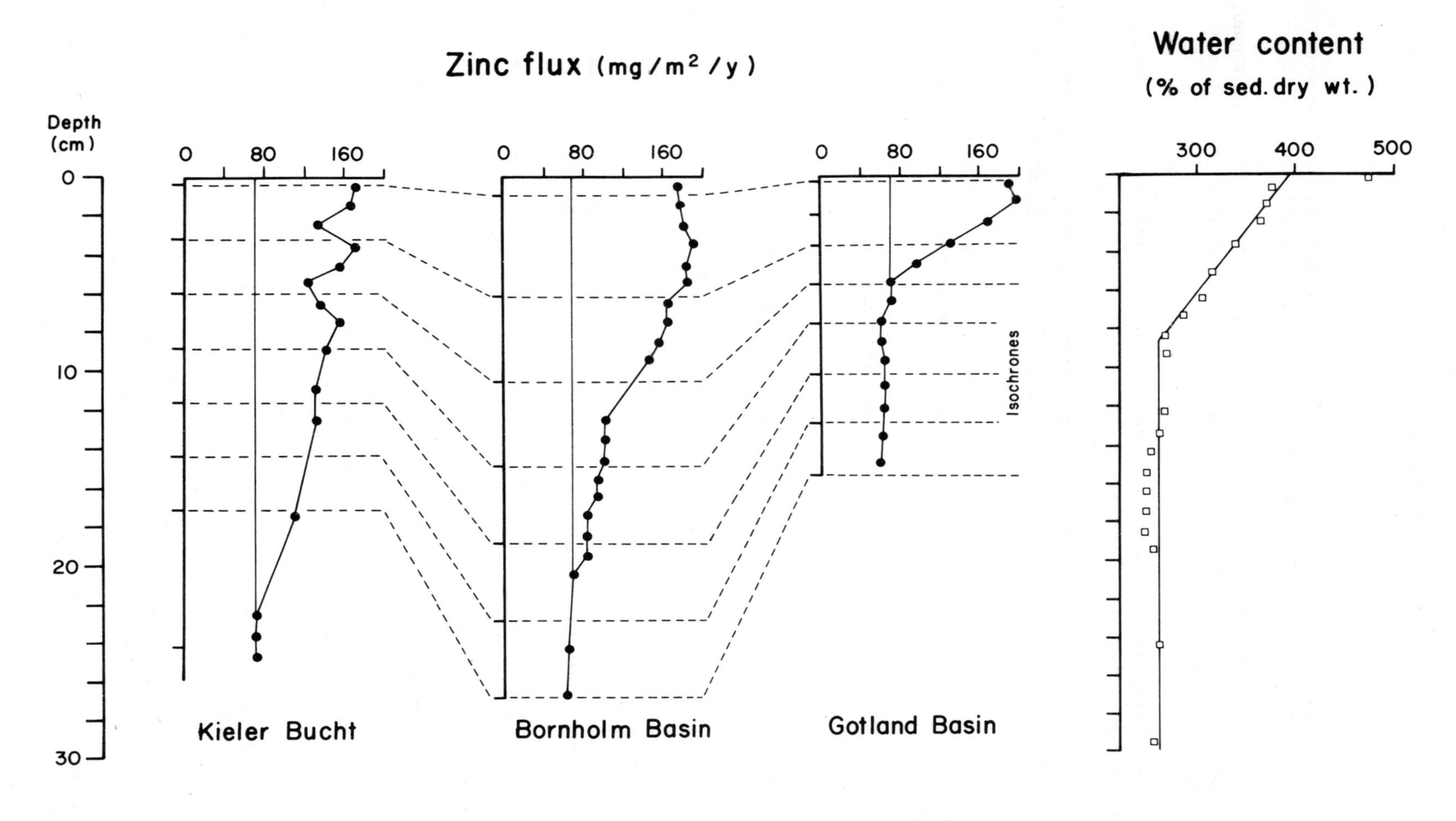


Fig. 8. Zn-flux to sediments of Baltic Sea basins; the vertical lines at about 75 mg $m^{-2}$ $y^{-1}$ separate natural from anthropogenic input; the isochrones were established by radiometric dating with the oldest indicating approximately the year 1840 A.D.; the water content used for flux calculations is shown for core from the Bornholm Basin; data from Suess and Erlenkeuser (1975).

## References

Aston, S.R.; Bruty, D.; Chester, R.; Padgham, R.C. 1973. Mercury in lake sediments : a possible indicator of technological growth. Nature, vol. 241, p. 450-1.

Banat, K.; Förstner, U.; Müller, G. 1972. Schwermetalle in Sedimenten von Donau, Rhein, Ems, Weser und Elbe im Bereich der Bundesrepublik Deutschland. Naturwissenschaften, vol. 59, p. 525-8.

Bertine, K.K.; Goldberg, E.D. 1971. Fossil fuel combustion and the major sedimentary cycle, Science., vol. 178, p. 233-5.

Bruland, K.W.; Bertine, K.; Koide, M.; Goldberg, E.D. 1974. History of metal pollution in Southern California coastal zone., Environm. Sci. & Techn., vol. 8, p. 425-31.

Chow, T.J.; Bruland, K.W.; Bertine, K.; Soutar, A.; Koide, M.; Goldberg, E.D. 1973. Lead pollution: records in Southern California coastal sediments. Science, vol. 181, p. 551-52.

Cordes, E. 1966. Aufbau und Bildungsbedingungen der Schwermineralseifen bei Skagen (Dänemark). Meyniana, vol. 16, p. 1-35.

Erlenkeuser, H. 1976. Environmental effects on radiocarbon isotope distribution in coastal marine sediments. Ninth internatl. radiocarbon conf., Los Angeles and San Diego, 1976.

Erlenkeuser, H.; Suess, E.; Willkomm, H. 1974. Industrialization affects heavy metal and carbon isotope concentrations in recent Baltic Sea sediments. Geochim. Cosmochim. Acta, vol. 38, p. 823-42.

Förstner, U.; Müller, G. 1974. Schwermetalle in Flüssen und Seen als Ausdruck der Umweltverschmutzung. Berlin, Heidelberg, New York, Springer-Verlag, p. 225.

Förstner, U.; Reineck, H.E. 1974. Die Anreicherung von Spurenelementen in den rezenten Sedimenten eines Profilkerns aus der Deutschen Bucht. Senkenbergiana marit., vol. 6, 175-84.

Gluskoter, H.J.; Lindahl, P.C. 1973. Cadmium: mode of occurrence in Illinois coal. Science, vol. 181, p. 264-5.

Hartmann, M.; Müller, P.J.; Suess, E.; van der Weijden, C.H. 1976. Chemistry of Late Quaternary sediments and their interstitial waters from the NW African continental margin. METEOR-Forsch. Ergebn., C 24, p. 1-67.

Krom, M.D. 1976. Chemical speciation and diagenetic reactions at the sediment-water interface in a Scottish Fjord. Abstract. Joint Oceanogr. Assembly, Edinburgh 1976, p. 89.

Linton, R.W.; Loh, A.; Natusch, D.S.F.; Evans, Jr., C.A.; Williams, P. 1976. Surface predominance of trace elements in airborne particles, Science, vol. 191, p. 852-4.

Lynn, D.C.; Bonatti, E. 1965. Mobility of manganese in diagenesis of deep-sea sediments. Marine Geol., vol. 3, p. 457-74.

Murozumi, M.; Chow, T.J.; Patterson, C. 1969. Chemical concentrations of pollutant lead aerosols, terrestrial dust and sea salts in Greenland and Antartic snow strata. Geochim. Cosmochim. Acta, vol. 33, p. 1247-92.

Peirson, D.H.; Cawse, P.A.; Salmon, L.; Cambray, R.S. 1973. Trace elements in the atmospheric environment. Nature, vol. 241, p. 252-6.

Price, N.B. 1967. Some geochemical observations on manganese-iron oxide nodules from different depth environments. Marine Geol., vol. 5/6, p. 511-38.

Price, N.B., 1973. Chemical diagenesis in sediments, WHOI-73-39, Techn. Report, Woods Hole, Mass., 73 p.

Smith, D.M.; Griffin, J.J.; Goldberg, E.D. 1973. Elemental carbon in marine sediments, a baseline for burning. Nature, vol. 241, p. 268-70.

Suess, E.; Djafari, D. 1976. Trace metal distribution in Baltic Sea ferromanganese concretions: Inferences on accretion rates. Manuscript submitted to Earth and Planetary Science Letters.

Suess, E.; Erlenkeuser, H. 1975. History of metal pollution and carbon input in Baltic Sea sediments, Meyniana, vol. 27, p. 63-75.

Tyler, G. 1972. Heavy metals pollute nature, may reduce productivity. Ambio, vol. 1 p. 52-9.

Vogel, J.C. 1972. Radiocarbon in the surface waters of the Atlantic Ocean. Proc. 8th internatl. conf. radiocarbon dating, Wellington, The Royal Society of New Zealand, p. 267-79.

Walger, E. 1966. Untersuchungen zum Vorgang der Transportsonderung von Mineralen am Beispiel von Strandsanden der westlichen Ostsee., Meyniana, vol. 16, p. 55-106.

Van Der Weijden, C.H. van der; Schuiling R.D.; Das, H.A. 1970. Some geochemical characteristics of sediments from the North Atlantic Ocean. Mar. Geol., vol. 9, p. 81-99.

Doyle, L.J.; Hopkins, T.L.; Betzer, P.R. 1976. Black magnetic spherule fallout in the Eastern Gulf of Mexico. Science, vol. 194, p. 1157-9.

Förstner, U. 1976. Lake sediments as indicators of heavy-metal pollution. Naturwissenschaften, vol. 63, p. 465-70.

Hansen, J.A.; Larsen, B.; Tjell, J.Ch. 1976. Tentative heavy metal budgets for the Baltic. In : Seminar on the recent development in the technological field with respect to prevention of pollution of the Baltic Sea area. Hanasaari Cultural Centre, Finland, 18-22 October 1976 (unpublished).

Johnson, D.B. 1976. Ultragiant urban aerosol particles, Science, vol. 194, p. 941-2.

# Some aspects of microbial growth in estuaries

P. J. le B. Williams*

Estuaries are regarded as areas of high biological and microbiological activity. It has long been recognized that estuarine sediments play an important part in these activities. There are certain factors which will facilitate (e.g. high organic content) , and other factors which may counter (e.g. oxygen deficiency) , the rates and extent of microbial activity in the sediments. Similarly, there are features of the water column which encourage or limit growth. As a consequence, certain estuaries are outwardly dominated by processes occurring in the sediments, whereas in others the bacterial role is less clear. It is thus profitable to consider some of the factors which affect or relate to microbial growth in the estuarine environment; the approach here will be more molecular and kinetic than descriptive.

While it is recognized that there is a risk in adopting this approach (relevant data are scarce, often non-existent and some of the concepts are not fully developed or are woolly), the advantages of the approach are threefold: at the present time, the precise measurement of processes within sediments, as opposed to those at the interface, presents formidable problems and the theoretical approach, based on existing knowledge of microbial processes can be justified; secondly, the descriptive aspects of microbiology are sometimes of limited value to the non-microbiologist, who may be more interested in the rates of processes; finally, these exercises are always valuable since they often more clearly define existent gaps in our knowledge.

## Equation of microbial growth

The equation conventionally used to describe microbial growth was proposed by Monod in 1942 and has since been found to be satisfactory for a majority of situations. The kinetics are robust and satisfactory results can be obtained even with mixed populations. The equation is :

$$\mu = \mu_{max} \frac{S}{S + K} ,$$

where $\mu$ = growth rate ($time^{-1}$), $\mu_{max}$ = maximum growth rate ($time^{-1}$), S = concentration of limiting substrate (mass $volume^{-1}$) and K = uptake constant (mass $volume^{-1}$).

The gross rate of utilization of the substrate is :

$$ds/dt = \frac{B}{Y} \mu_{max} \frac{S}{S + K} = V_{max} \frac{S}{S + K} ,$$

where B = microbial biomass (mass $volume^{-1}$), Y = growth efficiency (dimensionless) and $V_{max}$ = maximum rate of substrate uptake (mass $volume^{-1}$).

* Department of Oceanography, The University, Southampton, United Kingdom.

In general, kinetic data on bacteria isolated from aquatic systems, (although little information in known for populations from sediments) suggests that $\mu_{max}$ and K usually increase together. The growth efficiency, Y, has a value in the order of 0.5 for many situations and although there is little information on this value for natural populations, there are signs that it is not markedly lower than that for cultures.

Various factors may affect microbial growth in estuarine systems, a partial list of physico-chemical factors is as follows :

(1) Concentration of organic substrates; (2) pH; (3) oxygen concentration; (4) temperature; (5) rate of dispersion; (6) rate of molecular diffusion; and (7) presence of sediment particles. It is intended only to discuss here the last three on the list as many of the others are adequately covered by the other contributions to this volume.

## Rate of dispersion

A major difference between the pelagic and sedimentary environments, as far as micro-organisms are concerned, is the lack of dispersion in the sediments. Dispersion offers certain advantages to the microbial community; for example, for Southampton Water, it has been calculated that dispersion processes are more important in replacing the oxygen used by microbial heterotrophs than the immediate transfer from the atmosphere. On the other hand, dispersion may retard or inhibit the development of the planktonic community. The theory developed for the growth of micro-organisms in the continuous flow systems, such as a chemostat, is helpful here. If, for example, the rate of dispersion exceeds the rate of growth then the micro-organisms will be in a "wash-out" situation and the development of the population is prevented. The rates of microbial growth in natural systems are probably in the order of a day or so and will thus be within the range of dispersion rates encountered in estuaries. More information is needed on the growth kinetics of micro-organisms before it will be possible to use this approach to its fullest advantage but some calculations are made here for the purpose of illustration (see Table 1). Those organisms in Table 1 which were studied by Jannasch (1969) isolated from offshore water, whereas Herbert *et al.* (1956) studied *Aerobacter* in culture. The substrate concentration of 100μg/l which is assumed for the calculation, is intended to be on the high end of the range; it may be seen that the maximum growth rate for the organisms in Table 1 is 0.36 $day^{-1}$; the rate of dispersion in many estuaries would be much faster than this and would prohibit the development of a substantial population.

Similar calculations may be made for phytoplanktonic organisms and these are given in the lower part of Table 1. In this case, the nitrate concentration used in the calculation is possibly low for an estuary, but since it is very much greater than K, its exact value is not important. Faster growth rates are obtained for algae than for bacteria. At first it may seem suprising, however, this is probably largely a reflexion of the greater success achieved by the algologist in isolating strains from natural environments. Figure 1 illustrates that the phytoplankton bloom in an estuary is held back relative to that of an adjacent coastal water. Dispersion of the bloom would appear to be the most likely explanation for this; the development of the bloom in the estuary occur in midsummer when growth rates are highest and rates of flushing lowest.

## Presence of Surfaces

Under this heading we may consider and explore two related issues: molecular diffusion and the possible mode of action of extracellular enzymes.

Since the early work of Waksman, many marine microbiologists have held that surfaces play an important role in facilitating microbial processes in natural systems. Micro-organisms are often associated with particles and microbial activity is high in turbid systems. It is argued that sorption to the surface, or by slimes produced by the micro-organism, provide a microzone rich in organic substrates. These notions are not, however, very convincing from the physico-chemical standpoint.

The microzones around the micro-organism, however, does merit some further consideration; for example, rough calculations from a simple diffusion equation may be written thus :

$$C_s = C_\infty - \frac{Q}{4\pi Dr} ,$$

where $C_s$ is the substrate concentration adjacent to the cell, C is that at a distance from the cell, Q is the rate of substrate uptake, D is the diffusion constant and r is the radious of the cell. Calculations from data given in Table 2 suggest that Q/4xDr have a value in the region of 1µg/l. This type of analysis is crude and makes no allowance for feedback control between the substrate concentration and the activity of the organism. A more satisfactory approach is provided in the paper by Pasciak and Gavis (1974). They combine the Monod equation with the diffusion equation. The resultant quadratic equation :

$$C_s^2 + ((V_{max}/4\pi Dr) + K - C_\infty)C_s - KC_\infty = 0$$

is convenient to work with. Again, appropriate data for estuarine organisms are not available. From the data given in Table 3, it may be calculated that at substrate concentrations in the region of 50µg/l, the surface concentrations is 1 per cent below that of the surrounding medium, at 5µg/l it is only 2 per cent; one would normally expect substrate concentrations in sediments to be much greater than 50µg/l.

Thus from the data available, it may be concluded that molecular diffusion processes prevent a deficient zone from occurring round the microbial cell and the gradient required to sustain the inward diffusion of nutrients is hence very small.

The active dispersion of organic material by molecular diffusion raises the interesting problem of how micro-organisms manage their extracellular enzymes in natural aqueous environments. Micro-organisms are only able to take up freely diffusible material; particulate and plymeric material (from which must originate a significant part of their organic supply) must thus be hydrolysed outside the cell and this therefore requires the existence of extracellularly liberated enzymes. In order for the organism to achieve anything from this exercise it must efficiently retrieve the hydrolysis products formed as a result of this enzyme action - at least in excess of that energy invested in the synthesis and released of these enzymes. At the concentrations at which these enzymes will act (say 1-10 nM), the rate of dispersion is very rapid, having a time scale of $10^{-8}$ sec. Thus for the success of the operation it is important for the organism to maintain the enzyme close to itself and at the present it is difficult to be sure of the enzyme/organism relationship. It has been argued that this is perhaps achieved by the enzymes being sandwiched between cell surface and substrate (the latter may be on the surface itself or adsorbed to it). This is a subject which needs a great deal more thought and study since it is the key to the understanding of the decomposition of much of the material that enters the sediments. One suspects that there is probably sufficient information in the biochemical literature to construct first approximation models of the situation.

In conclusion, the rate of dispersion can be a severe hindrance to the development of micro-organisms in the water column. If substrate concentrations are as low as is believed, then it may be difficult to maintain growth rates of heterotrophic micro-organisms at a sufficient level to overcome the rates of dispersion encountered in well-flushed estuaries. This favours the estuarine sediments as a zone for microbial activity and although the lack of dispersion in the sediments will restrict the availability of oxygen and other material, molecular diffusion appears to be sufficiently active to prevent the micro-organisms from exhausting the substrate in the immediate environment. The importance of surfaces is not clear. Sorption processes, it would appear, do not facilitate growth by providing some substrate-rich zone. It is, however, possible that bacteria may exploit surfaces in order to achieve maximum recoveries from the products of their extracellular enzymes.

One point that has come out of the study is the apparent lack of appropriate kinetic data on estuarine micro-organisms. If the type of approach adopted in the present paper offers promise, then this deficiency will need remedying. The data are fairly easy to obtain, the main problem being in knowing how representative the isolates are of any commmunity as a whole. This is a familiar problem to the microbial ecologist.

TABLE 1 KINETIC DATA ON RATES OF GROWTH OF BACTERIA AND ALGAE

| Organisms | Substrate | K (µg/l) | $\mu_{max}$ ($day^{-1}$) | Calculated growth rate at 100µg/l substrate conc ($day^{-1}$) | Source of kinetic data |
|---|---|---|---|---|---|
| BACTERIA | | | | | |
| Achromobacter | Lactate | 1000 | 3.6 | 0.32 | Jannasch (1969) |
| | Glucose | 5000 | 9.0 | 0.16 | " " |
| Vibrio | Lactate | 900 | 3.6 | 0.36 | " " |
| Aerobacter | Glycerol | 10000 | 51 | 0.20 | Herbert et al. (1956) |
| ALGAE | | (µgN/l) | | | |
| Asterionella | Nitrate | 21 | 1.9 | 1.6 | Eppley & Thomas (1969) |
| Chaetoceros | Nitrate | 7 | 3.2 | 3.0 | " " |
| Skeletonema | Nitrate | 7 | 2.4 | 2.2 | " " |

TABLE 2 KINETIC DATA ON RATE OF SUBSTRATE UPTAKE BY BACTERIA

| Origin of Isolate | K ($\mu g/l^{-1}$) | $V_{max}$ ($\mu g$ $cell^{-1}$ $hr^{-1}$) | Source |
|---|---|---|---|
| Southampton Water | 25-100 | $10^{-8}$ | White (1969) |
| Pacific Ocean | 6-300 | $10^{-9}$ - $10^{-10}$ | Hamilton et al. (1966) |
| Lake Erken | 5 | $10^{-8}$ | Wright & Hobbie (1965) |

TABLE 3 CALCULATION OF DIFFUSION GRADIENT

SIMPLE CALCULATION :

$$C_s = C_\infty - \frac{Q}{4\pi Dr} ,$$

where $C_s$ = concentration at surface ($\mu g\ dm^{-3}$),

$C_\infty$ = concentration of distance ($\mu g\ dm^{-3}$),

$Q$ = rate of substrate uptake by cell $\sim 10^{-8} \mu g\ hr^{-1}$,

$D$ = diffusion constant $\sim 2 \times 10^{-4}\ dm^2\ hr^{-1}$, and

$r$ = radius of cell $\sim 0.5 \times 10^{-5}$ dm

$$\frac{Q}{4\pi Dr} = \frac{10^{-8}}{4\pi \times 2 \times 10^{-4} \times 0.5 \times 10^{-5}} \sim \mu g\ dm^{-3}$$

DETAILED CALCULATION

$C_s$ = Concentration at surface ($\mu g\ dm^{-3}$)

$C_\infty$ = Concentration at distance ($\mu g\ dm^{-3}$)

$D$ = Molecular diffusion constant $\sim 2 \times 10^{-4} dm^2 hr^{-1}$

$r$ = radius of cell $\sim 0.5 \times 10^{-5}$dm

$V_{max}$ = maximum uptake rate of cell $\sim 10^{-8} \mu g\ hr^{-1}$

$K$ = uptake constant $\sim 50$ g $dm^{-3}$

If $C_\infty = 50 \mu g\ dm^{-3}$ then $C_s = 49.5\ \mu g\ dm^{-3}$

$C_\infty = 5 \mu g\ dm^{-3}$ then $C_s = 4.9\ \mu g\ dm^{-3}$

$C_\infty = 0.5 \mu g\ dm^{-3}$ then $C_s = 0.48\ \mu g\ dm^{-3}$

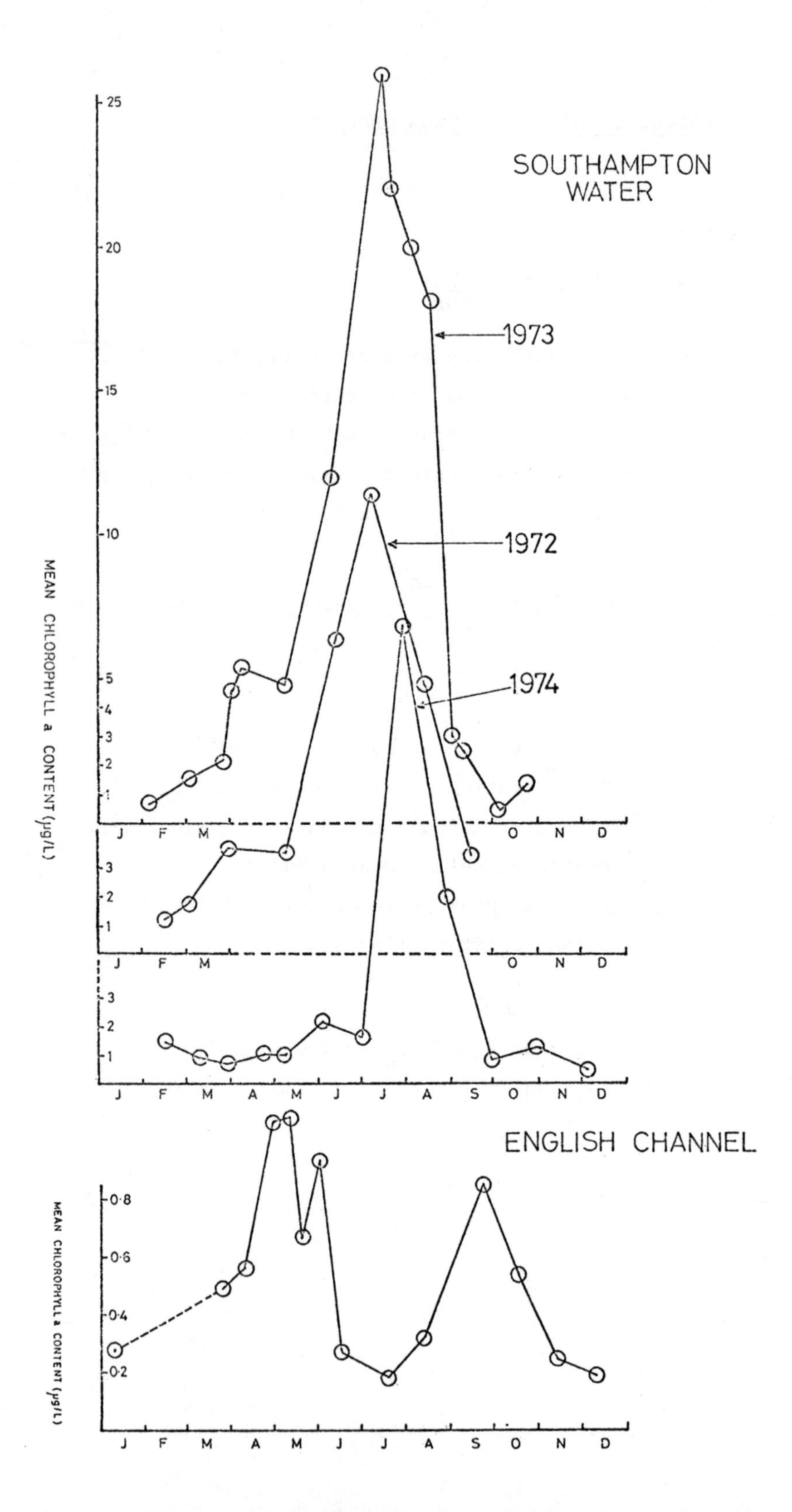


Fig. 1. Comparison of seasonal cycles of chlorophyll *a* in Southampton water and the English Channel (Station E,).

## References

Epply, R.W.; Thomas, W.H. 1969. J. Phycol., vol. 6, p. 375-9.

Hamilton, R.D.; Morgan, K.M.; Strickland, J.D.H. 1966. Can. J. Microbiol., vol. 12, p. 995-1003.

Herbert, D.; Elsworth, R.; Telling, R.C. 1956. J. gen. Microbiol., vol. 14, p.601-22.

Jannasch, H.W. 1969. J. Bacteriol., vol. 99, p. 156-60.

Pasciak, W.J.; Gavis, J. 1974. Limnol. Oceangr., vol. 19, p. 881-8.

White, B. 1969. M.Sc. Thesis, Southampton University.

Wright, R.T.; Hobbie, J.E. 1965. Ocean Sci. and Ocean Eng., vol. 1, p. 116-27.

# Means of determining natural versus anthropogenic fluxes to estuarine sediments

Kathe Bertine*

Influxes of material to estuarine sediments caused by man have been distinguished from their natural counterparts in several ways. The principal method that has been used by many investigators is to plot the varying concentrations of a metal or other material with depth for sedimentary profiles (see, for example, Bruland *et al.*, 1974). If a deposit has increased concentration of a suspected pollutant in its uppermost levels relative to its concentration in the deeper sediments, the differences in concentration are ascribed to anthropogenic sources. Such characteristic profiles for metals in Southern California Borderland sediments are shown in Fig. 1.

Several precautions must be taken before anthropogenic sources are unequivocally established. Firstly, many cores have oxidizing sediments overlying anoxic sediments; some metals too, such as manganese, are mobile under anoxic conditions and precipitate under oxidizing conditions (Li *et al.*, 1969). Freshly precipitated manganese oxides can co-precipitate other metals. Thus, the distribution of manganese must be determined in each core. If the manganese concentration is higher at the surface, an anthropogenic source for increased metal concentrations in these sediment levels may not be valid. Where there is a surface oxidized layer with deeper strata in an anoxic regime, such a situation could result. Notice that in Fig. 1 the Mn:Al ratio is constant, indicating that anoxic conditions probably extended to the surface sediments.

Increases in biological productivity can dilute natural and anthropogenic inputs of metals to a sediment (note the decreasing Al content in the upper San Pedro basin sediments). For this reason, the metal aluminium ratio is commonly plotted. Aluminium concentrations are primarily, if not completely, derived from weathering processes and thus, this ratio corrects for changes in biological productivity. For instance, if just the Zn content had been considered in San Pedro basin sediments (Fig. 1), no anthropogenic input would have been detected; however, when the Zn:Al ratio is plotted, a significant anthropogenic Zn influx be- me evident.

In interpreting the sediments described above, one assumes that sedimentary conditions at the site of the core have been uniform over the time period considered and there has been particle-by-particle deposition. In estuaries, such events as major storms or bioturbation can mix the sediment and add or erode layers of sediment so that the historical record is more difficult to interpret An example is given in Fig. 2 of Chesapeake Bay, where Hurricane Agnes in 1972 either placed a thick wedge of sediment on top or mixed the sediment already existing. Fortunately,

* Department of Geological Sciences, San Diego State University, San Diego, California 92182 U.S.A.

instances of non-particle-by-particle deposition can be identified through $^{210}Pb$ measurements. With particle-by-particle deposition, the unsupported $^{210}Pb$ activity should decrease exponentially with depth. In core CHSP-7505-1314, $^{210}Pb$ activity is constant to a depth of 30 cm, then decreases below that level. The metal concentrations exhibit similar patterns. Therefore, care must be taken during interpretation of the results. In general, $^{210}Pb$ activities should be measured to determine whether particle-by-particle deposition has occurred over the dating range of 100 years or so.

In most interpretations of sedimentary profiles, there is the assumption that the sediment has been derived from the same source over the length of time that is being considered. Such an assumption may not always be valid in estuarine systems. For instance, in the above example, sediment in a dam on the Susquehanna River was flushed out during Hurricane Agnes. If that sediment reached the core site and contained higher metal contents, it could account for the higher metal contents in the upper 30 cm of sediment. Alternatively, sedimentary material from that dam may not have reached the core site and Hurricane Agnes may just have homogenized the upper 30 cm of sediment. In that case, the higher metal concentrations in the upper sediments would be ascribed to those derived from man's influence.

Absolute concentrations of metals or metal:Al ratios are not sufficient to determine the amount of anthropogenic metal input at any site. Comparing Zn:Al ratios in San Pedro Basin sediments with Santa Barbara Basin sediments (Fig. 1) suggests that sediments from San Pedro Basin had received a far larger anthropogenic Zn influx to its uppermost sediments. However, the sedimentation rate at Santa Barbara Basin is far higher and dilutes the anthropogenic component. In reality, this Zn flux is higher in Santa Barbara sediments. Therefore, the flux of a metal is of far more importance than its concentration.

Identification and the singling out of a predominant source can be made by comparing anthropogenic fluxes in the sediments to the flux emanating from a particular source. These fluxes can be calculated by taking the concentration of a material at any depth, subtracting from it the "natural" concentration in the lower portions of the core, multiplying it by the sedimentation rate obtained by $^{210}Pb$, or any other method, and by the sediment density. For instance, metal fluxes originating from fossil fuel burning from the United Illuminating electric generating station in New Haven, Connecticut were shown to be one to two orders of magnitude less than the anthropogenic metal fluxes in sediments of one nearby reservoir (Lake Whitney). Thus, fossil fuel burning from that plant was not the primary cause of the increased metal contents in the sediments.

To pinpoint a source, not only the flux but also the time pattern generated by a particular flux must agree. For instance, the metal fluxes from coal burning at the United Illuminating generating station could account for most of the anthropogenic metal fluxes in another reservoir (Lake Saltonstall). However, United Illuminating switched to burning oil in the early 1960's. Oil burning released far less metals and a decrease in the anthropogenic metal fluxes in Lake Saltonstall sediments should have resulted. Such a decrease was not found in the sedimentary profile. Hence, even though the fluxes agree, coal burning at United Illuminating is probably not the source of the anthropogenic metals in Lake Saltonstall.

When both the time pattern and fluxes agree, this is evidence that the metals are from a particular source. Erlenkeuser *et al.* (1974) showed that the rate of increase in coal burning during the last 170 years around the Baltic reflects the rate of increase in metal concentration in Baltic sediments.

Isotopic ratios of Pb, C and O can be used to distinguish between natural and anthropogenic materials in sediments. The Pb isotopic ratios in sediment from Lake Whitney in New Haven, Connecticut are presented in Table 1. As the concentration of Pb increases in the sediment, the Pb isotopic ratios in the sediments tend toward the ratios measured in gasoline from Boston, Massachusetts and in coal from Pennsylvania. The Pb isotopic concentration was affected by man's activities as early as 1877. Since this time is well before the advent of the automobile, the increases have been ascribed to coal burning. In more recent times, lead tetraethyl added to gasoline has been the primary source of anthropogenic lead in these sediments. This is confirmed by the decrease in lead concentrations in the sediments after 1965, which resulted from the opening of a new highway which routed traffic away from the reservoir.

The scanning electron microscope (SEM) has been used to determine the morphological characteristics of particles in sediments. If differences exist between man-generated and natural particles, the source may be characterized. For instance, elemental carbon particles have been isolated chemically from sediments. From their morphology under the SEM and their chemical composition determined by energy dispersive X-ray, carbon from wood burning, coal, gasoline, and oil burning can sometimes be differentiated (Fig. 3). It is preferable to isolate the carbon particles manually since chemical methods remove the diagnostic trace metals (Pb for gasoline and V and Ni for oils). If the concentration of elemental carbon is also determined in the sediment, the amounts of metals that would be associated with the fossil fuel fly ash could be estimated from values in the literature.

Many other investigations involving morphology can be envisaged. For instance, heavy liquid fractionation of the sediment could be performed. The heavy liquid fraction particles could then be investigated under the SEM. The identity of individual particles such as pigments, metal fillings, etc., could then be determined using energy dispersive X-rays to obtain their composition.

Table 1

Lead Isotope Ratios in Lake Whitney Sediments

| Depth (cm. | Year | $Pb\frac{206}{204}$ | $Pb\frac{206}{207}$ | $Pb\frac{206}{208}$ | Pb conc. (µg/g) |
|---|---|---|---|---|---|
| 1 - 2 | 1974 | 18.91 | 1.2021 | 0.4901 | 707 |
| 10 - 11 | 1965 | 18.77 | 1.191 | 0.4869 | 1100 |
| 18 - 19 | 1956 | 18.64 | 1.179 | 0.4826 | 470 |
| 26 - 27 | 1948 | 18.78 | 1.190 | 0.4822 | 270 |
| 34 - 35 | 1912 | 18.93 | 1.202 | 0.488 | 170 |
| 38 - 39 | 1896 | 19.22 | 1.216 | 0.488 | 170 |
| 43 - 44 | 1877 | 19.77 | 1.241 | 0.4926 | 100 |
| 48 - 49 | 1858 | 20.79 | 1.302 | 0.5103 | 100 |
| 53 - 54 | 1839 | 20.90 | 1.305 | 0.5076 | 44 |
| 56 - 57 | 1827 | 20.50 | 1.291 | 0.5057 | 40 |
| Gasoline* | | | | | |
| Boston Mass. | 1964 | 18.45 | 1.179 | 0.4819 | - |
| Coal* | | | | | |
| Ashley, Pa. | | 18.64 | 1.189 | 0.4824 | 34 |
| Jeddo, Pa. | | 18.82 | 1.197 | 0.4850 | 11 |
| Treverton, Pa. | | 18.82 | 1.200 | 0.4849 | 10 |
| St. Nicholas, Pa. | | 18.79 | 1.199 | 0.4849 | 14 |

*(Chow *et al.*, 1975).

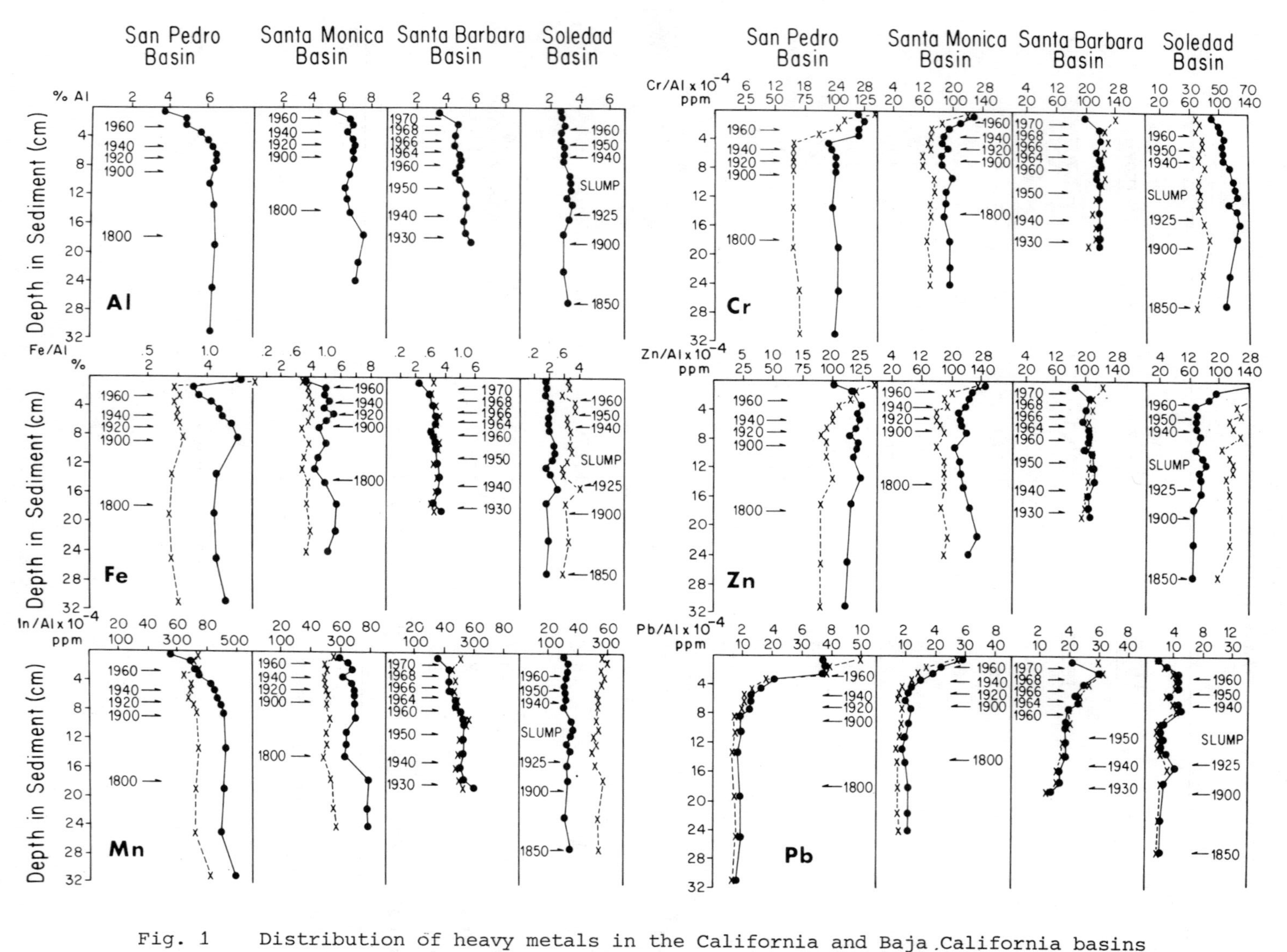


Fig. 1 Distribution of heavy metals in the California and Baja California basins as a function of depth and time. Filled circles and heavy curves refer to absolute concentrations in the sediments. Dashed lines and X's refer to the ratios of the species concentration to that of aluminium.

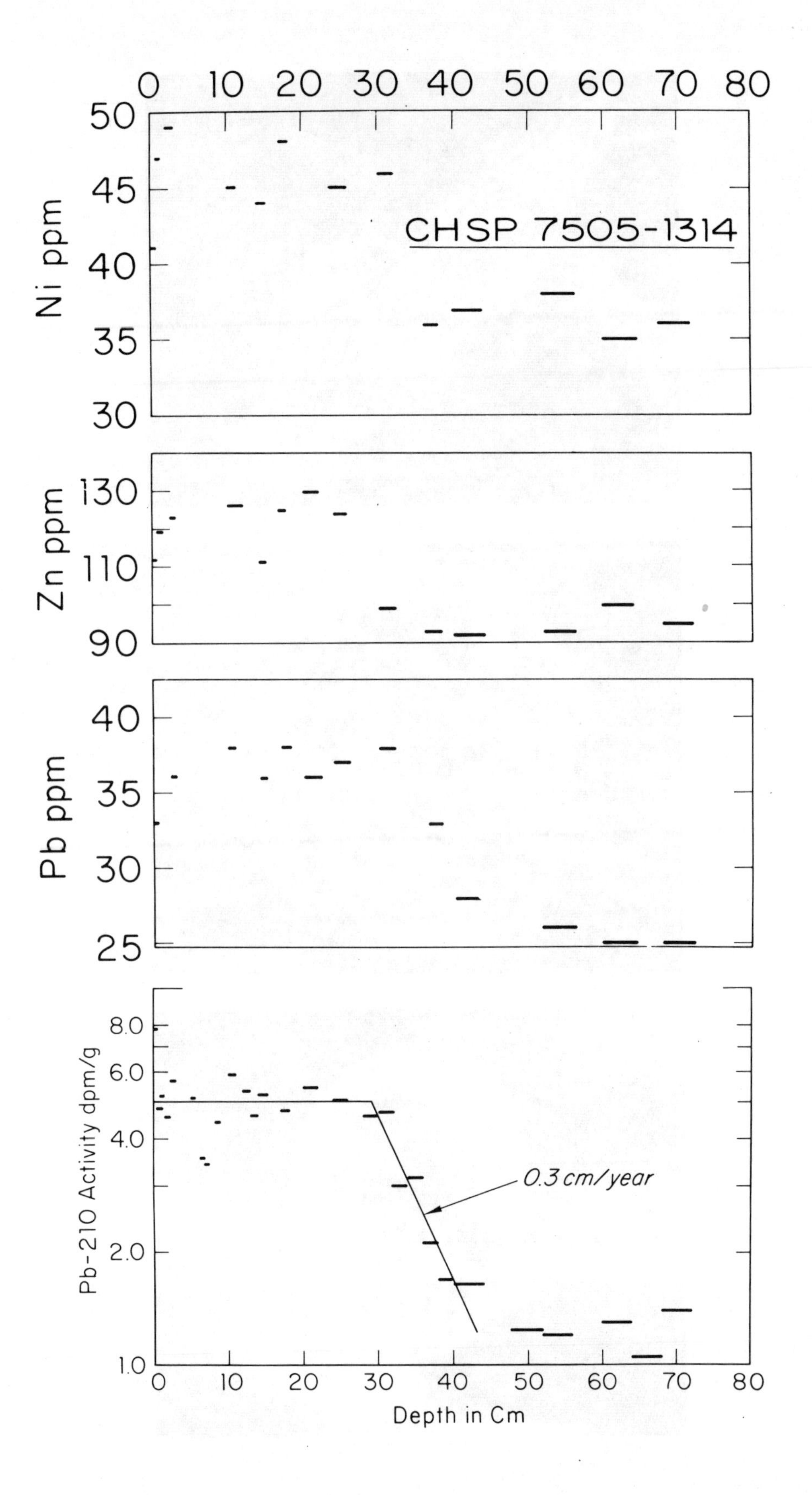


Fig. 2 $^{210}$Pb activities and metal concentrations as a function of depth in sediments from Chesapeake Bay (E.D. Goldberg, personal communication).

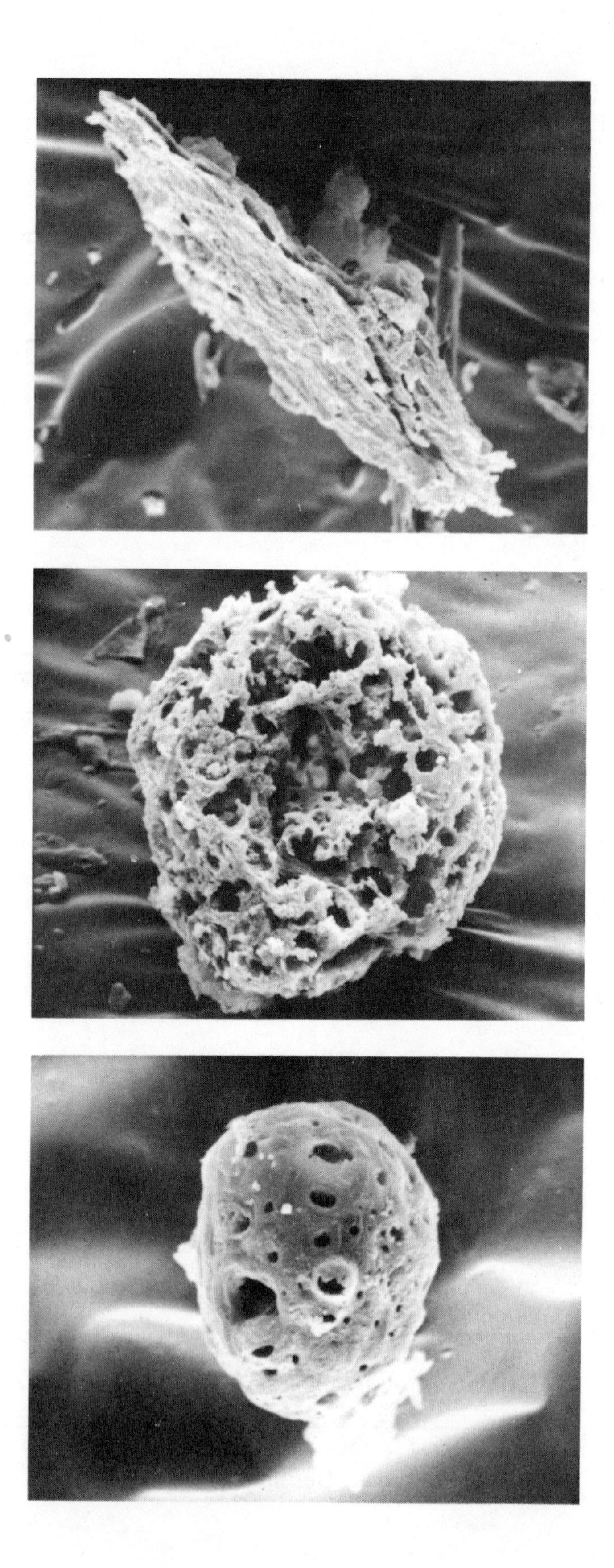

Fig. 3 SEM photographs of elemental carbon particles isolated from Lake Whitney sediments.

References

Bruland, K.W.; Bertine, K.; Koide, M.; Goldberg, E.D. 1974. History of metal pollution in Southern California coastal zone, Envir. Sci. Tech., vol. 8, p. 425.

Erlenkeuser, H.; Suess, E.; Willkomm, H. 1974. Industrialization affects heavy metals and carbon isotope concentrations in recent Baltic Sea sediments, Geochim. Cosmochim. Acta, vol. 38, p. 832.

Li, Y-H.; Bischoff, J.; Mathieu, G. 1969. The migration of manganese in the Arctic Basin sediment, Earth Planet. Sci. Lett., vol. 7, p. 265.

# The dependence of the various kinds of microbial metabolism on the redox state of the medium

G. Billen*

I. INTRODUCTION

From the point of view of microbial life, one of the most salient physico-chemical features of the recent sedimentary environment is the occurrence of a steep vertical gradient of oxidation-reduction conditions.

This gradient is caused by organic material within the sediment but it, in turn, influences the kind of microbiological metabolism likely to occur.

The characterization of the redox conditions and the knowledge of the relationship between redox conditions and microbial metabolism is therefore very important for an understanding of the biological processes within the sediments.

II. SIGNIFICANCE OF THE REDOX POTENTIAL CONCEPT AND MEASUREMENT IN NATURAL ENVIRONMENTS

In a medium where all the redox couples $Red_i/Ox_i$, reacting according to the equation,

$$Red_i + m_i\, H_2O \rightleftarrows Ox_i + l_i\, H^+ + n_i\, e^-,$$

are in thermodynamic equilibrium each with respect to the other, the redox potential ($E_h$) of this medium is given by the Nernst relation :

$$E_h = \frac{\Delta G°_i}{n_i F} + \frac{RT}{n_i F} \ln \frac{a(Ox_i)}{a(Red_i)} - \frac{RT l_i}{n_i F}\, pH \qquad \text{(for all i's)}.$$

For obvious practical reasons, this theoretical definition can be extended to the cases of media in metastable or partial equilibria. In such cases, the $E_h$ is defined with respect to the only couples which are reactive enough in the considered conditions to reach mutual equilibrium. (The term "conditions" includes the presence or absence of specific catalysts, such as micro-organisms).

Most natural media are dynamic systems subject to an energy flow. This energy flow is fed either by sunlight, used by photoautotrophs to syntesize highly unstable organic molecules, or by a direct input of these organic molecules from outside or from another part of the ecosystem. Heterotrophic and chemoautotrophic metabolisms dissipate this energy flow by tending to bring the system back to its thermodynamic equilibrium. The stationary state eventually reached in such dynamic systems cannot *a priori* be described by a thermodynamic equilibrium model, and the significance of the concept of redox potential in natural environments is therefore questionable.

* Université Libre de Bruxelles, Université Libre de Bruxelles, 50 ave. F. Roosevelt 1050 Brussels, Belgium

However, many observations concerning the distribution of chemical species in natural media displaying gradients of redox conditions and subjected to intense bacterial activity (Richards, 1965; Thorstenson, 1970; Billen 1975; Billen et al., 1976) show that an internal thermodynamic equilibrium is not far from being attained in the sub-systems formed by the main mineral redox couples involved in the energy metabolism of micro-organisms - such as $H_2O/O_2$ or $H_2O_2/O_2$ (Sato, 1960; Breck, 1972), $Mn^{2+}/MnO_2$, $NH_4^+/NO_3^-$, $Fe(OH)_3/FeCO_3$, $SO_4^{2}/HS^-$, $NH_4^+/N_2$, $CH_4/CO_2$ (Fig. 1).

The redox potential of natural environments can therefore be best defined with respect to this sub-system, considered, as a first approximation - be be at equilibrium. So defined, it does not take into account the presence of highly reduced organic matter, but only characterizes the availability of oxidants susceptible to use for the energy metabolism of micro-organisms.

By definition, the best way to determine the redox potential is to measure the concentration of the two redox forms of a particular couple $Red_i/Ox_i$. Thus, some couples which occur in most environments can be used as natural redox indicators e.g. $Mn^{2+}/MnO_2$; $Fe^{2+}/Fe(OH)_3$. The direct measurementof $E_h$ by the traditional platinum electrode, although it can provide valuable relative indications, must be interpretated with caution, amongst other reasons because of the different reactivity of some redox substances at the platinum electrode and in solution.

## III. OCCURRENCE OF VARIOUS REDOX MICROBIAL METABOLISMS IN NATURAL ENVIRONMENTS

Fermentative metabolisms result in reoganization of organic matter without modifying their gross redox level (see footnote). Therefore, they remain insensitive to the redox conditions of the medium. On the other hand, the respiratory metabolism, either organo- or chemo-lithotrophic, consists of the oxidation of a reduced substrate by an oxidant present in the medium. Not every combination of donor substrate with final acceptor is possible in all environments since the electron transfers have to be exoenergetic for its use as an energy-yielding metabolic system.

This allows us to define the general and ecological conditions for the existence of various respiratory metabolisms.

A necessary condition for the existence of a metabolic system is that the mean free energy change per electron during the oxidation of a donor couple by the acceptor couple must be negative under standard conditions (25°C, 1 atm. pressure, unit activity of all chemical species except at $aH^+ = aOH^- = 10^{-7}$). This free energy change can be directly calculated from a comparison of the standard redox potential at pH 7 of the couples concerned. Such a comparison is represented in Fig. 2 according to a scheme used by MacCarthy (1972). In this representation, the mean free energy change per electron is the intersection of the horizontal line, corresponding to the acceptor couple, with the oblique line, corresponding to the donor substrate. There are few metabolisms predicted possibly by this figure which do not occur in natural environments (no bacterium has as yet been isolated which oxidizes $NH_4^+$ to $N_2$, or $N_2$ to $NO_3^-$ as an energy-yielding process, although these would be thermodynamically conceivable).

Note: An exception to this is when molecular hydrogen is produced during the fermentation. This case presents some - more than a formal - analogy with respiratory metabolisms, of which it could represent a primitive form (Gray and Gest, 1965).

In a particular environment, the free energy change associated with the oxidation of a reduced substrate is a function of the availability of an oxidant, and thus of the $E_h$ of this environment as we have so defined it. Considering a natural medium in which the sub-system formed by the mineral redox couples is at thermodynamic equilibrium at a certain value of pH and $E_h$ :

$$E_h = E_i'^{\circ} + \frac{RT}{n_iF} \ln \frac{Ox_i}{Red_i} .$$

If a reduced substrate, $S_{red}$, organic or mineral, is brought into this medium, which is considered to be of infinite size, irrespective of which oxidant used, the maximal energy provided by the oxidation of the substrate will be given by :

$$\Delta G = nF\, E'^{\circ}_{Sred/Sox} + RT \ln Sox/Sred - nF\, E_h .$$

This relationship is graphically represented for some substrates in Fig. 3. From this it can be concluded that, within the range of $E_h$ encountered in natural media (Baas Becking et al., 1960), oxidation of organic compounds (i.e. chemo-organotrophic metabolisms) is always exoenergetic, although the energy yield is lower at lower $E_h$.

For chemolithotrophic metabolisms on the other hand, stricter limitations exist since the oxidation of mineral substrates is not exoenergetic over the whole $E_h$ range. Thus it can be predicted that Thiobacter are active only at pH and $E_h$ values higher than the upper limit of stability of the reduced sulphur compounds they metabolize. Observations by Baas Becking and Wood in 1955 entirely confirm this prediction (Fig. 4). Likewise, ferro-oxidizing bacteria are only active above the stability limit of iron carbonate, as shown by the data compiled by Baas Becking et al. (1960) in Fig. 5. Nitrification occurs only at $E_h$ values at which oxidation of ammonium into nitrate is exoenergetic, as shown both by observations on pure cultures by Zobell (1935) and by direct measurements of nitrifying activity in coastal marine sediments and in the Scheldt estuary by Billen (1976) and Somville (unpublished).

The preceding thermodynamic considerations only concern the possibility of reduced substrate oxidation not allowing the prediction, as a function of $E_h$, of which oxidant may be employed in a respiratory metabolism. However, many observations suggest that various oxidants are used successively, in order of their decreasing E°'; aerobic respiration will occur first and thus denitrification will not occur in the presence of oxygen or a $E_h$ values higher than + 250 mV (Fig. 6); sulphate-reduction is inhibited by oxygen (Wheatland, 1954) and by nitrate (Fig. 7); methane production begins only after sulphates are completely exhausted (Martens and Berner, 1974). Although direct experimental data are lacking concerning these metabolisms, by analogy, it can be predicted that utilization of Mn(IV) and Fe(III) as oxidants will occur respectively after aerobic respiration and after denitrification.

Although this sequence of utilization allows the best energy yield to be obtained from the oxidation of reduced substrates, no thermodynamic argument fully explains it. One explanation could involve the existence of a competition between micro-organisms for reduced substrates, with selection of the energetically more favourable metabolic system. However, such a reason could not explain the conditions in environments rich in organic deposits where reduced substrates are not limiting. Direct physiological effects do exist to prevent the use of one oxidant in the presence of a better one. These physiological effects include :

(i) A general, non-specific toxicity of some oxidants for oganisms not adapted to them. This is the case with oxygen in obligate anaerobic bacteria. Such an effect is, however, unlikely for poorly reactive oxidants such as nitrate or sulphate.

(ii) The specific action of an oxidant on the electron transport system of micro-organisms normally adapted to use another one, e.g. by "drainage" of the electrons towards the better oxidant; and

(iii) The regulation of the synthesis and activity of enzymes involved in redox metabolism. When an organism is adapted to use several energy-yielding metabolisms, regulatory mechanisms generally exist which allow the optimal one to be used for a given set of environmental conditions (e.g. the switch from fermentation to respiration when oxidants become available or from denitrification to aerobic respiration when the oxygen concentration rises).

## IV. CONCLUSION

The redox potential in natural systems, defined with respect to the sub-system formed by the main mineral couples involved in the energy-yielding metabolism of micro-organisms, is an important parametre from which the type of metabolism which will occur under a particular set of conditions can be deduced. In diagenetic models of recent sediments, it can therefore be a useful conceptual - as well as empirical - parametre for introducing vertical zoning with different types of microbiological activities occurring successively with increasing depth.

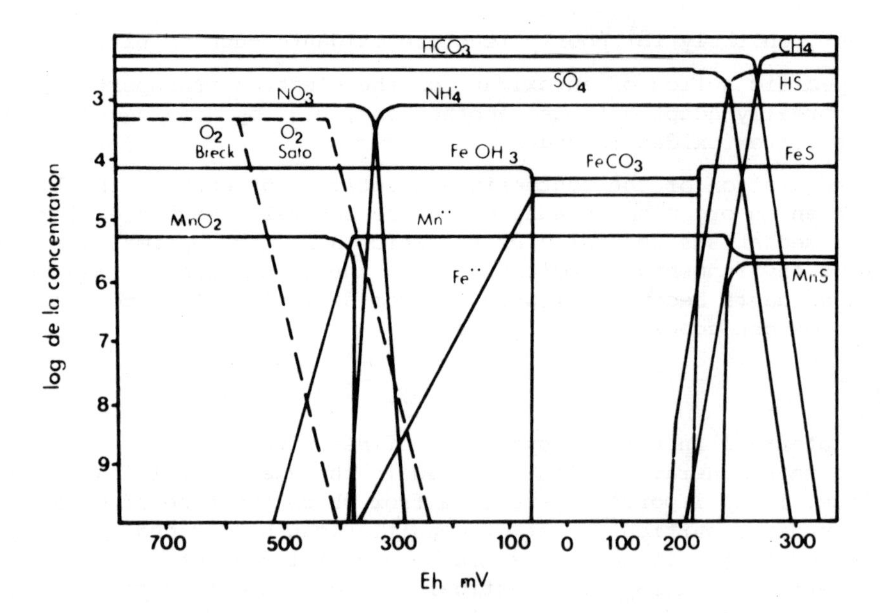


Fig. 1. Equilibrium diagram of the system $H_2O_2/O_2$, $Mn^{2+}/MnO_2$, $NH_4^+/NO_3^-$, $Fe(OH)_3/FeCO_3$, $SO_4^{2-}/HS^-$, $CH_4/CO_2$, in a typical freshwater at pH 7.5.

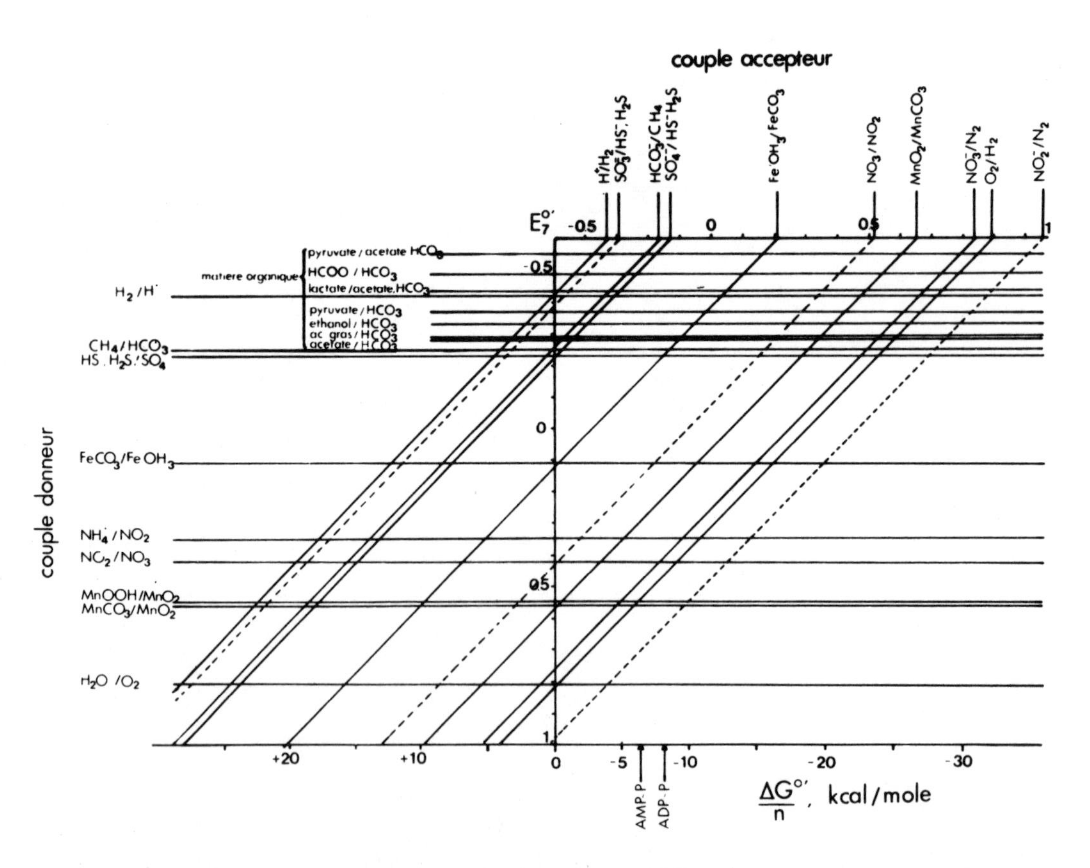


Fig. 2 Free energy change per electron during the oxidation of a donor couple at STP; (thermodynamic data from Garrels and Christ, (1965) and Edsall and Wymann (1958).

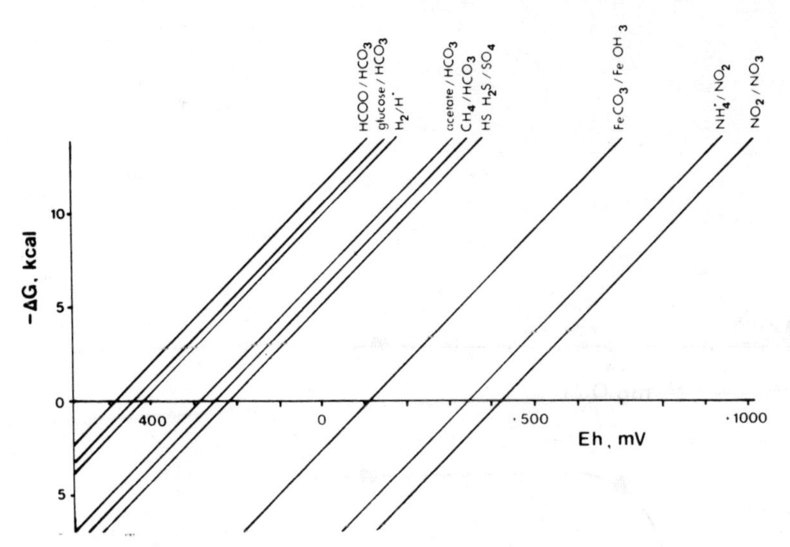


Fig. 3 Maximum free energy change per electron during some biological oxidations (organotrophic and lithotrophic metabolisms) as a function of the redox potential of the environment (at pH 7).

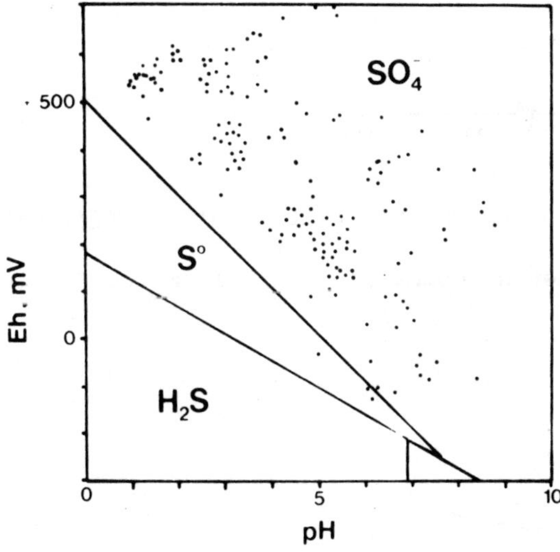


Fig. 4 $E_h$ and pH in active cultures of Thiobacillus on sulphur (from Baas Becking and Wood, 1955). Thermodynamic stability of sulphur with respect to sulphates.

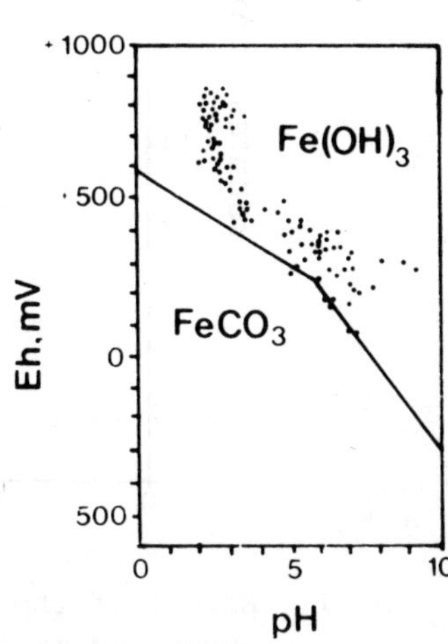


Fig. 5 Zone of activity of ferro-oxidizing bacteria in pure cultures and in natural environments (from Baas Becking *et al.*, 1960). Thermodynamic stability zone of ferric hydroxide and ferrous carbonate.

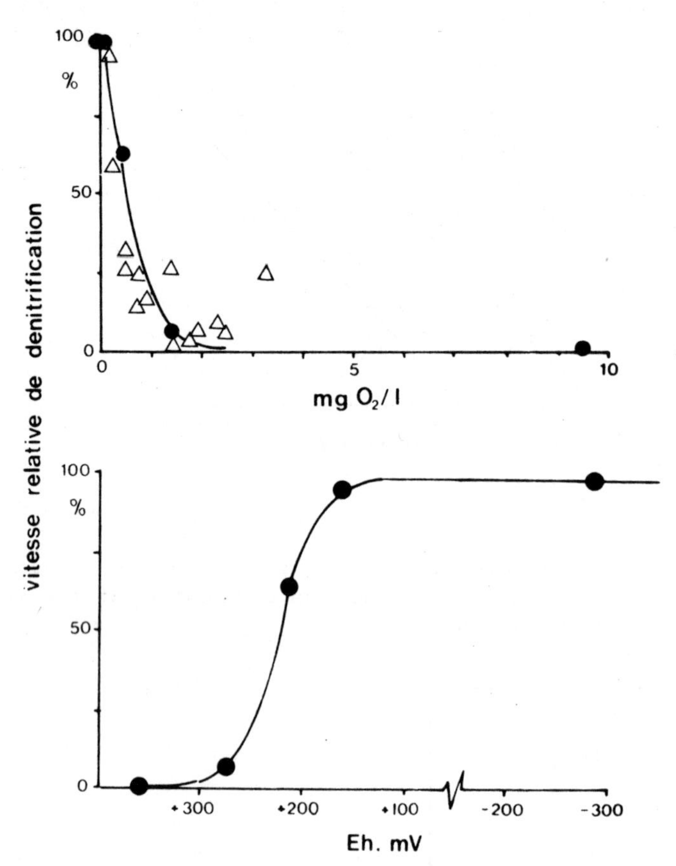


Fig. 6 Effect of oxygen concentration and $E_h$ on denitrification. (□) Data of Wheatland (1959).
(●) Experiment on a suspension of muddy sediments from the Sluice Dock at Ostend (Belgium).

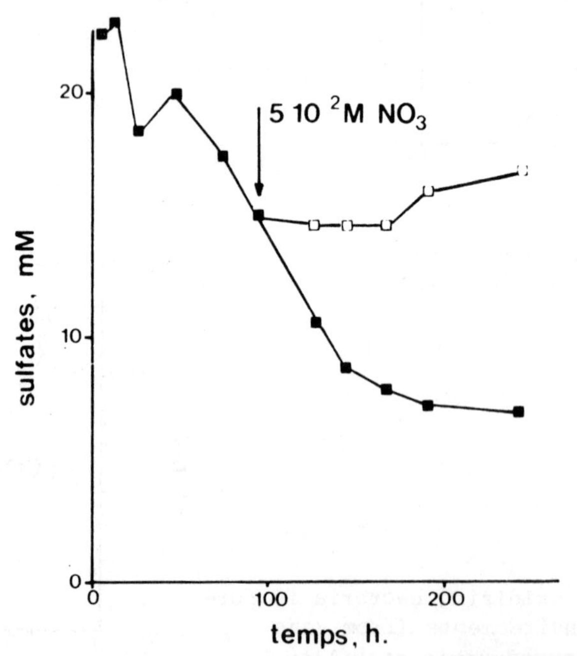


Fig. 7 Effect of the addition of nitrate on sulphate reduction in a suspension of muddy sediment from the Sluice Dock at Ostend.

## References

Baas Becking, L.G.M.; Wood, E.J.F. 1955. Biological processes in the estuarine environment. I. Ecology of the sulphur cycle. Proc. Konink. Nederland Adad. V. Wetenschappen, Ser.B., vol. 58, p. 160-81.

Baas Becking, L.G.M.; Kaplan, I.R., Moore, D. 1960. Limits of the natural environment in terms of pH and oxidation-reduction potentials. J. Geol., vol. 68, p.243-84.

Billen, G. 1975. Nitrification in the Scheldt estuary (Belgium and the Netherlands). Estuar. Coast. Mar. Sci., vol. 3, p. 79-89.

Billen, G. 1976. A method for evaluating nitrifying activity in sediments by dark $^{14}C$-bicarbonate incorporation. Water Res., vol. 10, p. 51-7.

Billen, G.; Smitz, J.; Somville, M.; Wollast, R. 1976. Dégradation de la matière organique et processus d'oxydo-reduction dans l'Estuaire de l'Escaut. In : R. Wollast; J.C.J. Nihoul, (eds.) : Modèle mathématique de la Mer du Nord. Rapport de Synthèse. (Vol. Estuaire de l'Escaut).

Breck, W.G. 1972. Redox potential by equilibration. J. Mar. Res., vol. 30, p. 121-39.

Gray, C.T.; Gest, H. 1965. Biological formation of molecular hydrogen. Science, vol. 148, p. 186-92.

MacCarthy, P.L. 1972. Energetics of organic matter degradation. In : R. Mitchell, (ed.) : Water Pollution Microbiology, p. 91-118, New York, Wiley.

Martens, C.S.; Berner, R.A. 1974. Methane production in intersitial water of sulphate-depleted marine sediments. Science, vol. 185, p. 1167-9.

Richards, F.A. 1965. Anoxic basins and fjords. In: S.P. Riley; G. Skirrow (eds.), Chemical oceanography., New York, Academic Press, vol. 1, p. 611-45.

Sato, m. 1960. Oxidation of sulphide ores bodies. I. Geochemical environment in terms of Eh and pH. Econ. Geol., vol. 55, p. 928-61.

Thorstenson, D.C. 1970. Equilibrium distribution of small organic molecules in natural waters. Geochim. Cosmochim. Acta, vol. 34, p. 745-70.

Wheatland, A.B. 1954. Factors affecting the formation and oxidation of sulphides in a polluted estuary., J. Hyg. Camb., vol. 52, p. 194-210.

Zobell, C. 1935. Oxidation-reduction potentials and the activity of marine nutrifers. J. Bact., vol. 29, p. 78.

# Zonation of reactions in sediments

William S. Reeburgh*

To date, most of our attention in this area has been given to sampling and analytical methods. While better resolution with depth is still needed, these measurements and methods have progressed to a point where typical depth distributions can be recognized and some of the controls on the distributions can be elucidated. Most of our attention has been focused on the obvious products of biological degradation processes such as sulphide, methane, and to some extent, carbon dioxide. We have a fair understanding of the further reactions and processes that these products are party to, such as the formation of iron sulphides (Berner, 1975) and escape of methane bubbles from the sediments (Reeburgh, 1969; Martens, 1976). Apart from sulphate, we know very little about reactants. We know from microbiology that the organisms responsible for sulphate reduction and methane production require simple substrates, yet we continue to refer simply to "organic matter", seldom measured and too complicated to characterize.

Far too few measurements of the various possible forms of carbon, nitrogen and sulphur have been made to permit construction of reliable mass or flux balances. Virtually no new chemical reactions have been identified in field measurements of sediments. Our present level of understanding has been reached by combining field measurements with thermodynamic and microbiological insights, and laboratory studies (Cappenberg, 1974, 1975; Martens and Berner, 1974) or mathematical models (Berner, 1975).

Thereis little question that the intersitial water reactions are mediated microbiologically, but little is known of the requirements and substrates of the bacteria. Culture work has indicated that the bacteria have rather specific substrate and environmental requirements and field evidence for zonation of reactions is beginning to appear. An example is anaerobic methane oxidation, a reaction that microbiological information indicated should not occur (Quale, 1972). This reaction occurs in nature in anoxic water columns and sediments (Reeburgh, 1976; Barnes and Goldberg, 1976) and appears to be a major sink both for methane as well as sulphate in marine systems (Reeburgh and Heggie, 1976), indicating that sulphate is required. Distributions of methane, sulphate, total carbon dioxide and $\delta^{13}CO_2$ (see Fig.) indicate that the reaction is located in a thin sub-surface zone, possibly where the product, $(SO_4^{2-})X(CH_4)$, is at a maximum (Reeburgh, submitted). This reaction appears to be diffusion-controlled, but the ultimate controls are probably the reactions which provide simple substrates from organic matter. Cappenberg (1974) indicates that acetate is a product of sulphate-reducers as well as a substrate for methane-producers. Similar interrelationships must be identified for other reaction pairs.

---

* Institute of Marine Science, University of Alaska, Fairbanks, Alaska 99701, U.S.A.

Future work should involve measurements of the concentrations and release or production rates of simple molecules like lactate, acetate, glucose and others of demonstrated importance in bacterial systems. These measurements should be used to produce depth distributions in sediments to identify reaction zones or sites. Continuous culture studies and use of specific inhibitors as shown by Cappenberg (1974, 1975), as well as direct rate measurements similar to those of Jorgenson and Fenchel (1974), should be employed. We should also fully exploit the potential offered by carbon and sulphur isotopes. The increase in one species of isotope is always accompanied by a corresponding decrease in another; the systems we have studied previously have not been constrained with enough care to identify these effects.

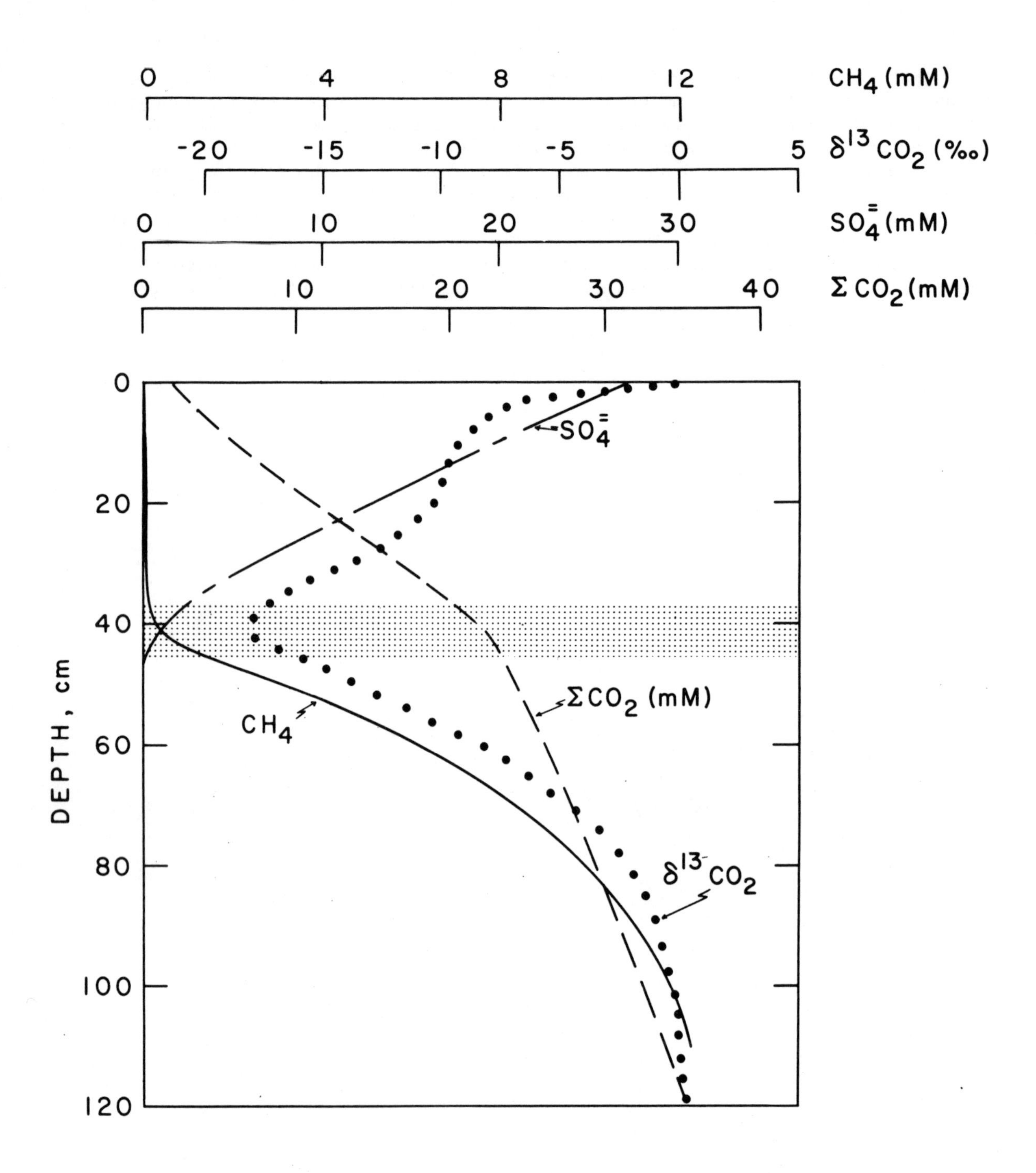


Fig. Schematic diagram (from Reeburgh, 1976, submitted) showing depth distributions of methane, sulphate, total carbon dioxide and carbon isotope ratio of carbon dioxide in interstitial waters of marine sediments. All the distributions show breaks or slope changes in the stippled area, which shows the presumed location of the anaerobic methane oxidation zone. The $\delta^{13}CO_2$ minimum at this same depth suggests injection of isotopically light carbon dioxide by anaerobic methane oxidation.

## References


Barnes, R.O.; Goldberg, E.D. 1976. Methane production and consumption in anoxic marine sediments. Geology, vol. 4, p. 297-300.

Berner, R.A. 1975. Diagenetic models of dissolved species in interstitial waters of compacting sediments. Am. J. Sci., vol. 275- p. 88-96.

Cappenberg, Th.E. 1974. Interrelations between sulphate-reducing and methane-producing bacteria in bottom deposits of a freshwater lake. II. Inhibition experiments. Antonie van Leeuwenhoek, vol. 40, p. 297-306.

Cappenberg, Th.E. 1975. A study of mixed continuous cultures of sulphate-reducing and methane-producing bacteria. Microbial Ecology, vol. 2, p. 60-72.

Jorgenson, B.B.; Fenchel, T. 1974. The sulphur cycle of a marine sediment model system. Mar. Biol., vol. 24, p. 189-201.

Martens, C.S.; Berner, R.A. 1974. Methane production in the interstitial waters of sulphate-depleted marine sediments., Science, vol. 185, p. 1167-9.

Martens, C.S. 1976. Control of methane sediment-water bubble transport by macroinfaunal irrigation in Cape Lookout Bight, North Carolina, Science, vol. 192, p.998-1000.

Quale, J.R. 1972. The metabolism of one-carbon compounds by micro-organisms. Advances in Microbial Physiology, vol. 7, p. 119-203.

Reeburgh, W.S. 1969. Observations of gases in Chesapeake Bay sediments., Limnol. Oceanogr., vol. 14, p. 368-75.

Reeburgh, W.S. 1976. Methane consumption in Cariaco Trench waters and sediments. Earth Planet. Sci. Lett., vol. 28, p. 337-44.

Reeburgh, W.S.; Heggie, D.T. 1976. Microbial methane consumption reactions and their effect on methane distributions in freshwater and marine environments. Limnol. Oceanogr. (in press).

Reeburgh, W.S. 1976. A major sink and flux control for methane in marine sediments: anaerobic consumption. (submitted) In : R.A. Fanning and F.T. Manheim, (eds.), Joint Oceanographic Assembly Benthic Boundary Layer Volume. (in press).

# Some of the chemical consequences of microbially mediated degradation of organic materials in estuarine sediments

Christopher S. Martens *

## Biogeochemical depth zonation in estuarine sediments

Bacteria obtain energy for growth and cell maintenance from a series of dehydrogenation or coupled oxidation-reduction reactions utilizing specific organic or inorganic molecules as electron acceptors for coenzyme reoxidation (e.g. Mechalas, 1974). Many different metabolic types may be present, each in its own ecological niche and depending on factors such as substrate availability, micronutrient supply, trace metal concentrations and physical restrictions imposed by the environment. Organisms which obtain the greatest metabolic energy from existing substrates and electron acceptors are observed to dominate (Claypool and Kaplan, 1974; Mechalas, 1974). The removal of various oxidized chemical species and their subsequent replacement by reduced forms imposes a further selective pressure which may lead to a temporal or spatial succession of different bacterial types.

Two general types of metabolism are generally found: (1) processes utilizing inorganic substances as electron acceptors and (2) fermentative processes in which organic molecules are employed as electron acceptors. In the absence of competition for the same substrates and electron acceptors, both types should occur simultaneously in the same location.

The well-documented sequence of microbially mediated reactions and associated reactants and products in estuarine sediments is schematically represented in Fig. 1. Organic matter deposited in estuarine sediments moves downwards with time through the series of biogeochemical zones whose vertical dimensions change in response to variations in the organic matter composition. Once oxygen is exhausted a sequence of successive anaerobic respiratory proceeses occurs including nitrate, sulphate and bicarbonate reduction. Goldhaber and Kaplan (1975) have compared the free energy yield of reaction between hydrogen (assumed to represent a constant organic carbon source) and the inorganic electron acceptors involved in this sequence (Table 1). Fermentative processes in which complex organic materials are degraded to simple organic acids accompany sulphate and bicarbonate reduction.

## Chemical consequences of biodegradation

Chemical reactants and products of the respiratory processes utilizing the major electron acceptors are summarized briefly in Fig. 1. Following the complete removal of oxygen, nitrate reduction leads to production of $NO_2^-$ and $N_2$. In most estuarine sediments, production of these species has not been observed due to the low initial concentration of $NO_3^-$ and the correspondingly rapid transition from aerobic respiration to sulphate reduction. Vanderborght and Billen (1975) have shown that diffusion and denitrification control nitrate concentration in the organic-rich, muddy sediments of an artificial Belgian lagoon, while nitrification is an additional important process in the sandy sediments. Barnes et al. (1975) have observed a signi-

*Marine Sciences Program, University of North Carolina, Chapel Hill, North Carolina 27514 USA

ficant increase in dissolved $N_2$ in the interstitial waters of surface sediments from the Santa Barbara Basin which result from additional nitrogen sources, including ammonia diffusing from below into the zone of nitrate reduction (Fig. 1). Such reactions may occur in estuarine sediments yet be masked by relatively rapid transport processes and/or occurrence in a relatively thin sediment layer.

Sulphate reduction follows nitrate reduction and, in estuarine sediments and elsewhere, is carried out primarily by bacteria of the genus Desulfovibrio (Goldhaber and Kaplan, 1975). It is important to note that, under laboratory conditions, sulphate reducers are dependent on a sufficient supply of short three-and four-carbon chain carboxylic acids which are generated by other heterotrophic bacteria through hydrolysis and anaerobic fermentation. The primary end product of sulphate reduction is acetate. Sulphate reducers have a hydrogenase system that allows the coupling of sulphate reduction with $H_2$; however, no cell growth can occur in the absence of the carbon-containing substrates (Mechalas, 1974).

Methane production by a symbiotic community of two or more organisms (Toerien and Hattingh, 1969) is the result of stepwise degradation of simple organic compounds, such as ethanol, to $H_2$ and acetic acid followed by the reduction of $CO_2$ :

$$CO_2 + 4\ H_2 \rightarrow CH_4 + 2\ H_2O.$$

In addition to $CO_2$, only acetate and formate appear to be used directly for methane production (Bryant et al., 1967; Goldhaber and Kaplan, 1975; Mechalas, 1974) though the mechanism for direct breakdown of acetate is not fully understood.

Methane does not appear to be produced in significant quantities in marine sediments until sulphate reduction is complete, even though acetate is the end product of sulphate reduction (Martens and Berner, 1974, 1977). Laboratory studies by Cappenberg (1975) indicate that this results from the toxicity of $H_2S$ to methanogenic bacteria. An alternative explanation is that sulphate reducers successfully compete (see Table 1) for $H_2$ utilizing their hydrogenase system as discussed above. It is known that the accumulation of $H_2$ is inhibitory to hydrogen-producing populations (Bryant et al., 1967; Mechalas, 1974). Thus large populations of $H_2$ producers may be initially linked with sulphate reducers and only subsequently support $CO_2$ reduction by methanogenic bacteria following exhaustion of available $SO_4^{2-}$.

During oxidation of organic materials in marine sediments, nitrogen and phosphorus are generally released in proportion to the organic C:N:P ratios (Sholkovitz, 1973; Hartmann et al., 1973). Within or below the zone of sulphate reduction, or under conditions of slow nitrification (e.g. in muddy sediments - see above), the nitrogen builds up as ammonia. Stoichiometric models (Redfield et al., 1963; Richards, 1965), often utilized to model nutrient regeneration utilizing a variety of electron acceptors, may not be directly applicable because of complications arising from preferential release of phosphorus and nitrogen with respect to carbon (e.g. see Sholkovitz, 1973) and adsorption of ammonia and phosphorus (Berner, 1976).

Changes in the stoichiometry of ammonia and phosphorus release into interstitial water during sulphate reduction have been observed at offshore and harbour stations in Long Island Sound by Martens et al. (1977). A comparison of these stations (Table 2) indicates a shoreward increase in the amount of ammonia and phosphorus generated per sulphate reduced probably reflecting (a) more rapid deposition (and thus less preferential stripping of N and P from organic matter prior to burial in harbour stations) and/or (b) selective stripping of N and P compounds in the heavily bioturbated upper 10 cm of sediment at offshore locations. Adsorption has also been shown to be an important control for ammonia in these sediments (Rosenfeld and Berner, 1976).

Rates of bacterial degradation

Clear evidence exists that the rate of degradation of organic material is dependent on both the amount present and its chemical nature (Berner, 1970; Sweeney and Kaplan, 1973; Goldhaber and Kaplan 1975). Associated factors include the state of complexing of the organic material, the environments of deposition and decomposition and the rate of sedimentation. Additionally, temperature ranges typical of estuarine sediments (5-25°C) have been shown to have a three- to five-fold effect on rates of sulphate reduction (Kaplan and Rittenberg, 1964; Goldhaber et al., 1977).

Indirect relationships between the amount of organic matter and degradation rates are provided through correlation with sedimentary pyrite content (Berner, 1970; Goldhaber and Kaplan, 1975; Sweeney and Kaplan, 1973). Correlations between the amount of soluble organic carbon and its degradation rate (Ramm and Bella, 1974) and changes in rates associated with the electron donor (Kaplan and Rittenberg, 1964) directly indicate the importance of the nature of the organic material. Relatively rapid complexation of organic matter in recent sediment to humics and kerogen (Nissenbaum and Kaplan, 1972) render it less susceptible to bacterial degradation, invalidating direct application of models based solely on the amount of organic material to all marine environments. Degrading cellular materials apparently go through a series of reactions leading to fulvic, humic, and finally kerogen substances.

In nearshore environments on the basis of stable carbon isotope studies (Nissenbaum and Kaplan, 1972; Schultz and Calder, 1976) and amino-acid and sugar distributions (Degens and Mopper, 1975) it has been demonstrated that there is a rapid spatial transition (within a few kilometres and near river mouths) from terrestrial lignin-derived substances to plankton-derived material with the former polyaromatic substances (Ghessemi and Christman, 1968) being less susceptible to bacterial degradation (Degens and Mopper, 1975). Over an 18 km transect in the White Oak (North Carolina, U.S.A.) river estuary, Martens and Goldhaber (1977) have been able to demonstrate the probable influence of such a transition in sedimentary organic materials on generation rates of $\Sigma CO_2$, nutrients and methane (Figs. 2-4).

Effects of deposition and accumulation on rates of degradation of organic materials in anoxic sediments have been illustrated by Zsolnay (1971) in oxic and anoxic Baltic basins and by Sholkovitz (1973) through comparisons of slope and bottom basin sediments of the Santa Barbara Basin.

Rates of degradation of sedimentary organic materials utilizing various electron acceptors have been quantified through both direct measurements (e.g. decreases in concentrations of electron acceptors with time in laboratory experiments) and through kinetic models (Berner, 1964, 1974; Goldhaber and Kaplan, 1975). The kinetic models have dealt primarily with sulphate reduction (see Goldhaber and Kaplan, 1975) and are thus of considerable interest for estuarine sediments. A range in rates of more than 3 orders of magnitude were observed in nearshore environments.

Observed vertical variations in rates in estuarine sediments appear to be controlled by (1) the nature of available substrates, (2) the concentrations of electron acceptors and nutrients, and (3) the size and density of bacterial populations. Observed vertical variations in a given environment can thus be envisaged as resulting from depletion of substrate and thus decreases in numbers with depth (e.g. Oppenheimer, 1960; Sorokin, 1962) and possibly changes in the mode of respiration (Claypool and Kaplan, 1974). Observed profiles of $\Sigma CO_2$ in the White Oak (Fig. 3) do not appear to indicate a rate change and switch to methanogenesis upon depletion of $SO_4^{2-}$ (Fig. 2).

The value of kinetic models in determining both the nature and rates of chemical reactions resulting from degradation of organic matter can be illustrated for methane consumption during sulphate reduction in marine sediments. This is undetectable over short periods of time in laboratory experiments (Martens and Berner, 1974). However, Barnes and Goldberg (1976) and Reeburgh (1976) demonstrate that concave-up distributions of methane in Santa Barbara Basin and Cariaco Trench sediments could be explained only through such consumption. A similar methane profile in Long Island Sound sediments has been modeled by Martens and Berner (1977) utilizing the following steady-state kinetic model which include the influences of diffusion, consumption and compaction :

$$D_s \frac{\partial^2 c}{\partial z^2} - w \frac{\partial c}{\partial z} - k_1 c = 0.$$

The methane profile could not be fit by zero order consumption but could be well described with a $k_1 = 8 \times 10^{-9}$ $sec^{-1}$ (see Fig. 5) and the predicted slow rate of methane consumption in these sediments would not be observed over ordinary laboratory time scales. Recently 420 day laboratory experiment with anoxic sediments collected from Cape Lookout Bight, North Carolina, Martens and Sansone (1977) have observed decreases in methane concentration during sulphate reduction which are well described by a first order model with a $k_1$ of approximately $1 \times 10^{-7}$ $sec^{-1}$.

## Influence of physical mixing and molecular diffusion

Chemical distributions and rates of processes in estuarine sediments resulting from the degradation processes discussed above can be significantly modified by physical mixing processes, such as current stirring (e.g. Hammond et al., 1975) and biological irrigation (e.g. Goldhaber et al., 1977), and by molecular diffusion below (or seasonally, both in and below) the mixing zone. Sulphate and nutrient concentrations in the interstitial waters of the upper 10-20 cm of Long Island Sound sediments are controlled by molecular diffusion in winter months and by irrigation during the summer and late fall months (Goldhaber et al., 1977). In Cape Lookout Bight, N.C., irrigating macrofauna indirectly control methane production and subsequent bubble ebullition by pumping sulphate-rich overlying waters into surface sediments (Martens, 1976). In the bight interior where macrofauna are absent and sulphate is exhausted in the upper 10-15 cm, direct methane bubble transport to overlying waters occurs. Near the bight entrance where extensive irrigation takes place, linear vertical sulphate concentration profiles occur in spite of active sulphate reduction as indicated by $H_2S$ and black iron monosulphides, with low resultant methane concentrations.

It is clear that the chemistry of estuarine sediments can be controlled directly by both transport and redistribution of chemical species, and indirectly through transport of electron acceptors such as $O_2$ and $SO_4^{2-}$, thereby altering the vertical distribution of the microbially mediated chemical zones discussed above.

## Acknowledgements

Support provided by National Science Foundation Oceanography Section Grant DES 75-06199 is acknowledged. The writer has been particularly benefited from discussions with Martin B. Goldhaber of the U.S. Geological Survey, Denver, who also reviewed the manuscript.

Table 1. Free energy of reaction between hydrogen and various inorganic electron acceptors (from Goldhaber and Kaplan, 1975).

| Reaction | $\Delta G_r°$ Kcal. $mole^{-1}$ |
|---|---|
| $3\ O_2 + 6\ H_2 \rightarrow 6\ H_2O$ | -340.2 |
| $2\ NO_3^- + 6\ H_2 \rightarrow 6\ H_2O + N_2$ | -287.4 |
| $1.5\ SO_4^{2-} + 6\ H_2 \rightarrow 4.5\ H_2O + 1.5\ HS^- + 1.5\ OH^-$ | - 42.1 |
| $1.5\ HCO_3^- + 6\ H_2 \rightarrow 3\ H_2O + 1.5\ CH_4 + 1.5\ OH^-$ | - 34.0 |

Table 2. Ratio of regenerated sulphate, ammonia and reactive phosphate ($\Delta SO_4$: $\Delta NH_4$:$\Delta \Sigma P$) in interstitial waters of nearshore Long Island Sound surface sediments (from Martens _et al._, 1977).

| Location | No. Cores | mean $\Delta SO_4:\Delta NH_4:\Delta \Sigma P$ |
|---|---|---|
| Station TH (2 km offshore) | 6 | -53:4.8:0.37 |
| Station BS (harbour) | 2 | -53:19:3.7 |
| Richards (1965) stoichiometric model | - | 53:16:1 |

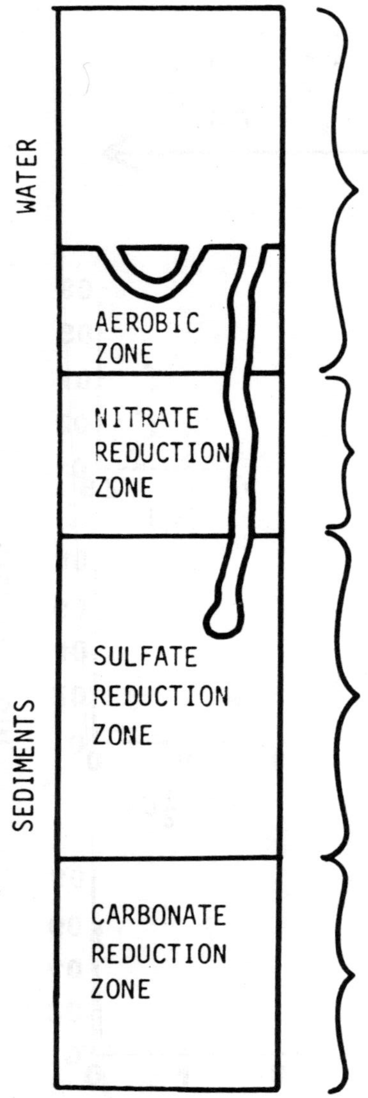

WATER

SEDIMENTS

AEROBIC ZONE

$$(CH_2O)_{106}(NH_4)_{16}H_3PO_4 + 106\ O_2 \rightarrow 106\ CO_2 + 16\ NH_3 + H_3PO_4 + 106\ H_2O$$

$$NH_4^+ + 1.5\ O_2 \rightarrow NO_2^- + H_2O + 2\ H^+$$

$$NO_2^- + 0.5\ O_2 \rightarrow NO_3^-$$

$$CH_4^+ + 2\ O_2 \rightarrow 2\ H_2O$$

NITRATE REDUCTION ZONE

$$(CH_2O)_{106}(NH_3)_{16}H_3PO_4 + 84.8\ NO_3^- \rightarrow 106\ CO_2 + 42.4\ N_2 + 16\ NH_3 + H_3PO_4 + 148.4\ H_2O$$

$$5\ NH_4^+ + 3\ NO_3^- \rightarrow 4\ N_2 + 9\ H_2O + 2\ H^+$$

SULFATE REDUCTION ZONE

$$(CH_2O)_{106}(NH_3)_{16}\ H_3PO_4 + 53\ SO_4^{2-} \rightarrow 106\ CO_2 + 53\ S^{2-} + 16\ NH_3 + H_3PO_4 + 106\ H_2O$$

$$CH_4 + SO_4^{2-} \rightarrow HCO_3^- + HS^- + H_2O$$

$$2\ CH_3CHOHCOOH + SO_4^{2-} \rightarrow 2\ CH_3COOH + 2\ HCO_3^- + H_2S$$

CARBONATE REDUCTION ZONE

$$CH_3COOH \rightarrow CH_4 + CO_2$$

$$CO_2 + 4\ H_2 \rightarrow CH_4 + 2\ H_2O$$

Fig. 1 Observed sequence of microbially mediated reactions in estuarine sediments including stoichiometric decomposition equations.

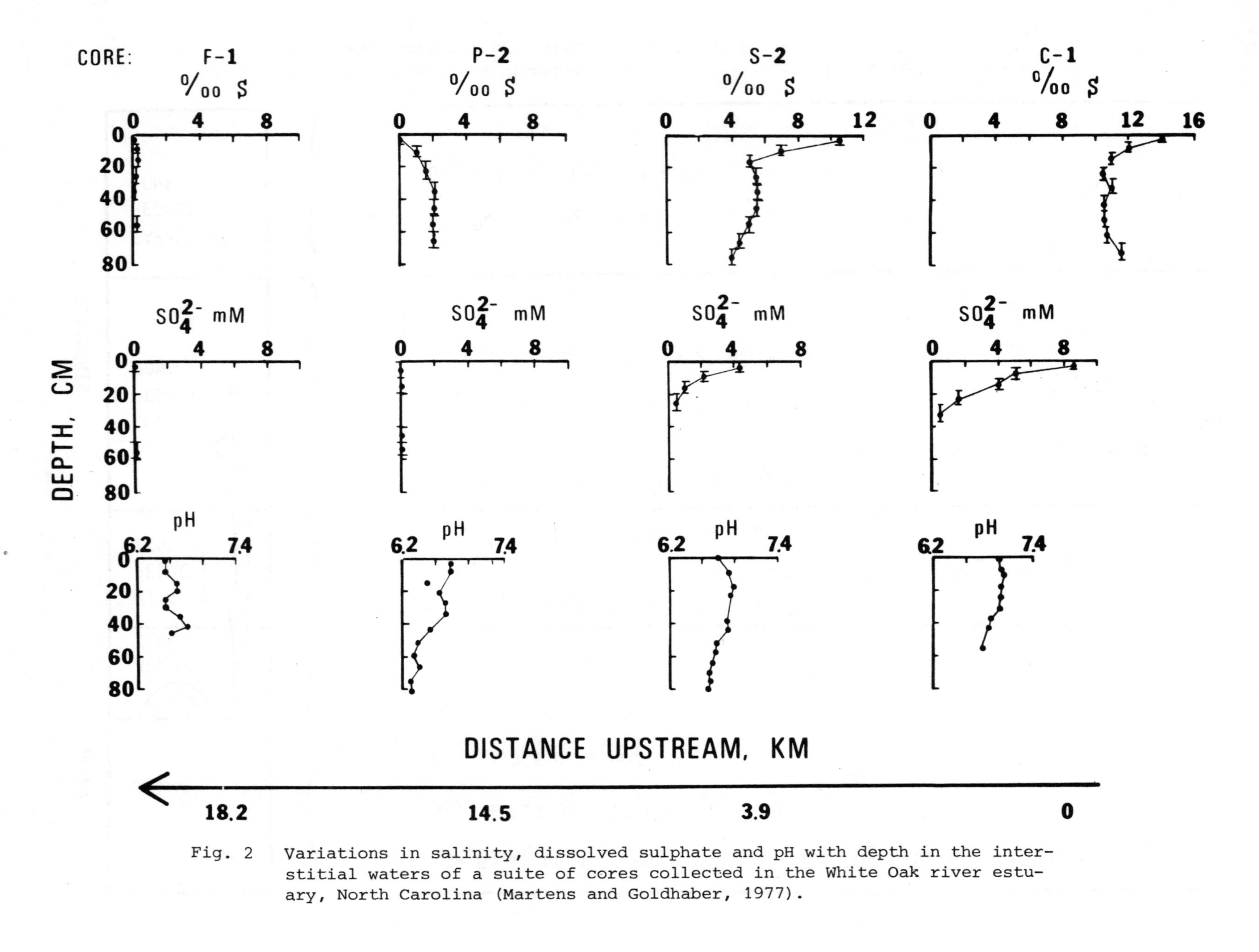


Fig. 2 Variations in salinity, dissolved sulphate and pH with depth in the interstitial waters of a suite of cores collected in the White Oak river estuary, North Carolina (Martens and Goldhaber, 1977).

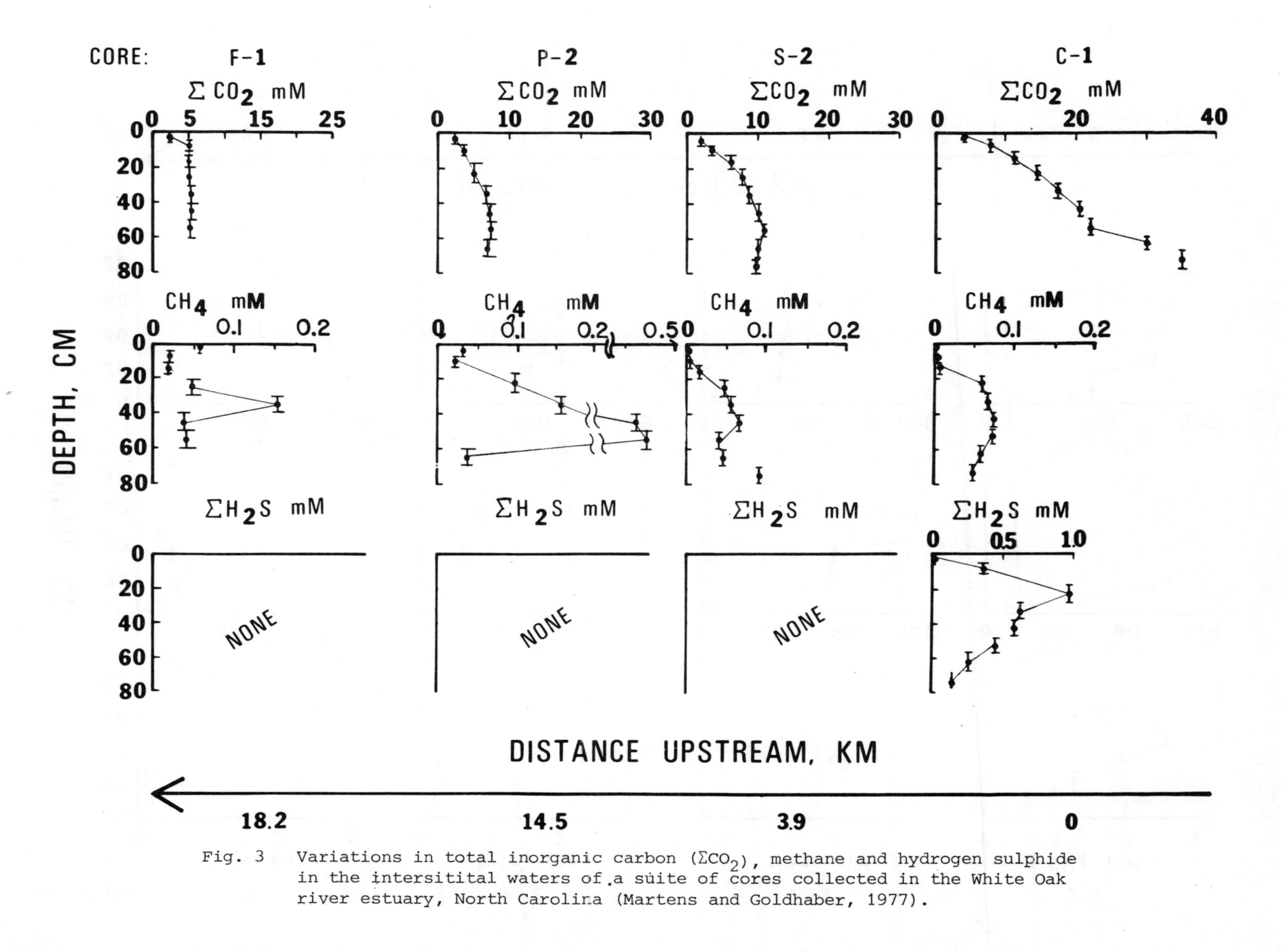


Fig. 3 Variations in total inorganic carbon ($\Sigma CO_2$), methane and hydrogen sulphide in the intersititial waters of a suite of cores collected in the White Oak river estuary, North Carolina (Martens and Goldhaber, 1977).

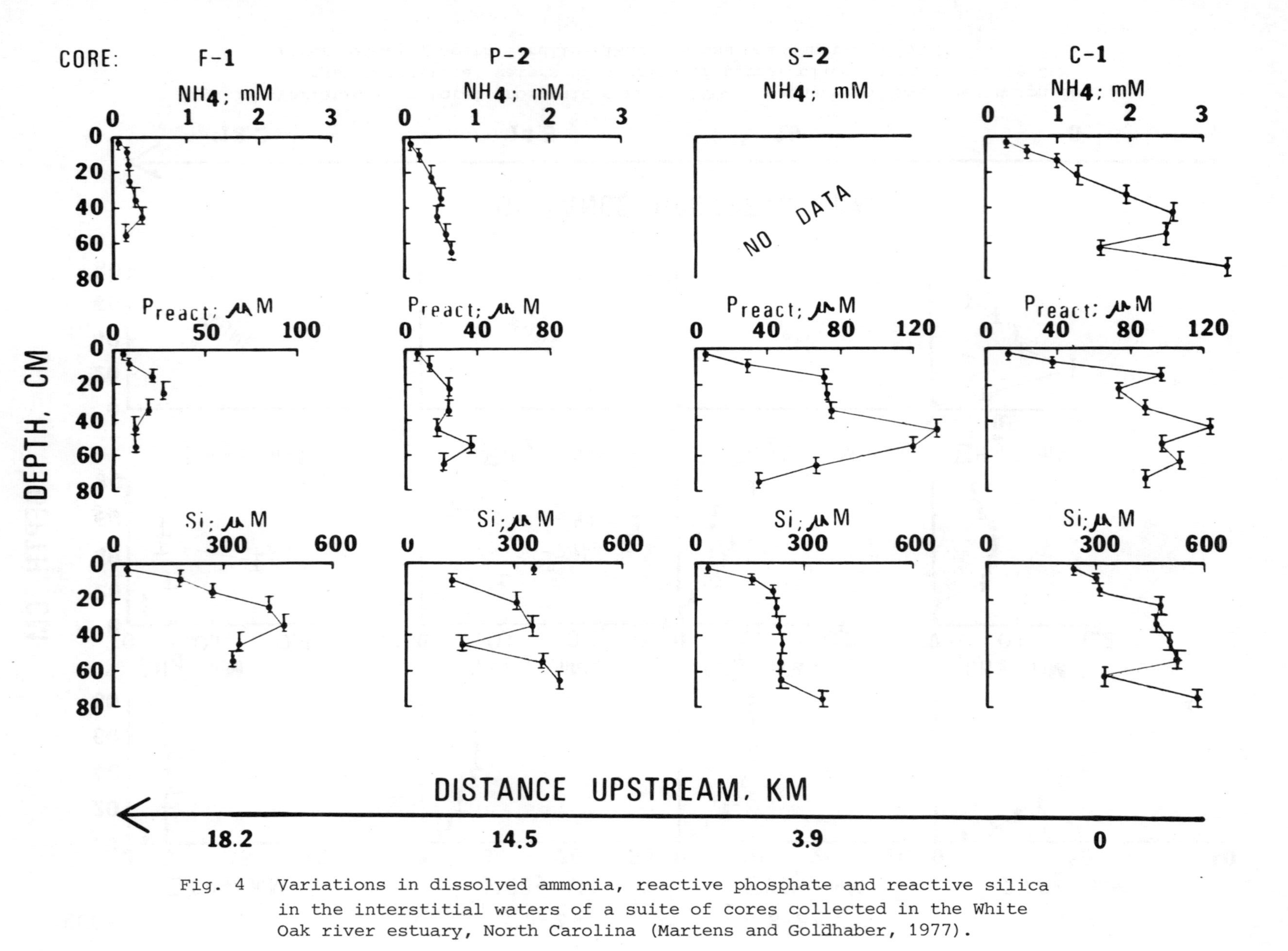


Fig. 4 Variations in dissolved ammonia, reactive phosphate and reactive silica in the interstitial waters of a suite of cores collected in the White Oak river estuary, North Carolina (Martens and Goldhaber, 1977).

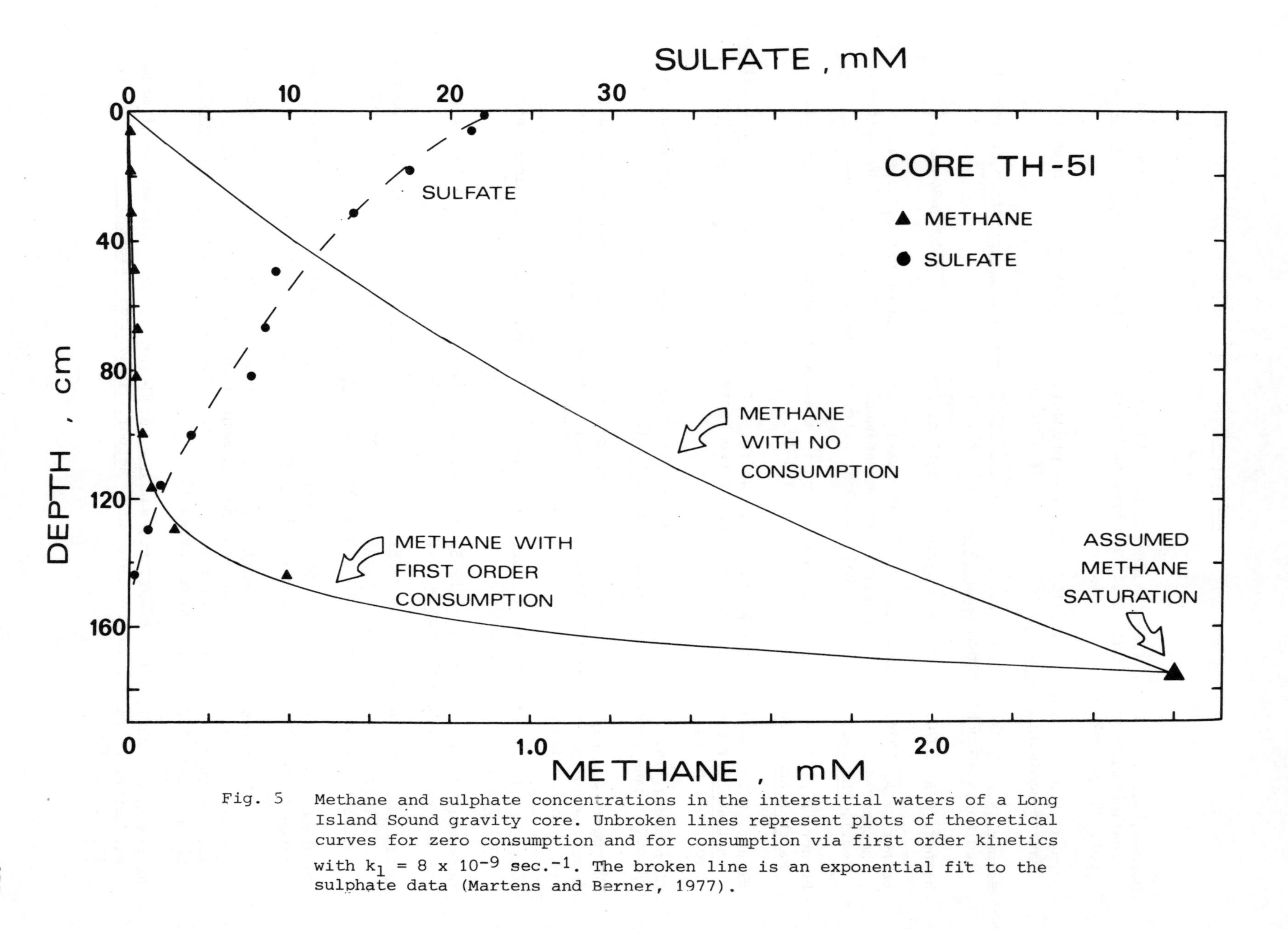


Fig. 5 Methane and sulphate concentrations in the interstitial waters of a Long Island Sound gravity core. Unbroken lines represent plots of theoretical curves for zero consumption and for consumption via first order kinetics with $k_1 = 8 \times 10^{-9}$ sec.$^{-1}$. The broken line is an exponential fit to the sulphate data (Martens and Berner, 1977).

## References

Barnes, R.O.; Bertine, K.K.; Goldberg, E.D. 1975. $N_2$:Ar, nitrification and denitrification in Southern California borderland basin sediments. Limnol. oceanogr., vol. 20, p. 962-70.

Barnes, R.O.; Goldberg, E.D. 1976. Methane production and consumption in anoxic marine sediments. Geology, vol. 4, p. 297-300.

Berner, R.A. 1964. An idealized model of dissolved sulphate distribution in recent sediments. Geochim. Cosmochim. Acta, vol. 28, p. 1497-503.

Berner, R.A. 1970. Sedimentary pyrite formation. Am. J. Sci., vol. 268, p. 1-23.

Berner, R.A. 1974. Kinetic models for the early diagenesis of nitrogen, sulphur, phosphorus and silicon in anoxic marine sediments, p. 347-61. In : E.D. Goldberg (ed.), The Sea, vol. 5, New York, London, Wiley and Sons.

Berner, R.A. 1976. Inclusion of adsorption in the modelling of early diagenesis. Earth Planet. Sci. Lett., vol. 29, p. 333-40.

Bryant, M.P.; Wolin, E.A.; Wolin, M.J.; Wolfe, R.S.; Mandel, M. 1967. Methanobacillus omelianskii, a symbiotic association of two bacterial species. Bacteriol. Proc., vol. 67, p. 19.

Cappenberg, Th.E. 1975. A study of mixed continuous cultures of sulphate reducing and methane producing bacteria. Microb. Ecol., vol. 2, p. 60-72.

Claypool, G.; Kaplan, I.R. 1974. The origin and distribution of methane in marine sediments, p. 99-139. In : I.R. Kaplan (ed.), Natural Gases in Marine Sediments. New York, Plenum.

Degens, E.T.; Mopper, K. 1975. Early diagenesis of organic matter in marine soils. Soil Sci., vol. 119, p. 65-72.

Ghessemi, M.; Christman, R.F. 1968. Properties of the yellow organic acids of natural waters., Limnol. oceanogr., vol. 13, p. 585-97.

Goldhaber, M.B.; Aller, R.C.; Cochran, J.K.; Rosenfeld, J.K.; Martens, C.S.; Berner, R.A. 1977. Sulphate reduction, diffusion, and bioturbation in Long Island Sound sediments: Report of the FOAM group. Amer. J. Sci., vol. 277, p. 193-237.

Goldhaber, M.B.; Kaplan, I.R. 1975. Controls and consequences of sulphate reduction rates in recent marine sediments. Soil Sci., vol. 119, p. 42-5.

Hammond, D.E.; Simpson, H.J.; Mathieu, G. 1975. Methane and Radon-222 as tracers for mechanisms of exchange across the sediment-water interface in the Hudson River estuary, p. 119-32. In : T. Church (ed.), Marine Chemistry in the Coastal Environment. A.C.S. Symposium Series 18.

Hartmann, M.; Muller, P.; Suess, E.; van der Weijden, C.H. 1973. Oxidation of organic matter in recent marine sediments. Meteor. Forschungs. Ergebnisse, Reihe C., 12, p. 74-86.

Kaplan, I.R.; Rittenberg, S.C. 1964. Microbiological fractionation of sulphur isotopes. J. Gen. Microbiol., vol. 34, p. 195-212.

Martens, C.S. 1976. Control of methane sediment-water bubble transport by macroinfauna irrigation in Cape Lookout Bight, North Carolina. Science, vol. 192, p.998-1000.

Martens, C.S.; Berner, R.A. 1974. Methane production in the interstitial waters of sulphate depleted marine sediments. Science, vol. 185, p. 1167-9.

Martens, C.S.; Berner, R.A. 1977. Interstitial water chemistry of anoxic Long Island Sound sediments: 1. Dissolved gases. Limnol. oceanogr., vol. 22, p.10-25.

Martens, C.S. Berner, R.A.; Rosenfeld, J.K. 1977. Interstitial water chemistry of anoxic Long Island Sound sediments: 2. Nutrient regeneration and phosphate removal. Limnol. oceanogr.,

Martens, C.S.; Goldhaber, M.B. 1977. Early diagenesis in transitional sedimentary environments of the White Oak river-estuary, North Carolina. Limnol. oceanogr. (in press).

Martens, C.S.; Sansone, F.J. 1977. A methane budget for Cape Lookout Bight, North Carolina. (in preparation).

Mechalas, B.J. 1974. Pathways and environmental requirements for biogenic gas production in the ocean, p. 11-25. In : I.R. Kaplan (ed.), Natural Gases in Marine sediments., New York, Plenum.

Nissenbaum, A.; Kaplan, I.R. 1972. Chemical and isotopic evidence for the in situ origin of marine humic substances. Limnol. oceanogr., vol. 17, p. 570-82.

Nissenbaum, A.; Presley, B.J.; Kaplan, I.R. 1972. Early diagenesis in a reducing fjord, Saanich Inlet, British Columbia-I. Chemical and isotopic changes in major components of interstitial water. Geochim. Cosmochim. Acta, vol. 36, p. 1007-27

Oppenheimer, C.H. 1960. Bacterial activity in sediments of shallow marine bays. Geochim. Cosmochim. Acta, vol. 19, p. 244-60.

Ramm, A.E.; Bella, D.A. 1974. Sulphide production in anaerobic microcosms. Limnol. oceanogr., vol. 19, p. 110-260.

Redfield, A.C. Ketchum, B.J.; Richards, F.A. 1963. The influence of organisms on the consumption of sea water, p. 26-77. In : M.N. Hill (ed.), The Sea, vol. 2, New York, London, Wiley-Interscience.

Reeburgh, W.S. 1976. Methane consumption in Cariaco Trench waters and sediments. Earth Planet. Sci. Lett., vol. 15, p. 337-44.

Richards, F.A. 1965. Anoxic basins and fjords, p. 611-45. In : J.P. Riley and G. Skirrow (eds.), Chemical Oceanography, vol. 1, New York, London, Acadamic Press.

Rosenfeld, J.K.; Berner, R.A. 1976. Ammonia adsorption in nearshore anoxic sediments. Proc. Geol. Sic. Am. (Abstract).

Schultz, D.J.; Calder, J.A. 1976. Organic carbon $^{13}C/^{12}C$ variations in estuarine sediments. Geochim. Cosmochim. Acta, vol. 40, p. 381-5.

Sholkovitz, E. 1973. Interstitial water chemistry of the Santa Barbara Basin sediments. Geochim. Cosmochim. Acta, vol. 37, p. 2043-73.

Sorokin, Yu.I. 1962. Experimental investigations of bacterial sulphate reduction in the Black Sea using $S^{35}$. Mikrobiologiya, vol. 31, p. 402-10.

Sweeney, R.E.; Kaplan I.R. 1973. Pyrite framboid formation: Laboratory synthesis and marine sediments. Econ. Geol., vol. 68, p. 618-34.

Torien, D.F.; Hattingh, W.H.J. 1969. Anaerobic digestion I. The microbiology of anaerobic digestion. Water Res., vol. 3, p. 385-416.

Vanderborght, J-P.; Billen, G. 1975. Vertical distribution of nitrate concentration in interstitial water of marine sediments with nitrification and denitrification. Limnol. oceanogr., vol. 20, p. 953-61.

Zsolnay, V.A. 1971. Diagenesis as a function of redox conditions in nature: A comparative survey of certain organic and inorganic compounds in an oxic and anoxic Baltic basin. Kieler Meersforschungen, vol. 27, p. 135-65.

# Biogeochemical properties of an estuarine system: the River Scheldt

R. Wollast and J. J. Peters*

## Introduction

The hydrographic basin of the Scheldt estuary (Fig. 1) covers one of the most heavily populated regions of Europe and supports highly diversified industrial activity. Since most of the waste discharges are uncontrolled, large amounts of domestic and industrial wastes are carried by the river. The tidal range varies along the estuary from 3.7 m at the mouth, increasing to 5 m near Antwerp, and decreasing to 2 m in Gent. The mean river discharge amount to 120 $m^3$/sec at the mouth, or 5 x $10^6$ $m^3$ during one tidal period, while the volume of seawater flowing up the estuary during the flood tide is about $10^9$ $m^3$.

The Scheldt may be considered a well mixed estuary with only small and local vertical salinity gradients. The mixing zone of fresh and salt water extends over a distance of 70 to 100 km. A comparison of the river water composition at the head to that of the brackish water at the mouth of the estuary reveals that important physical, chemical and biological processes occurring in the mixing zone strikingly modify the transport of pollutants out to the sea.

This intense sedimentation, typical of slightly stratified estuaries, is one of the most effective processes by which pollutants are removed from the surface water. On the other hand, the intense activity of heterotrophic bacteria, which is related to the high organic load, is responsible for overall control of the oxidation-reduction conditions which, in turn, affect the speciation and behaviour of various elements.

Finally, the downstream zone of the estuary supports intense photosynthetic activity during spring and summer, characteristic of the eutrophication of that region. This phenomenon affects mainly the distribution of nutrients and is responsible for the synthesis of large amounts of organic matter.

We present here briefly the main processes affecting the behaviour of the elements in the estuarine zone of the Scheldt and their influence on the mass-transfer, the accumulation and the transformation of some typical elements.

## Mechanisms of mud deposition and the accumulation of pollutants in sediments

One of the most important characteristics for the transport of pollutants in an estuarine system is the usually long residence time of the freshwater masses. In the case of the Scheldt estuary, the average cross-sectional ebb or flood currents are about 0.7 m/sec, with instantaneous maximum values reaching 1.5 m/sec. However,

* Université de Bruxelles, 50 ave. F. Roosevelt, 1050 Brussels, Belgium

the residual currents averaged over one complete tidal cycle in a cross-section drop from 0.08 m/sec at 100 km, to 0.01 m/sec at 50 km. The total residence time in the brackish water zone, which extends over 100 km, is between two and three months. From an environmental standpoint, this implies a high accumulation of persistent pollutants and intense modifications of the chemically or biologically active substances in the estuarine region. On the other hand, mixing of fresh and salt water induces complicated water movements and influences the physico-chemical behaviour of both suspended and dissolved species.

The measurements of vertical profiles of salinity, temperature, and currents, permit the distinction of two zones with different hydrodynamic characteristics. The lower one, extending from the sea to 50 km, is characterized by well-defined flood and ebb channels which contribute to the intense mixing; vertical stratification is generally small. The upper zone extending from 50 km to the freshwater zone 100 km is characterized by a single and narrower channel; the vertical stratification is greater, chiefly in the area of the harbour of Antwerp, where average salinity gradients of about 0.2‰ $m^{-1}$ are observed. Despite their relatively low values, these salinity gradients sufficiently influence the vertical distribution of the currents to modify markedly the residual currents averaged over one complete tidal cycle. The density currents slacken the ebb movement and accelerate the flood movement near the bottom. A reverse effect in the surface layer compensates for this bottom movement. Consequently, the residual currents near the bottom are orientated upstream in the lower zone. They are, however, orientated downstream in the freshwater zone and the two opposed movements cancel out in the area of the harbour of Antwerp, which is thus a highly favourable zone for the accumulation of sediments.

The existence of a vertical gradient of turbidity, associated with these water movements, creates a zone of maximum turbidity, which is well demonstrated in the case of the Scheldt (Fig. 2). This region also corresponds roughly to the transition from fresh to brackish water. The suspended matter transported by the river water is mainly composed of colloidal particles which flocculate as the salinity increases. Laboratory experiments carried out with suspended matter of the Scheldt show that an intense flocculation occurs as soon as the salinity reaches 2‰, and is completed at a salinity of 5‰. A further salinity increase produces larger flocs but with lower densities and, finally, lower sedimentation velocities. The optimum values of salinity for flocculation occur in the zones favourable to sedimentation and accumulation, leading to the intense shoaling of mud in a restricted area (Fig. 3). The influence of salinity on the removal of suspended matter from the surface waters is well demonstrated in Fig. 4. The restricted accumulation of mud in the region of the harbour of Antwerp is also depicted in Fig. 5, which shows the organic content of the sediment along the estuary. The high values near Terneuzen are due to a local input of highly polluted waters transported by the Terneuzen Canal (see Fig. 1).

Taking into account the physical characteristics of the Scheldt, the estuary was divided into two zones : an upper one from 100 km to 55 km and a lower one from 55 km to the mouth. Four times a year at fixed stations situated at the boundaries of these regions, hourly samples for 5 days were taken at three different depths and continuous measurements of the profile of the currents along a vertical plane were taken. A longitudinal monthly survey was also carried out, following the low tide from the mouth to 130 km. Approximately 50 surface samples were collected during each survey. Observations over three years enabled us to estimate annual mass balances of input, transport and accumulation by sedimentation of various pollutants in the two estuarine regions.

The mass balances were established for each compartment by considering the net flow due to river discharge for both suspended and dissolved compounds, their longitudinal turbulent dispersion (estimated from the salinity profile), their sedimentation process and their lateral input from tributaries and sewers.

In Fig. 6 is shown the mass balance obtained for suspended solids. From a total load of $1.52 \times 10^6$ tons/yr discharged in the first zone, $1.2 \times 10^6$ tons/yr are deposited in this upper compartment and $3.2 \times 10^5$ tons/y are transported to the second, where the sedimentation is much less, and finally only $1.2 \times 10^5$ tons/y reach the North Sea. Coarse sand is also transported upwards by the strong bottom currents. The contribution of this process was evaluated by comparing the chemical composition of suspended matter carried by freshwater and by the bottom currents to the sediments deposited in that region. The estimation of $2 \times 10^6$ tons/yr of solid deposited in the upper region is in close agreement with an estimation of the shoaling in this region, which amounts to $10^7$ $m^3$/yr of mud (with an 80 per cent water content by weight).

As one may expect, pollutants discharged in the river strongly affect the composition of suspended matter, especially organic compounds and heavy metals, such as Pb, Zn and Cu. The accumulation of sediments in the upper zone acts thus as a very efficient removal process which prevents further transport of the sediments out to the sea.

## The evolution of organic matter and nutrients

The evolution of the organic matter in the Scheldt may be easily represented schematically by the longitudinal profile of the permanganate "oxidisability" of the water (Fig. 7). During the winter, only a rapid removal of the organic matter is observed, mainly in the upper part of the estuary, due to sedimentation and microbial degradation. During spring and summer, reaching a maximum in June, there is an increase of organic matter in the lower part of the estuary, due to an intense primary production which may reach 250 mg C/$m^3$ per day. The mean annual primary productivity may be estimated at 180 g C/$m^2$/yr for the lower part of the estuary (between 0 and 60 km).

The rate of biodegradation of organic matter also strongly varies seasonally, depending particularly on the temperature of the water. The heterotrophic activity measured by $^{14}C$-labelled bicarbonate incorporation in the dark indicates that the rate of biodegradation varies from 10 to 250 mg C/$m^3$/hr in the upper part of the estuary.

In the lower part however, there is no significant increase of the heterotrophic activity during the summer, which suggests that the zooplankton plays a more important role in the turnover of the organic matter.

The evaluation of the mass balance in the upper zone, given in Fig. 8, indicates that the removal of organic matter by sedimentation and biodegradation is significant; only 20 per cent of the total organic input is transferred downwards.

The intense activity of heterotrophic bacteria is responsible for the existence of a large anaerobic region in the estuary, especially during the summer (see Fig. 2). The re-aeration in the upper zone, which was estimated at around $60 \times 10^3$ tons $O_2$/yr is insufficient to provide the oxidants necessary to account for the bacteriological activity. Other oxidants like $NO_3^-$, $MnO_2$, $Fe_2O_3$, and $SO_4^{2-}$ can be used, and a complete oxidation-reduction budget is necessary to describe correctly the evolution of the chemical composition of water under microbiological influences (Billen *et al.*, 1976).

In the case of the upper Scheldt estuary, the degradation of the organic matter in the freshly deposited sediments is essentially related to sulphate reduction, producing measurable amounts of iron sulphides (greigite, pyrite). A tentative mass-balance for oxygen is given in Fig. 8. It does not include the oxidation-reduction reactions for nitrogen, iron and manganese which are, however, important for description of the behaviour of these elements.

In the anaerobic zone $Mn^{4+}$, $NO_3^-$ and $Fe^{3+}$ are successively reduced, and in the downward zone, they are regenerated in the reverse order as the redox potential increases (Fig. 9).

As one might expect, the activity of autotrophic and heterotrophic organisms also influences markedly the longitudinal profile of nutrients in the estuary. Typical distributions of the dissolved species of N, P and Si, which, in turn affect the input of nutrients into the North Sea, are shown in Fig. 10.

The behaviour of silica is of particular interest. From May to early October, dissolved silica carried by freshwater is rapidly consumed in a limited region situated between 10 and 50 km. During the winter, dissolved silica behaves conservately. The seasonal variation suggests that this phenomenon is related to the activity of the diatoms which abound in the estuarine waters. The restricted area of their activity may be explained by the fact that they become active only when the turbidity is low enough to permit photosynthesis to occur. This hypothesis is confirmed by primary productivity measurements.

The large amounts of nitrogen and phosphorus discharged in the Scheldt persist in the lower part of the estuary, which may be considered as partially eutrophied. The supply of dissolved silica to the North Sea is practically nil during the summer, 95 per cent of the dissolved silica carried by the river water being consumed in the estuary itself. The same phenomenon occurs in the Rhine (Van Bennekom et al., 1974), and the North Sea is actually deprived of an important source of a major nutrient, here too in contrast with nitrogen and phosphorus, which are discharged in large quantities. This disequilibrium between the nutrients may affect the planktonic populations of the North Sea.

## The longitudinal profile of heavy metal concentrations

A large fraction of heavy metals are introduced in the river as solid compounds or are rapidly precipitated if discharged in a soluble form. The intense deposition of sediments in the upper zone thus again constitutes an efficient removal process. It should also be noted that the ratio of solute-to-suspended matter increases for heavy metals as salinity increases. A careful investigation presently in progress of the longitudinal concentration profiles indicates that copper (Fig. 11) and, to a lesser extent Zn, dissolve or become desorbed as the salinity increases. This may possibly be related to the redox potential and the presence of even minute levels of $H_2S$ in the anaerobic zone. The degradation of organic matter on the other hand, may release complexed or adsorbed heavy metals.

Finally, the uptake of these elements by living organisms may play an important role in the transfer from the dissolved to particulate states. More information however, is needed on the importance of these various mechanisms in order to estimate their relative importance. Tentative mass-balances for copper, zinc and lead in the Scheldt estuary are given in Fig. 12.

---

This paper is a contribution to the Belgian Program of Research and Development in the Physical and Biological Environment, sponsored by the Dept. of Science Policy, Office of the Prime Minister, Belgium.

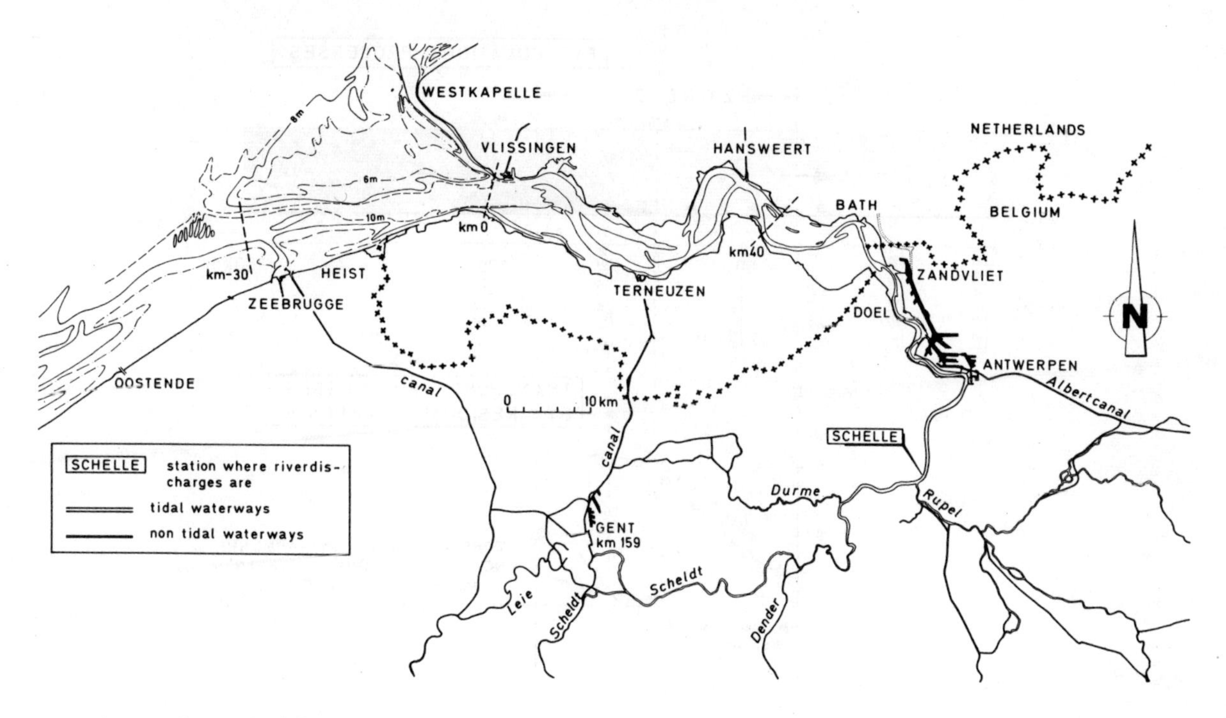


Fig. 1. The Scheldt estuary.

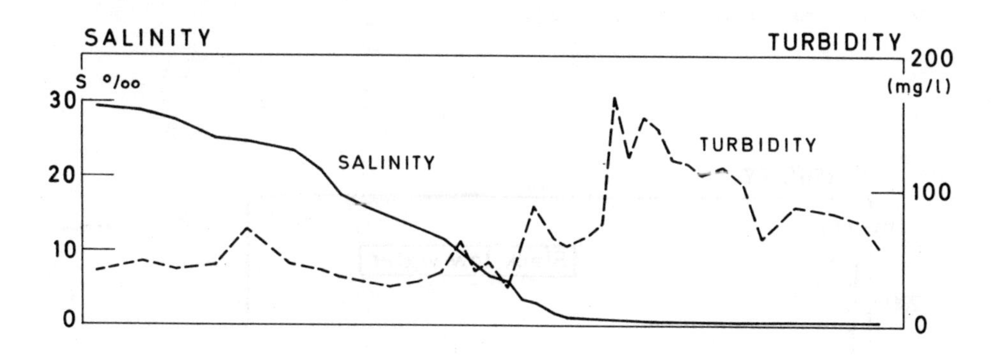


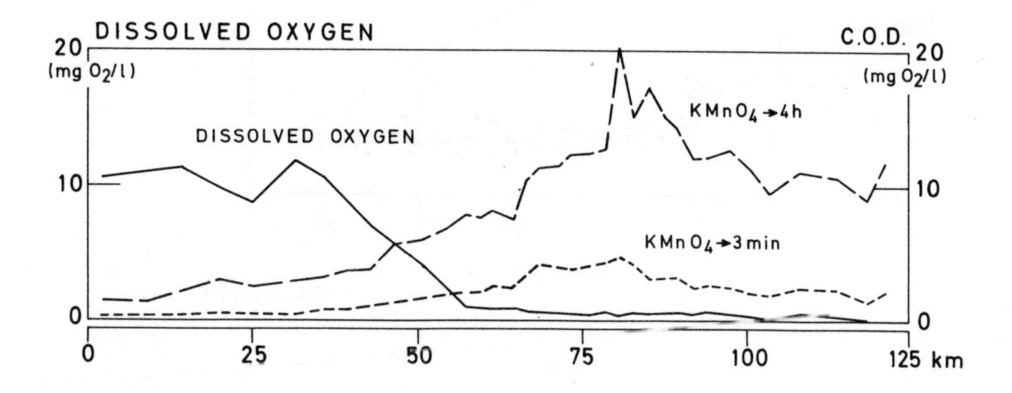


Fig. 2. Longitudinal profile of salinity, turbidity, chemical oxygen demand and dissolved oxygen content (January, 1973).

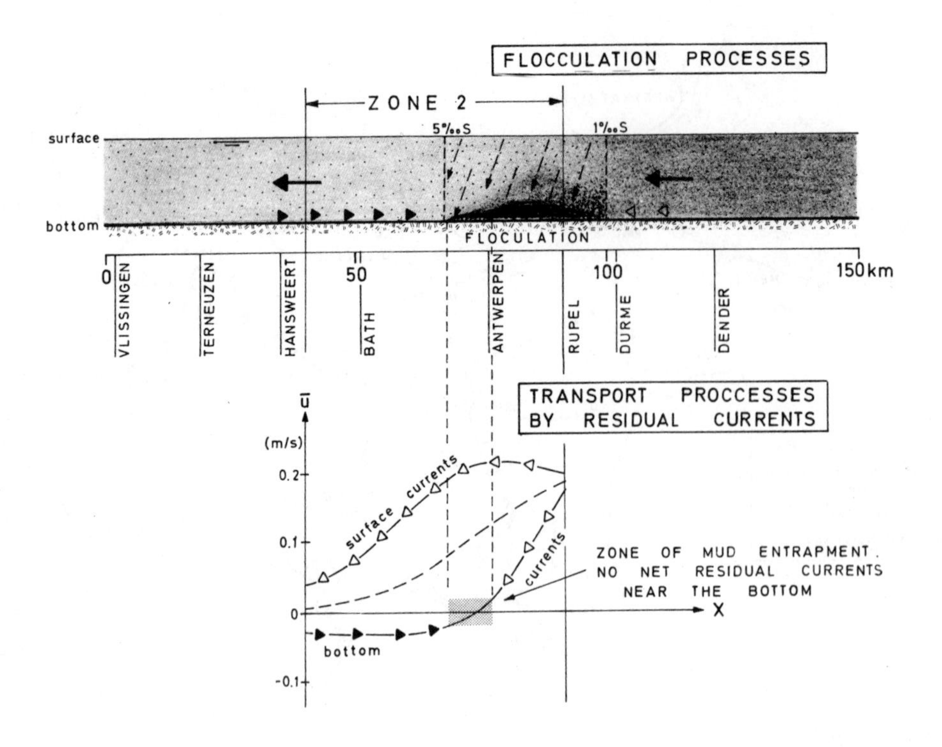


Fig. 3. Mechanisms of mud deposition.

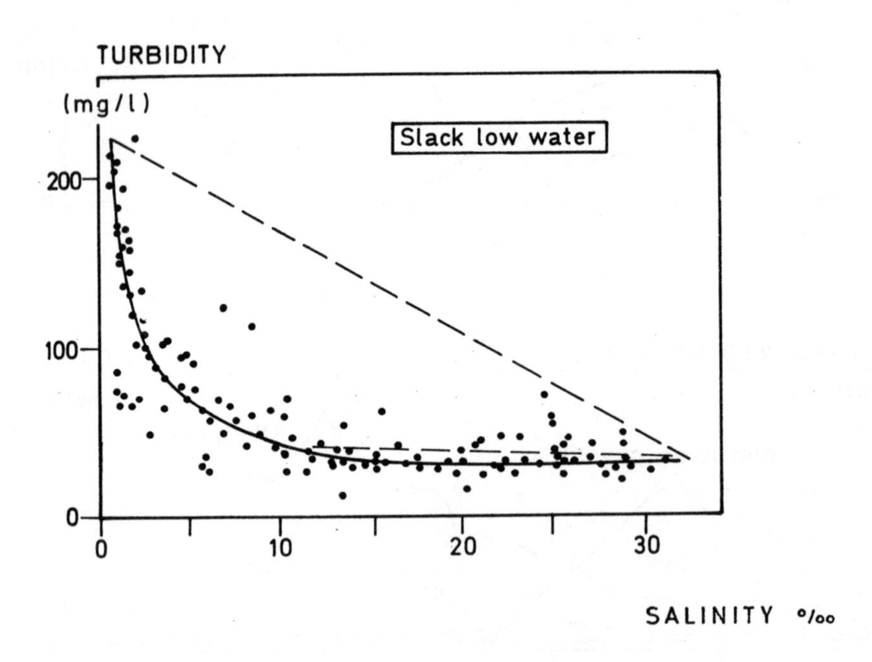


Fig. 4. Evolution of turbidity as a function of salinity in the Scheldt estuary. (The broken line represents the evaluation from simple mixing of freshwater with seawater).

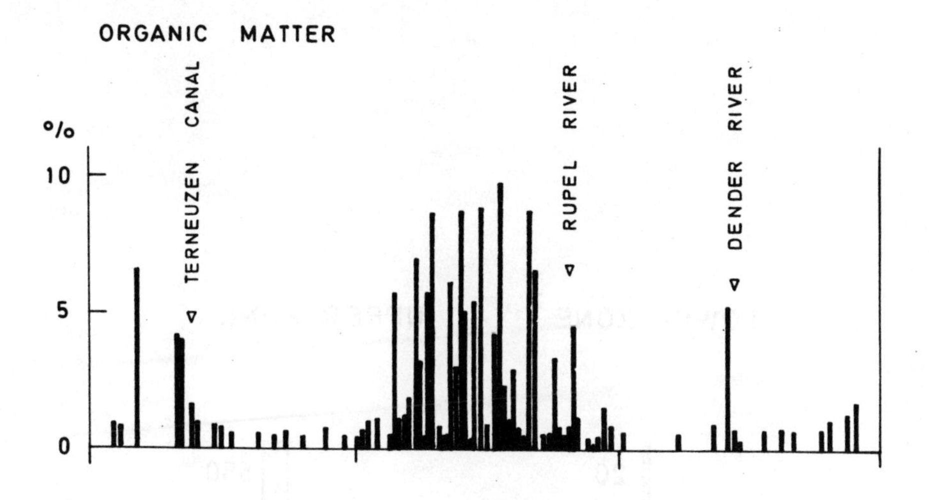


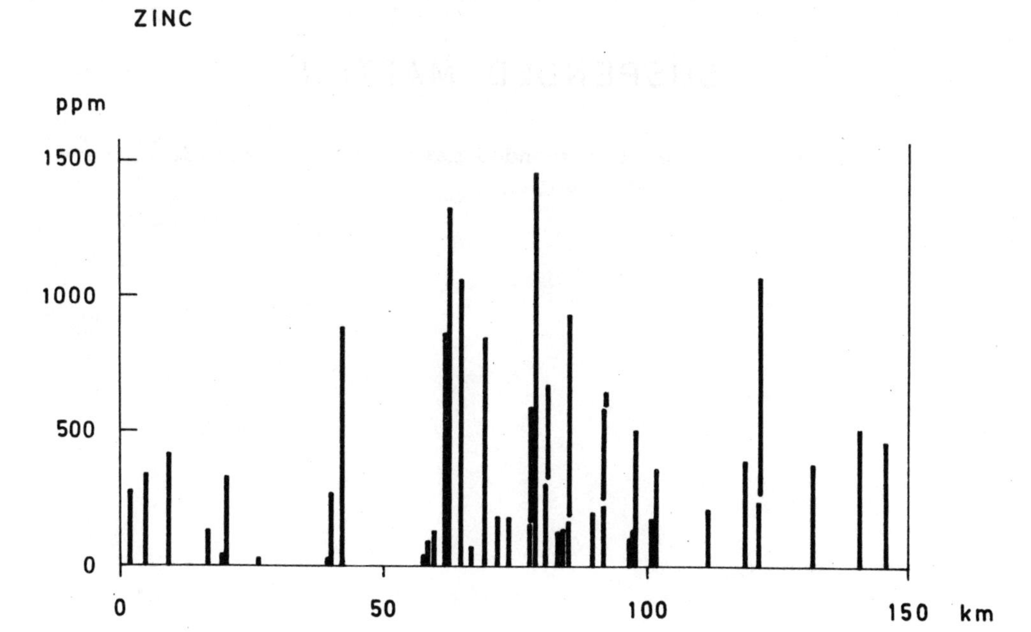


Fig. 5. Organic matter and zinc in sediments as function of the distance to the sea.

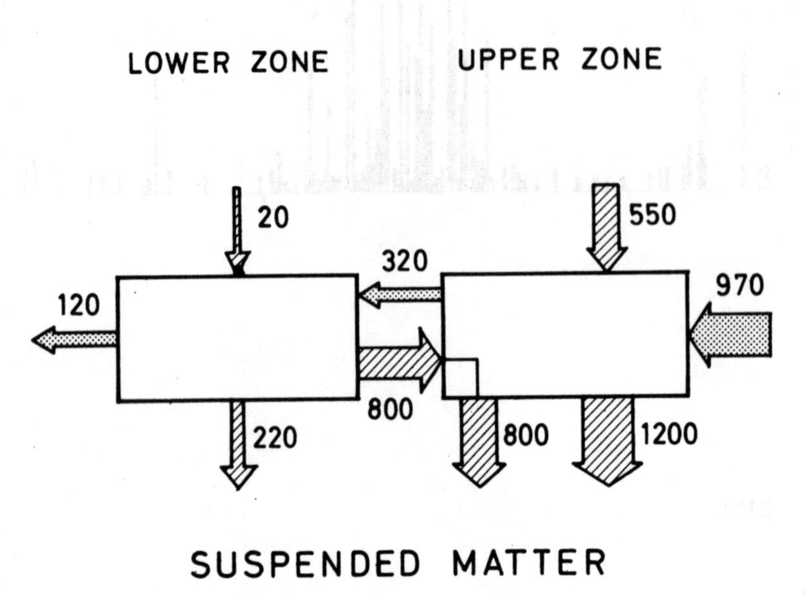


Fig. 6. Mass balance of suspended matter in the Scheldt estuary.

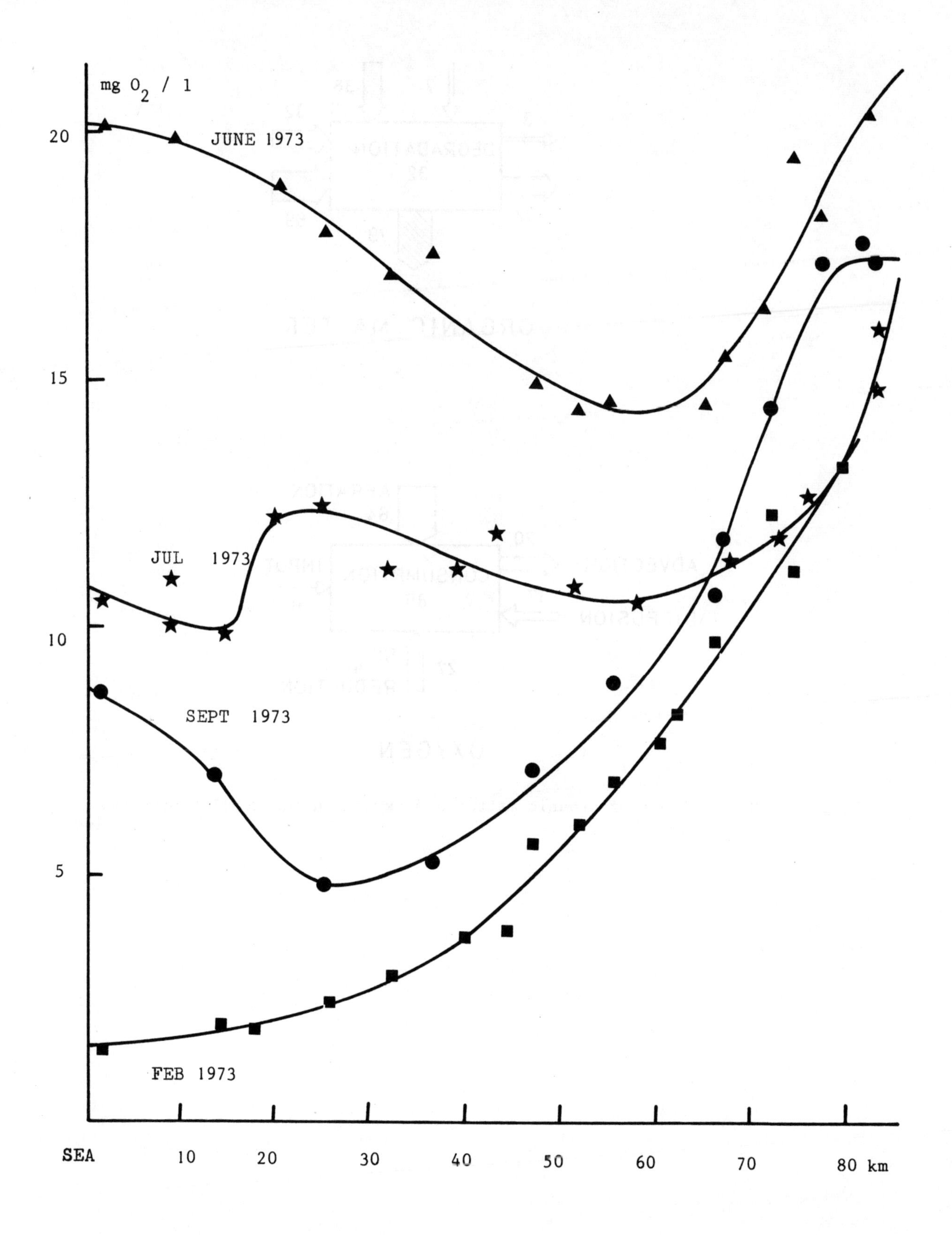


Fig. 7. "Oxidisability" by $KMnO_4$ (4 hours).

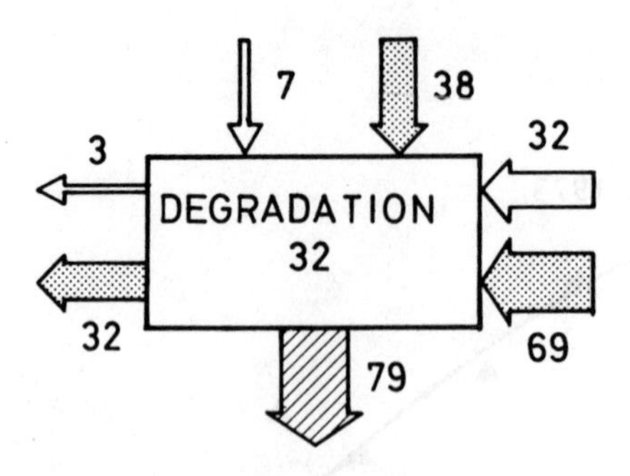


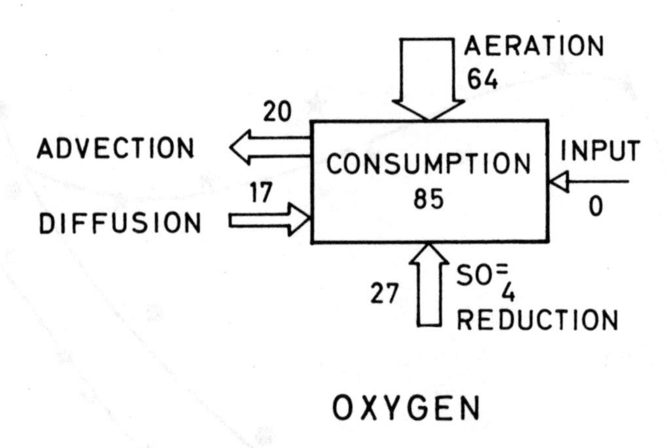


Fig. 8. Mass balance of organic matter and oxygen in the Scheldt estuary.

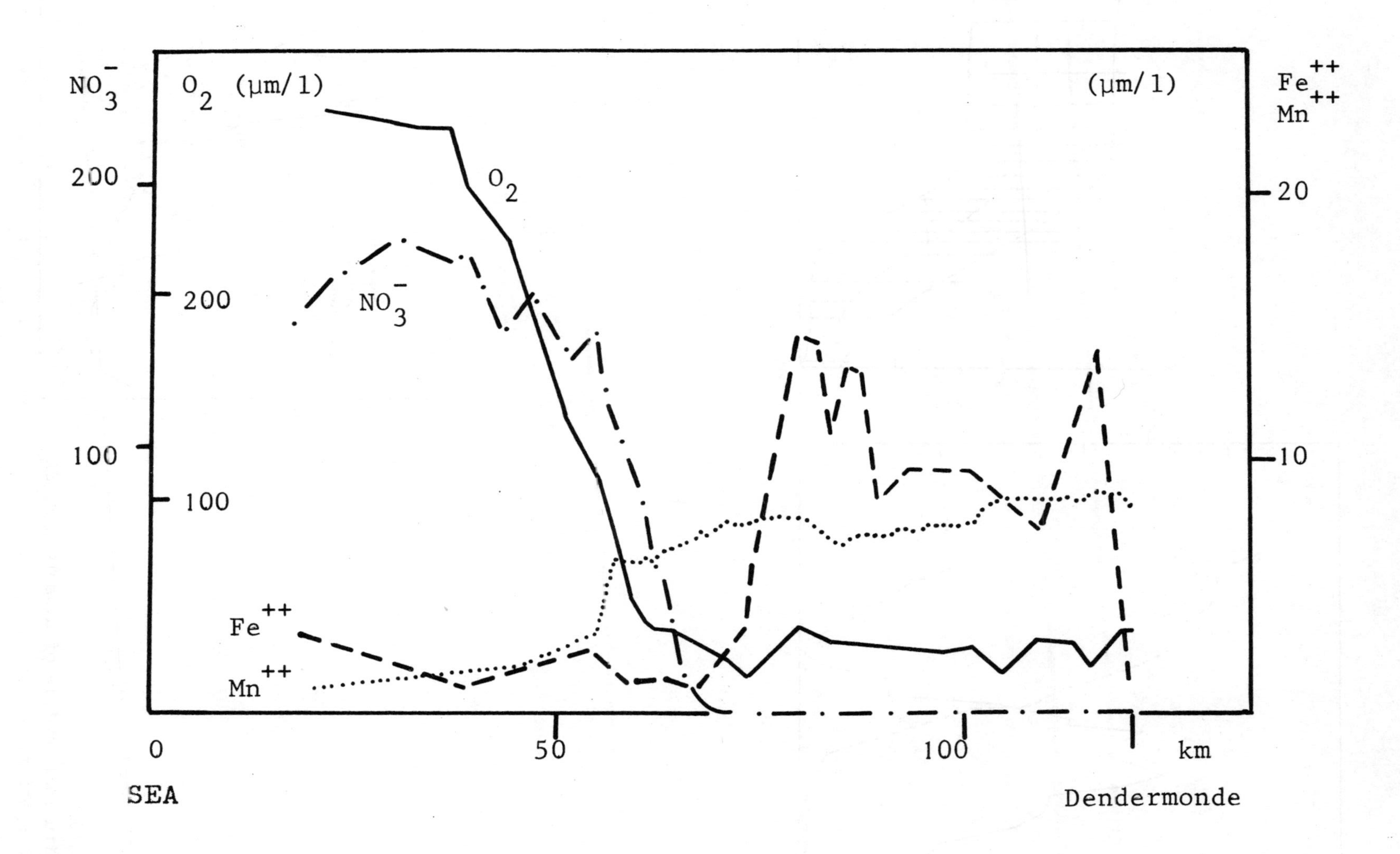


Fig. 9. Evolution of redox couples in the estuary.

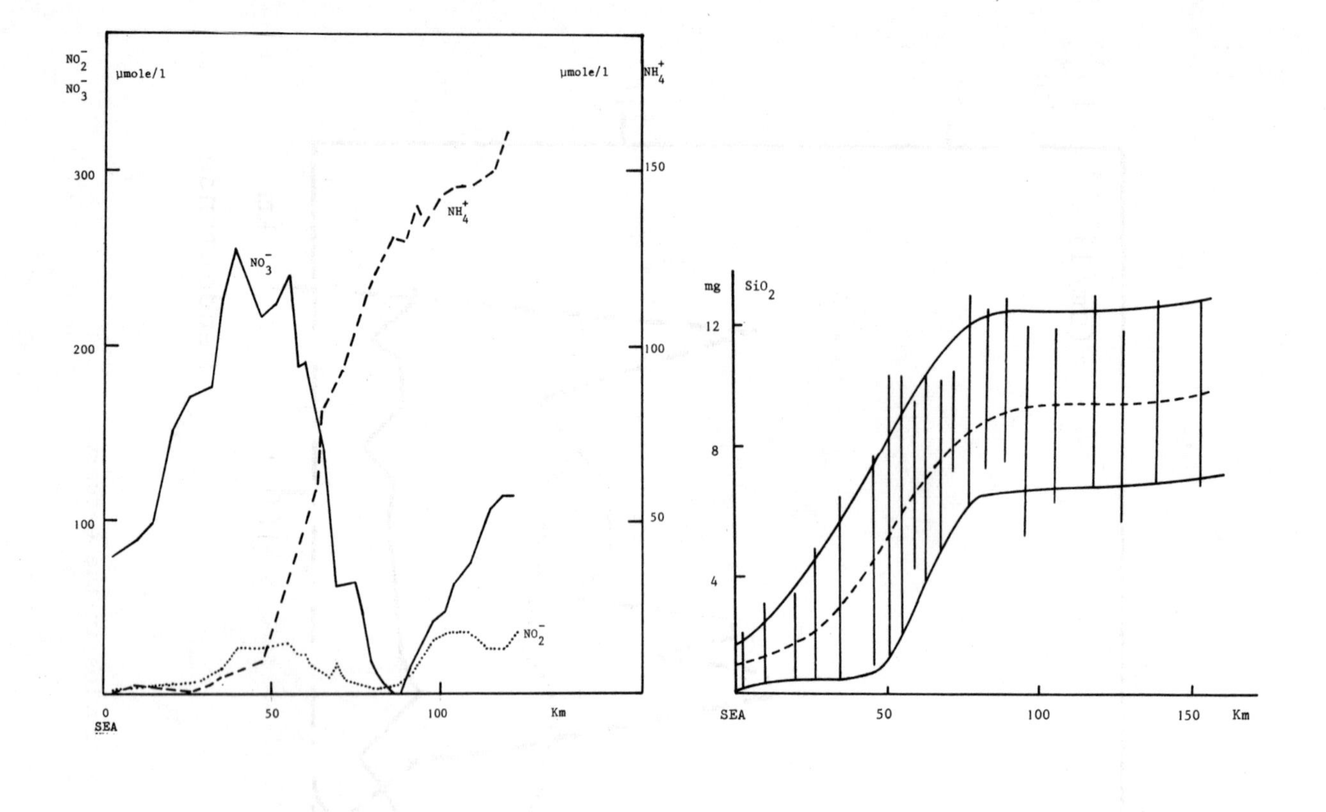


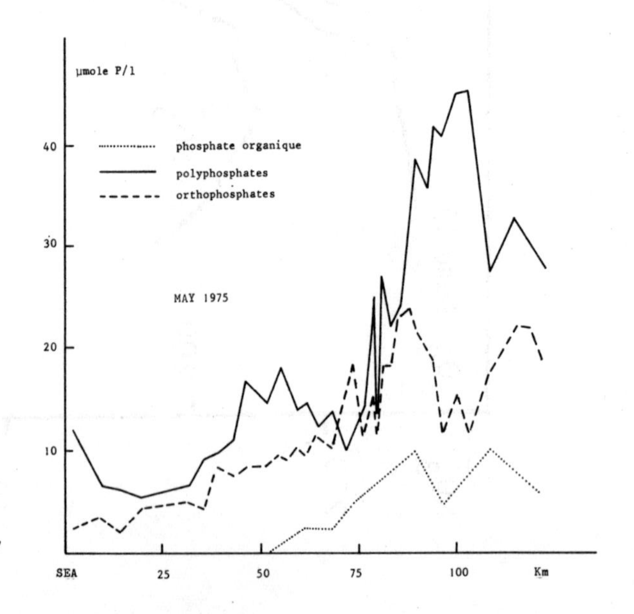


Fig. 10.
Longitudinal profile of dissolved N, Si, and P species.

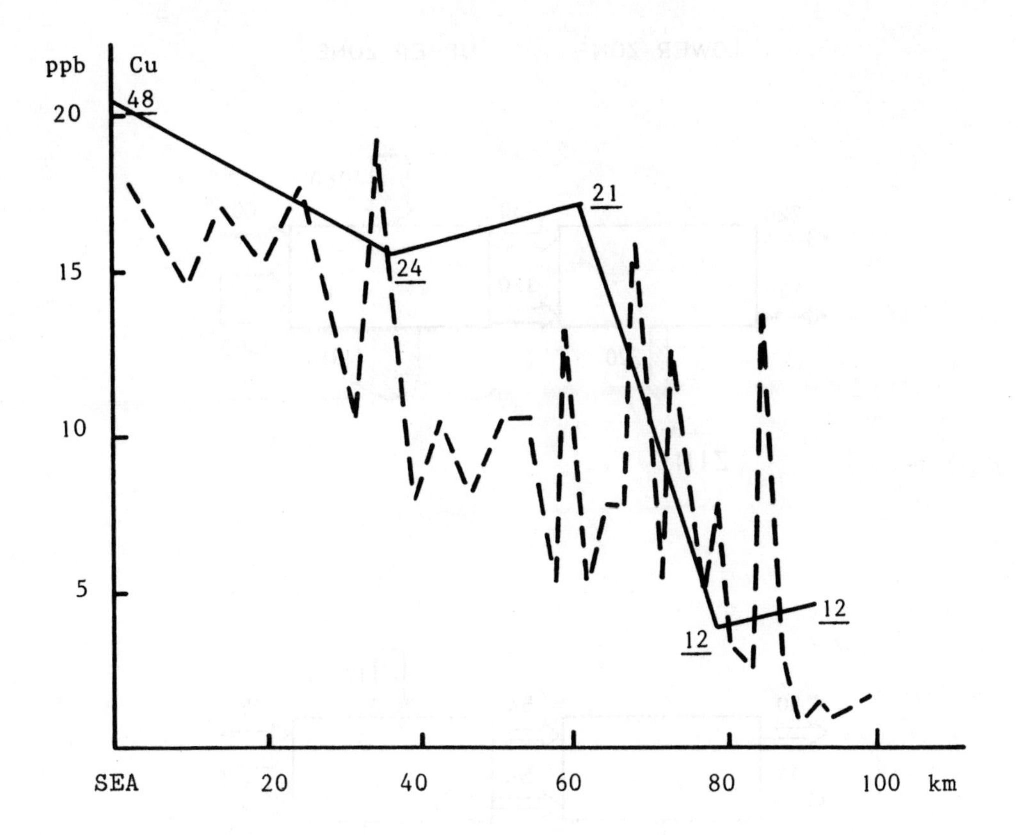


Fig. 11. Longitudinal profile of dissolved copper.
Broken line : instantaneous profile:
Continuous line : annual mean concentrations.
(the figures refer to the number of samples).

LOWER ZONE UPPER ZONE

1060
220 350 60
100 320
220 1810 1360

ZINC

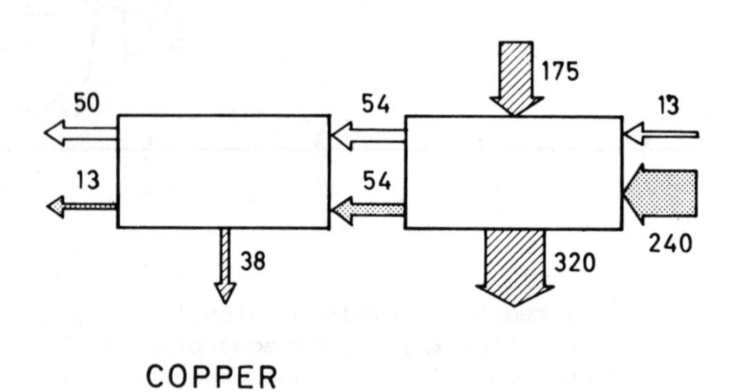


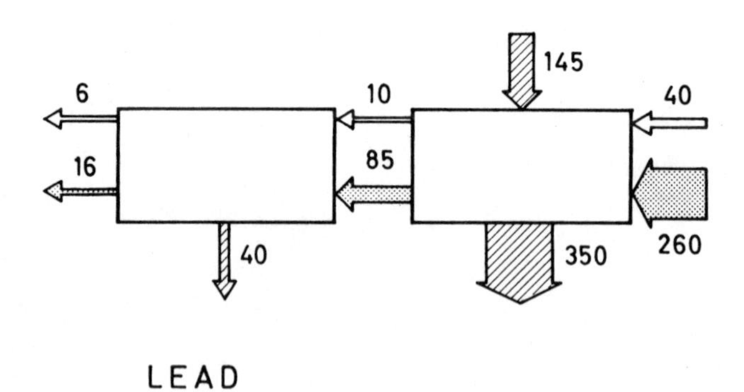


DISSOLVED MATTER SUSPENDED MATTER SEDIMENT

Fig. 12. Mass balances of zinc, copper and lead in the Scheldt estuary.

REFERENCES

Billen, G.; Smitz, J.; Somville, M.; Wollast, R. 1976. Dégradation de la matière organique et processus d'oxydo-réduction dans l'estuaire de l'Escaut. In : J.C.J. Nihoul and R. Wollast (eds.), Rapport Final - Projet Mer. Programme National de recherche et de Développement Environement - Eau., vol. 10, L'Estuaire de l'Escaut. Services du Premier Ministre. Belgique.

Van Bennekom, A.J.; Krijgsman-van Hartingsveld E.; Van der Veer, G.C.M.; Van Voorst, H.F.J. 1974. The seasonal cycles of reactive silicate and suspended diatoms in the Dutch Wadden Sea. Neth. J. of Sea Res., vol. 8, N°. 2-3, p. 174-207.

[B.] SC.78/D-102/A